Recent Trends in Ergodic Theory and Dynamical Systems

S. G. Dani

CONTEMPORARY MATHEMATICS

631

Recent Trends in Ergodic Theory and Dynamical Systems

International Conference in honor
of S. G. Dani's 65th Birthday
Recent Trends in Ergodic Theory and Dynamical Systems
December 26–29, 2012
Vadodara, India

Siddhartha Bhattacharya
Tarun Das
Anish Ghosh
Riddhi Shah

Editors

American Mathematical Society
Providence, Rhode Island

2010 *Mathematics Subject Classification.* Primary 22D40, 28D20, 37A15, 37A17, 37A20, 37A30, 37A35, 37B05, 37E35, 60B15.

Library of Congress Cataloging-in-Publication Data

Recent trends in ergodic theory and dynamical systems : international conference in honor of S.G. Dani's 65th birthday, December 26–29, 2012, Vadodara, India / Siddhartha Bhattacharya, Tarun Das, Anish Ghosh, Riddhi Shah, editors.
    pages cm. – (Contemporary mathematics ; volume 631)
    Includes bibliographical references.
    ISBN 978-1-4704-0931-9 (alk. paper)
    1. Dani, S. G.   2. Differentiable dynamical systems–Congresses.   3. Ergodic theory–Congresses. 4. Mathematical analysis–Congresses.   I. Bhattacharya, Siddhartha, 1971–   II. Das, Tarun, 1964– III. Ghosh, Anish, 1979–   IV. Shah, Riddhi.

QA614.8.R45   2014
515′.48—dc23
                                                         2014021630

Contemporary Mathematics ISSN: 0271-4132 (print); ISSN: 1098-3627 (online)

DOI: http://dx.doi.org/10.1090/conm/631

Dedicated to S. G. Dani on the occasion of his 65th birthday

Contents

CONTENTS

Preface

A seven day instructional workshop followed by a four day discussion meeting were held, respectively, December 18–24 and 26–29, 2012, at the Maharaja Sayajirao University of Baroda, Vadodara, India, on the topic "Recent trends in ergodic theory and dynamical systems". The organising committee comprised C. S. Aravinda, T. Das, R. Rao, N. Shah and R. Shah. Speakers in the discussion meeting included J. Aaronson, S. G. Dani, A. Gorodnik, E. Lindenstrauss, M. S. Raghunathan, B. Weiss and other leading mathematicians. At the inaugural session of the meeting, Professor S. G. Dani was felicitated on the occasion of his 65th birthday. Taking into account the rich academic content of the lectures, it was decided to publish the proceedings of the conference and the present editorial committee was set up. These proceedings are dedicated to S. G. Dani.

Dani is a leader in dynamical systems with interests spanning large parts of the modern theory of ergodic theory and dynamical systems as well as Diophantine approximation. Accordingly, the volume comprises original articles across the spectrum of the subject including homogeneous dynamics, Diophantine approximation, Teichmüller dynamics, ergodic theory, dynamics of iterated function systems, random walks and the ergodic theory of $\mathbb{Z}^d$ actions. Additionally, the editors commissioned two surveys on Dani's work. The survey by D. Morris discusses Dani's many contributions to homogeneous dynamics while the survey by F. Ledrappier and R. Shah discusses his work on probability measures on groups. An invited article by M. S. Raghunathan, who was Dani's PhD supervisor, contains some personal reminiscences.

It is a pleasure to thank the International Centre for Theoretical Sciences of the Tata Institute of Fundamental Research for providing financial support for the workshop and discussion meeting. It is also a pleasure to thank the Maharaja Sayajirao University for excellent local hospitality and especially the local organisers for their untiring efforts in making both the events an unqualified success.

Contemporary Mathematics
Volume **631**, 2015
http://dx.doi.org/10.1090/conm/631/12591

S. G. Dani as I have known him

M. S. Raghunathan

I am happy to have this opportunity to say a few words about Shrikrishna Dani as I have known him. I make no attempt here at describing his mathematics (for which I have the greatest regard, even while I cannot claim to have an in-depth understanding). I am afraid that at my age writing about any one I know would necessarily mean some nostalgia on display. In fact I will dwell almost entirely on my early association with Dani; my contacts with him in those days (the seventies and early eighties) were closer than they were in later years; and in any case happenings in our interactions of that time are more interesting than those in our maturer years.

I have known Shrikrishna Dani as well as his wife Jyotsna now for more than four decades—ever since he joined the Tata Institute of Fundamental Research (TIFR) as a graduate student in 1969 (Jyotsna was not yet his wife; incidentally no one that I know addresses or even refers to him as Shrikrishna (or Krishna); it is always "Dani"). That year I gave a graduate course on Discrete Subgroups of Lie Groups, which was rather advanced for a beginning student like Dani who was busy with the mandatory first year courses. I did not have much direct contact with him during that year. However my colleagues who were teaching the first year graduate students told me that he was one to look out for. Jyotsna was a student at Bombay University but was taking courses at TIFR and she was considered an equally promising student by my colleagues.

I came to know that he was associated with some left wing activists then operating in Mumbai. I myself have always had some mild left leanings and that was an additional reason—apart from his mathematical talent—for me to begin by being well-disposed towards him; I do not think though that he was aware of this then or later.

Over the next few years I got to know him well both professionally and personally. At some point he asked me if I would be willing to be his thesis adviser. I was more than happy to be his "mentor". By that time he had in fact more or less decided that he would specialize in Ergodic Theory, more specifically Homogeneous Dynamics. In fact he had already proved an interesting result in collaboration with Jyotsna which belonged there. I was no expert in Ergodic theory and was in fact learning some at that time and so my role as Thesis Advisor was necessarily rather limited. He studied a paper of Sinai's and came up with a solution of a problem suggested by that work. I was only vaguely aware that he was reading Sinai's paper when he came up with the solution of the problem.

When the thesis had been written up I went through the manuscript (and learnt some Ergodic Theory in the process). I was impressed with the extent of scholarship he had picked up along the way and it was all done without any significant help from me. The new ideas he had introduced too impressed me greatly. It was a beautiful piece of work in which he had shown that the an ergodic translation action of a semisimple element of a semisimple Lie group G on a homogeneous space of finite Haar measure is a Kolmogorov automorphism and he had also computed its entropy.

A few weeks after he had handed me the manuscript, he came one evening to my house with Jyotsna. He was in a somewhat disturbed state of mind as he had discovered that Sinai had in an Appendix to his paper broadly outlined a method of attack on the problem which was indeed the approach Dani had devised on his own. He had begun to think that his efforts had come to nought. I did not share his pessimism, but was not able to quite reassure him. After some discussion we found a possible way to recast the preliminary parts of the manuscript that would enhance the value of the work. Dani was quick to carry out the programme.

On my suggestion Dani submitted the paper to the American Journal of Mathematics. A couple of months later he got back the manuscript without any covering letter. He was naturally upset at this and so was I, the more so as I was convinced that the paper was an outstanding piece of work. Happily, a letter from the journal arrived a few days later accepting the paper as it stood! Incidentally Dani's exposition of mathematics in all his papers is uniformly excellent. Dani went on to study these semisimple translations in greater depth. He showed in later work that they were all Bernoulli shifts.

Around this time Jyotsna wrote a paper on the density of certain subsets of the $\mathbb{R}$-points of a vector space over a finite dimensional central division algebra over $\mathbb{Q}$. During some discussions about the paper with Dani, I realised that the question dealt with had a larger context and I suggested that one should perhaps look at orbits of horocycles in homogeneous spaces of finite Haar measure of semisimple Lie groups. He came up soon enough with density theorems for orbits in compact homogeneous spaces and then went on to prove theorems in the non-compact case as well. I was in constant touch with him mathematically, but acquired no in-depth knowledge of his work. This contact however is what led me to what came to be known as "Raghunathan Conjecture" about unipotent flows. It was on the one hand Dani's theorems and on the other hand Margulis's first results on unipotent 1-parameter flows (viz. that they cannot have non-compact closed orbits) that suggested the orbit-density statement (which is the conjecture). I had mentioned this to Dani in some conversation and it was he who put it down in one of his papers as my conjecture. He formulated a measure theoretic conjecture in the same paper which is closely related to the orbit-density conjecture and this formulation which was a result of his insights in ergodic theory that eventually enabled Ratner to prove my conjecture. Thus Dani's conjecture was crucial in the resolution of my conjecture. I was unhappy when Dani's formulation was labelled "Raghunathan Measure Conjecture" by Ratner (and had in fact protested to her). Fortunately the ergodic theory community is now aware that the "Measure Conjecture" is due to Dani.

Dani had also done some very interesting work on unipotent one parameter flows providing new insights into their behaviour improving on Margulis's first results. His interest shifted to questions connected with the Oppenheim conjecture soon after it was settled by Margulis. He has here a lot of interesting work about which again I have only a passing familiarity.

I have some (but not really in-depth) acquaintance with Dani's work on homogeneous flows, but am much less familiar with his extensive work on probability measures on locally compact groups. My mathematical contacts with him had become far and few between after he embarked on research in this area.

Over the years Dani has involved himself in the promotion of mathematics in the country. He has had numerous PhD students and continues to have some even now. Many of his students are now teachers in different university departments and their presence has no doubt enhanced the standing of these places. Nimish Shah worked with Dani for his PhD; technically I was Nimish's guide but it was Dani with whom he worked.

His contribution to the advancement of mathematics in India goes well beyond his own researches and the guidance of students. He has shouldered many administrative responsibilities over the years and has executed them diligently and with a great sense of responsibility. He has served on numerous national and international bodies engaged in the promotion of mathematics. He served as the chair for the Commission for Development and Exchange of the International Mathematical Union during 2006 -2010. For several years he worked closely with me as member (when I was the chair) of the National Board for Higher Mathematics. Later he became the chair while I continued as a member. He was the vice-chair of the Organizing Committee of the International Congress of Mathematicians held in Hyderabad in 2010. In these bodies he was an immense source of strength and could be depended upon to take on any responsibility whenever a need arose. I have often benefitted from informal consultations with him on a variety of issues relating to promotion of mathematics in the country.

In recent years Dani has taken a serious interest in the History of Mathematics and has brought his incisive thinking to research in this area. He has played a leading role in debunking the spurious claims of the so called "Vedic Mathematics".

Dani is of course a first rate mathematician: his contributions to mathematics speak for themselves. Those of us who have had more than just professional contacts with him know him to be a fine human being. He has many collaborators which indicates that he is easy to get along with. His political views continue to remain left of centre. He is a man of few words, is soft spoken and rather self effacing; but he is firm in his beliefs and unyielding in advocating causes in which he believes. He is a keen supporter of feminist rights.

It is indeed a matter of great pleasure for me and my wife Ramaa to have known him and Jyotsna and also their daughters Varsha and Pallavi. I take this opportunity to wish Dani and his family many more happy and fruitful years.

DEPARTMENT OF MATHEMATICS, INDIAN INSTITUTE OF TECHNOLOGY BOMBAY, MUMBAI 400076, INDIA

E-mail address: msr@math.iitb.ac.in

Contemporary Mathematics
Volume **631**, 2015
http://dx.doi.org/10.1090/conm/631/12592

Cubic averages and large intersections

V. Bergelson and A. Leibman

Dedicated to Professor S. G. Dani on the occasion of his 65th birthday

ABSTRACT. We discuss the nature of the phenomenon of "large limits along cubic averages" and obtain some rather general new results. We also prove a general version of Furstenberg's correspondence principle that applies to uncountable amenable discrete groups and utilize it to derive combinatorial corollaries of ergodic results involving large limits.

0. Introduction

Let $(X, \mathcal{B}, \mu, T)$ be an invertible probability measure preserving system. The classical Poincaré recurrence theorem (see [**Po**], Theorem I) states that for any $A \in \mathcal{B}$ with $\mu(A) > 0$ there exists $n \in \mathbb{N}$ such that $\mu(A \cap T^n A) > 0$. For mixing systems one has $\lim_{n \to \infty} \mu(A \cap T^n A) = \mu(A)^2$, and it is natural to ask whether for any $(X, \mathcal{B}, \mu, T)$, any $A \in \mathcal{B}$, and any $\delta > 0$ one can find $n \neq 0$ such that

$$\mu(A \cap T^n A) \geq \mu(A)^2 - \delta,$$

and, if so, how "large" the set

$$R_\delta(A) = \left\{ n \in \mathbb{Z} : \mu(A \cap T^n A) \geq \mu(A)^2 - \delta \right\}$$

can be. These questions were addressed in [**Kh**], where the following result, often quoted as "Khintchine's recurrence theorem", was obtained.

THEOREM 0.1. *For any invertible probability measure preserving system* $(X, \mathcal{B}, \mu, T)$, *any* $A \in \mathcal{B}$, *and any* $\delta > 0$, *the set* $R_\delta(A)$ *is syndetic.*[1]

(We are taking the liberty to formulate Khintchine's result for powers of a single measure preserving transformation. Khintchine's paper dealt with continuous measure preserving flows.)

In [**Kh**], Theorem 0.1 was derived from the following stronger result. (Again, we formulate it for $\mathbb{Z}$- rather than for $\mathbb{R}$-actions.)

2010 *Mathematics Subject Classification.* Primary: 28D15; Secondary: 05D10.

Key words and phrases. Measure preserving group actions, multiple recurrence, Furstenberg's correspondence principle.

Partially supported by NSF grants DMS-0901106 and DMS-1162073.

[1]A set S in a topological group G is said to be *(left) syndetic* if there exists a compact set $F \subseteq G$ such that $G = FS$. If G is a discrete group, F has to be finite.

THEOREM 0.2. *For any invertible probability measure preserving system* $(X, \mathcal{B}, \mu, T)$ *and any* $A \in \mathcal{B}$, *one has*

$$\lim_{N-M\to\infty} \frac{1}{N-M} \sum_{n=M}^{N-1} \mu(A \cap T^n A) \geq \mu(A)^2.$$

By iterating Khintchine's theorem, one immediately observes that for any $n \in R_\delta(A)$ there exists a syndetic set of $m \in \mathbb{Z}$ such that

$$(0.1) \quad \mu\big((A \cap T^n A) \cap T^m (A \cap T^n A)\big) = \mu(A \cap T^n A \cap T^m A \cap T^{n+m} A) > \mu(A)^4 - 2\delta.$$

This leads to the natural question of whether, for any $\delta > 0$, the set of pairs (n, m) satisfying (0.1) is syndetic in $\mathbb{Z}^2$. The affirmative answer is provided by the following theorem.

THEOREM 0.3 ([**B1**]). *For any invertible probability measure preserving system* $(X, \mathcal{B}, \mu, T)$ *and any* $A \in \mathcal{B}$, *one has*

$$\lim_{N-M\to\infty} \frac{1}{(N-M)^2} \sum_{n,m=M}^{N-1} \mu(A \cap T^n A \cap T^m A \cap T^{n+m} A) \geq \mu(A)^4.$$

COROLLARY 0.4. *For any invertible probability measure preserving system* $(X, \mathcal{B}, \mu, T)$, *any* $A \in \mathcal{B}$, *and any* $\delta > 0$, *the set*

$$\big\{(n,m) \in \mathbb{Z}^2 : \mu(A \cap T^n A \cap T^m A \cap T^{n+m} A) > \mu(A)^4 - \delta\big\}$$

is syndetic.

Theorem 0.3 was extended in [**HK1**] and [**HK2**] to multiparameter expressions of the form $\mu\big(\bigcap_{\epsilon_1,\ldots,\epsilon_k\in\{0,1\}} T^{\epsilon_1 n_1 + \ldots + \epsilon_k n_k} A\big)$, $n = (n_1, \ldots, n_k) \in \mathbb{Z}^k$, which in turn implies, for any $\delta > 0$, the syndeticity of the set

$$\Big\{n \in \mathbb{Z}^k : \mu\Big(\bigcap_{\epsilon_1,\ldots,\epsilon_k\in\{0,1\}} T^{\epsilon_1 n_1 + \ldots + \epsilon_k n_k} A\Big) > \mu(A)^{2^k} - \delta\Big\}.$$

The proofs of these results in [**HK1**] and [**HK2**] utilize in a rather crucial way knowledge about *characteristic factors*[2] responsible for the limiting behavior of the expressions $\frac{1}{(N-M)^k} \sum_{n_1,\ldots,n_k=M}^{N-1} \prod_{\epsilon_1,\ldots,\epsilon_k\in\{0,1\}} T^{\epsilon_1 n_1 + \ldots + \epsilon_k n_k} f$, and it is of interest to attempt to establish the results on large intersections in the situations where there is (so far) no sufficient knowledge on the structure of the corresponding characteristic factors. In particular, one would like to know if, given commuting invertible measure preserving transformations $T_1, \ldots, T_k$ of a probability space $(X, \mathcal{B}, \mu)$ and a set $A \in \mathcal{B}$ with $\mu(A) > 0$, the set

$$(0.2)$$
$$R_\delta^{(k)}(A) = \Big\{(n_1,\ldots,n_k) \in \mathbb{Z}^k : \mu\Big(\bigcap_{\epsilon_1,\ldots,\epsilon_k\in\{0,1\}} T_1^{\epsilon_1 n_1} \ldots T_k^{\epsilon_k n_k} A\Big) > \mu(A)^{2^k} - \delta\Big\}$$

is syndetic.

The goal of this short paper is to show that the phenomenon of large limits along cubic configurations has a very general scope and hinges on a simple "Fubini principle" (Lemma 1.1 below), which, roughly, states that the limits of uniform multi-parameter Cesàro averages can be replaced by iterated limits of uniform Cesàro

---

[2]A factor Y of X is *characteristic* if the limit, as $N - M \to \infty$, of the expressions under consideration depends not on f, but only on $E(f|Y)$ – the conditional expectation of f with respect to Y (the orthogonal projection of f on $L^2(Y) \subseteq L^2(X)$).

averages. In what follows we will establish a general scheme that can be viewed as a machine for obtaining various results on large limits of cubic averages. This machine has an input and an output. The input consists of two convergence statements: (i) a general multiparameter limiting theorem that stipulates the existence of certain uniform Cesàro limits (but does not require any description of these limits); and (ii) an ergodic limiting result pertaining to single or multiple recurrence that guarantees a "large" uniform Cesàro limit. An example of such a result is given by Theorem 0.2. (We will provide below a few more examples of various degrees of generality.) Then the machine's output is a theorem on large limits of cubic averages. An example of such a theorem is Theorem 0.3, but, as we will see, significantly more general theorems hold true as well.

To help the reader get a feeling about the general results which we obtain in this paper, let us give a sketch of the proof of the following fact, which is a rather special case of Theorem 0.8 below.

THEOREM 0.5. *Let $T_1, \ldots, T_k$ be invertible transformations of a probability measure space $(X, \mathcal{B}, \mu)$, which commute (or, more generally, generate a nilpotent group). Then for any $A \in \mathcal{B}$,*

$$\lim_{N-M \to \infty} \frac{1}{(N-M)^k} \sum_{M \leq n_1, \ldots, n_k \leq N-1} \mu \left(\bigcap_{\epsilon_1, \ldots, \epsilon_k \in \{0,1\}} T_1^{\epsilon_1 n_1} \cdots T_k^{\epsilon_k n_k} A \right) \geq \mu(A)^{2^k},$$

and hence the set $R_\delta^{(k)}(A)$ defined in (0.2) is syndetic.

PROOF SKETCH. The input statement of the form (i) that is needed for the proof is the following fact (see [**W**], Theorem 5.1):

- *For any $A \in \mathcal{B}$, the limit*

$$\lim_{N-M \to \infty} \frac{1}{(N-M)^k} \sum_{M \leq n_1, \ldots, n_k \leq N-1} \mu \left(\bigcap_{\epsilon_1, \ldots, \epsilon_k \in \{0,1\}} T_1^{\epsilon_1 n_1} \cdots T_k^{\epsilon_k n_k} A \right)$$

exists.

The input statement of type (ii) is Theorem 0.2 above:

- *For any $A \in \mathcal{B}$, $\lim_{N-M \to \infty} \frac{1}{N-M} \sum_{n=M}^{N-1} \mu(A \cap T^n A) \geq \mu(A)^2$.*

Take $k = 2$. Replacing the uniform double Cesàro limit by the iterated uniform Cesàro limits (this is an instance of the Fubini principle alluded to above), we have:

$$\lim_{N-M \to \infty} \frac{1}{(N-M)^2} \sum_{M \leq n_1, n_2 \leq N-1} \mu \left(A \cap T_1^{n_1} A \cap T_2^{n_2} A \cap T_1^{n_1} T_2^{n_2} A \right)$$

$$= \lim_{N_1 - M_1 \to \infty} \frac{1}{N_1 - M_1} \sum_{n_1 = M_1}^{N_1 - 1} \lim_{N_2 - M_2 \to \infty} \frac{1}{N_2 - M_2} \sum_{n_2 = M_2}^{N_2 - 1}$$

$$\mu \left((A \cap T_1^{n_1} A) \cap T_2^{n_2} (A \cap T_1^{n_1} A) \right)$$

$$\geq \lim_{N_1 - M_1 \to \infty} \frac{1}{N_1 - M_1} \sum_{n_1 = M_1}^{N_1 - 1} \mu(A \cap T_1^{n_1} A)^2$$

$$\geq \left(\lim_{N_1 - M_1 \to \infty} \frac{1}{N_1 - M_1} \sum_{n_1 = M_1}^{N_1 - 1} \mu(A \cap T_1^{n_1} A) \right)^2 \geq (\mu(A)^2)^2 = \mu(A)^4.$$

Inductively repeating this argument for $k = 3, 4, \ldots$, we obtain the desired result. $\square$

In order to formulate our results in proper generality we have to introduce some notation. For a (left) amenable group G with left Haar measure τ and a mapping $g \mapsto v_g$, $g \in G$, from G to a Banach space V, let us write UC-$\lim_{g \in G} v_g = v$ if for every left Følner net $(\Phi_\alpha)^3$ in G one has

$$\lim_{\alpha \to \infty} \frac{1}{\tau(\Phi_\alpha)} \int_{\Phi_\alpha} v_g \, d\tau(g) = v.$$

(In the case G is discrete, and thus τ is the counting measure, the above limit takes the form $\lim_{\alpha \to \infty} \frac{1}{|\Phi_\alpha|} \sum_{g \in \Phi_\alpha} v_g$. If G is σ-compact (that is, is a countable union of compact sets), Følner nets can be replaced by Følner sequences; see [**P**].)

In what follows, let $(X, \mathcal{B}, \mu)$ be a probability measure space. We will prefer to deal with functions from $L^2(X)$ instead of subsets of X; to obtain results about sets $A \in \mathcal{B}$ one takes $f = 1_A$. Notice also that to get a left action of a non-commutative group G on $L^2(X)$ we have to assume that G acts on X from the right; of course, this does not affect the results. (In Section 3 we pass back to a left action of G on X.) In Section 2 we will prove the following:

THEOREM 0.6. *Let $G_1, \ldots, G_k$ be locally compact amenable groups and, for each $i = 1, \ldots, k$, let $T_{i,1}, \ldots, T_{i,r_i}$ be (not necessarily homomorphic or continuous) mappings from G_i to the set of measure preserving transformations of X. Assume that the following two conditions are satisfied:*
(i) *for any collection $f_{j_1, \ldots, j_k}$, $1 \leq j_1 \leq r_1$, $\ldots$, $1 \leq j_k \leq r_k$, of functions from $L^\infty(X)$, the limit*

$$\operatorname*{UC-lim}_{\substack{(g_1, \ldots, g_k) \in G_1 \times \cdots \times G_k}} \int_X \prod_{\substack{1 \leq j_1 \leq r_1 \\ \vdots \\ 1 \leq j_k \leq r_k}} \Big(T_{1,j_1}(g_1) \cdots T_{k,j_k}(g_k) f_{j_1, \ldots, j_k} \Big) d\mu$$

exists;
(ii) *for any non-negative function $f \in L^2(X)$ and any $i \in \{1, \ldots, k\}$,*

$$\operatorname*{UC-lim}_{g_i \in G_i} \int_X \prod_{j_i = 1}^{r_i} \Big(T_{i,j_i}(g_i) f \Big) d\mu \geq \Big(\int_X f \, d\mu \Big)^{r_i}.$$

Then for any non-negative function $f \in L^\infty(X)$,
(a)

$$\operatorname*{UC-lim}_{\substack{(g_1, \ldots, g_k) \in G_1 \times \cdots \times G_k}} \int_X \prod_{\substack{1 \leq j_1 \leq r_1 \\ \vdots \\ 1 \leq j_k \leq r_k}} \Big(T_{1,j_1}(g_1) \cdots T_{k,j_k}(g_k) f \Big) d\mu \geq \Big(\int_X f \, d\mu \Big)^{r_1 \cdots r_k};$$

---

^{3}A *left Følner net* in a group G is a family $(\Phi_\alpha)_{\alpha \in \Lambda}$ of compact non-null subsets of G, indexed by a directed set Λ, such that $\lim_\alpha \tau(\Phi_\alpha \triangle (g\Phi_\alpha))/\tau(\Phi_\alpha) = 0$.

(b) *for any $\delta > 0$, the set*

$$\Big\{(g_1,\ldots,g_k) \in G_1 \times \cdots \times G_k :$$

$$\int_X \prod_{1 \le j_1 \le r_1} \Big(T_{1,j_1}(g_1) \cdots T_{k,j_k}(g_k)f\Big)d\mu > \Big(\int_X f\,d\mu\Big)^{r_1 \cdots r_k} - \delta\Big\}$$

$$\vdots$$
$$1 \le j_k \le r_k$$

is syndetic in $G_1 \times \cdots \times G_k$.

In order to apply this theorem we only need to verify, in concrete situations, whether conditions (i) and (ii) are satisfied. The following recent generalization by Zorin-Kranich of a theorem of Walsh ([**W**]) provides a convenient general result guaranteeing the fulfillment of condition (i):

THEOREM 0.7 ([**Z**]). *Let G be a locally compact amenable group and let $S_1, \ldots, S_k$ be continuous polynomial mappings[4] from G to a nilpotent group of measure preserving transformations of X. Then, for any collection of functions $f_1, \ldots, f_l \in L^\infty(X)$, the limit $\text{UC-lim}_{g \in G} \prod_{i=1}^{l} S_i(g)f_i$ exists in $L^2(X)$.*

This theorem will allow us to obtain two kinds of applications. First, we will get a general result (Theorem 0.8 below) on convergence and recurrence. One can then use a variant of Furstenberg's corresponcence principle for combinatorial actions (see [**BF**]) to obtain some combinatorial applications. There is however a much more interesting vista that Theorem 0.7 opens up. Namely, it applies, in particular, to discrete uncountable amenable groups, which – via an appropriate generalization of Furstenberg's correspondence principle – will allow us to obtain somewhat unexpected combinatorial applications (see Section 3).

As for condition (ii), it holds, for example, in the following situations:

(**1**) Let T be a measure preserving action on X of a locally compact amenable group G. For any $f \in L^2(X)$, by the classical von Neumann ergodic theorem, adapted to limits along Følner nets, one has $\text{UC-lim}_{g \in G} T_g f = Pf$, where P is the orthogonal projection onto the space $\{f \in L^2(X) : T_g f = f \text{ for all } g \in G\}$. For $f \ge 0$ we have $Pf \ge 0$ and $\int_X Pf\,d\mu = \int_X f\,d\mu$. Hence,

$$\text{UC-lim}_{g \in G} \int_X f \cdot T_g f\,d\mu = \int_X f \cdot Pf\,d\mu = \int_X Pf \cdot Pf\,d\mu = \|Pf\|^2$$

$$\ge \Big(\int_X Pf\,d\mu\Big)^2 = \Big(\int_X f\,d\mu\Big)^2.$$

Thus, condition (ii) is satisfied for $T_{i,1} = \text{Id}$ and $T_{i,2} = T$. This gives us the following result:

THEOREM 0.8. *Let $G_1, \ldots, G_k$ be locally compact amenable groups and let T_i, $i = 1, \ldots, k$, be (continuous) homomorphisms from G_i to a nilpotent group of measure preserving transformations of X. Then for any non-negative function*

---

[4]A mapping P from a group G to a (nilpotent) group H is said to be *polynomial* if there is $d \in \mathbb{N}$ such that for any $h_1, \ldots, h_d \in G$ one has $D_{h_1} \cdots D_{h_d} P = 1$, where the "derivative" D_h is defined by $(D_h P)(g) = P(g)^{-1} P(gh)$.

$f \in L^\infty(X)$,
(a)

$$\underset{(g_1,\ldots,g_k)\in G_1\times\cdots\times G_k}{\text{UC-lim}} \int_X \prod_{\epsilon_1,\ldots,\epsilon_k\in\{0,1\}} \left(T_1(g_1)^{\epsilon_1}\cdots T_k(g_k)^{\epsilon_k} f\right) d\mu \geq \left(\int_X f\,d\mu\right)^{2^k};$$

(b) *for any $\delta > 0$, the set*

$$\Big\{(g_1,\ldots,g_k)\in G_1\times\cdots\times G_k :$$
$$\int_X \prod_{\epsilon_1,\ldots,\epsilon_k\in\{0,1\}} \left(T_1(g_1)^{\epsilon_1}\cdots T_k(g_k)^{\epsilon_k} f\right) d\mu > \left(\int_X f\,d\mu\right)^{2^k} - \delta\Big\}$$

is syndetic in $G_1 \times \cdots \times G_k$.

(2) Let F be a (discrete) field and let V be a finite dimensional vector space over F. We call a mapping T from V to a group $\mathcal{T}$ *polynomial* if $T = S \circ P$, where P is a polynomial (in the usual sense) mapping from V to a finite dimensional vector space W over F and S is a homomorphism from W to $\mathcal{T}$. The following theorem, extending a result of Larick ([**La**]), can be found in [**BLM**]:

THEOREM 0.9. *Let V be a finite dimensional vector space and let U be a poly-nomial unitary action of V on a Hilbert space $\mathcal{H}$. Then for any $f \in \mathcal{H}$, the limit UC-$\lim_{v \in V} U(v)f$ exists and is the orthogonal projection of f on the space of $U(V)$-invariant vectors.*

(In [**BLM**], this theorem is stated and proved for the case where the base field F is countable, but it remains true, with the same proof, when F is uncountable.)

It follows that if T is a polynomial measure preserving action on X of a finite dimensional vector space V, then for any non-negative $f \in L^2(X)$, UC-$\lim_{v \in V} \int_X f \cdot T(v)f\,d\mu \geq \left(\int_X f\,d\mu\right)^2$. Based on this fact, Theorem 0.6 acquires the following form:

THEOREM 0.10. *Let $V_1,\ldots,V_k$ be finite dimensional vector spaces over a field F and let T_i, $i = 1,\ldots,k$, be polynomial mappings from V_i to a nilpotent group of measure preserving transformations of X. Then for any non-negative function $f \in L^\infty(X)$,*
(a)

$$\underset{(v_1,\ldots,v_k)\in V_1\oplus\ldots\oplus V_k}{\text{UC-lim}} \int_X \prod_{\epsilon_1,\ldots,\epsilon_k\in\{0,1\}} \left(T_1(v_1)^{\epsilon_1}\cdots T_k(v_k)^{\epsilon_k} f\right) d\mu \geq \left(\int_X f\,d\mu\right)^{2^k};$$

(b) *for any $\delta > 0$, the set*

$$\Big\{(u_1,\ldots,u_k)\in V_1\oplus\cdots\oplus V_k :$$
$$\int_X \prod_{\epsilon_1,\ldots,\epsilon_k\in\{0,1\}} \left(T_1(v_1)^{\epsilon_1}\cdots T_k(v_k)^{\epsilon_k} f\right) d\mu > \left(\int_X f\,d\mu\right)^{2^k} - \delta\Big\}$$

is syndetic in the (discrete) group $V_1 \oplus \cdots \oplus V_k$.

(3) Let us say that polynomials $p_1,\ldots,p_r$ are *essentially linearly independent* if the polynomials $p_1 - p_1(0),\ldots,p_r - p_r(0)$ are linearly independent. The following theorem can be found in [**FK**]:

THEOREM 0.11. *Let T be a totally ergodic measure preserving transformation of X and let $p_1, \ldots, p_r$ be integer valued, essentially linearly independent polynomials of one or several integer variables. Then for any $f_1, \ldots, f_r \in L^\infty(X)$, UC-$\lim_{n \in \mathbb{Z}} \prod_{j=1}^r T^{p_j(n)} f_j = \prod_{j=1}^r \int_X f_j \, d\mu$.*

(In [**FK**], this theorem was stated and proved for the case of polynomials of one variable only, but the same argument works in the case of polynomials of several variables as well.)

Via Theorem 0.6, this result implies the following:

THEOREM 0.12. *Let $T_1, \ldots, T_k$ be totally ergodic measure preserving transformation of X generating a nilpotent group, and let for each $i = 1, \ldots, k$, $p_{i,j}$, $j = 1, \ldots, r_k$, be essentially linearly independent polynomials $\mathbb{Z}^{d_i} \longrightarrow \mathbb{Z}$. Then for any non-negative function $f \in L^\infty(X)$,*
(a)

$$\operatorname*{UC-lim}_{(n_1, \ldots, n_k) \in \mathbb{Z}^{d_1} \times \cdots \times \mathbb{Z}^{d_k}} \int_X \prod_{\substack{1 \leq j_1 \leq r_1 \\ \vdots \\ 1 \leq j_k \leq r_k}} \left(T_1^{p_{1,j_1}(n_1)} \cdots T_k^{p_{k,j_k}(n_k)} f \right) d\mu \geq \left(\int_X f \, d\mu \right)^{r_1 \cdots r_k};$$

(b) *for any $\delta > 0$, the set*

$$\left\{ (n_1, \ldots, n_k) \in \mathbb{Z}^{d_1} \times \cdots \times \mathbb{Z}^{d_k} : \right.$$
$$\left. \int_X \prod_{\substack{1 \leq j_1 \leq r_1 \\ \vdots \\ 1 \leq j_k \leq r_k}} \left(T_1^{p_{1,j_1}(n_1)} \cdots T_k^{p_{k,j_k}(n_k)} f \right) d\mu > \left(\int_X f \, d\mu \right)^{r_1 \cdots r_k} - \delta \right\}$$

is syndetic in $\mathbb{Z}^{d_1 + \cdots + d_k}$.

(**4**) Theorem 0.11, stated for $\mathbb{Z}$-actions, implies a similar result for $\mathbb{R}$-actions (see [**BLMo**]). Moreover, for $\mathbb{R}$-flows the total ergodicity is no longer an issue ($\mathbb{R}$-flows are totally ergodic if ergodic, and the ergodic decomposition allows us to give up the condition of ergodicity as well). We therefore also get the following theorem:

THEOREM 0.13. *Let $(T_1)_{t \in \mathbb{R}}, \ldots, (T_k)_{t \in \mathbb{R}}$ be flows of measure preserving transformations of X generating a nilpotent group, and let, for each $i = 1, \ldots, k$, $p_{i,j}$, $j = 1, \ldots, r_k$, be essentially linearly independent polynomials $\mathbb{R}^{d_i} \longrightarrow \mathbb{R}$. Then for any non-negative function $f \in L^\infty(X)$,*
(a)

$$\operatorname*{UC-lim}_{(t_1, \ldots, t_k) \in \mathbb{R}^{d_1} \times \cdots \times \mathbb{R}^{d_k}} \int_X \prod_{\substack{1 \leq j_1 \leq r_1 \\ \vdots \\ 1 \leq j_k \leq r_k}} \left(T_1^{p_{1,j_1}(t_1)} \cdots T_k^{p_{k,j_k}(t_k)} f \right) d\mu \geq \left(\int_X f \, d\mu \right)^{r_1 \cdots r_k};$$

(b) *for any $\delta > 0$, the set*

$$\Big\{ (t_1, \ldots, t_k) \in \mathbb{R}^{d_1} \times \cdots \times \mathbb{R}^{d_k} :$$

$$\int_X \prod_{1 \leq j_1 \leq r_1} \Big(T_1^{p_{1,j_1}(t_1)} \cdots T_k^{p_{k,j_k}(t_k)} f \Big) d\mu > \Big(\int_X f \, d\mu \Big)^{r_1 \cdots r_k} - \delta \Big\}$$

$$\vdots$$

$$\scriptstyle 1 \leq j_k \leq r_k$$

is syndetic in (the topological) group $\mathbb{R}^{d_1 + \cdots + d_k}$.

The fundamental nature of cubic averages is also manifested by the fact that the multiple recurrence results, such as Theorem 0.8, lead to new sharp combinatorial applications involving large sets in uncountable amenable groups. For example, one has the following result (see Theorem 3.2 in Section 3 for a more general statement):

THEOREM 0.14. *Let $d, k \in \mathbb{N}$ and let m be an invariant mean[5] on the group $\mathbb{R}^d$, considered as a discrete group. Then for any $E \subseteq \mathbb{R}^k$ with $m(1_E) > 0$ and any $\delta > 0$, the set*

$$R_\delta = \Big\{ (u_1, \ldots, u_k) \in (\mathbb{R}^d)^k :$$

$$m \Big(\bigcap_{\epsilon_1, \ldots, \epsilon_k \in \{0,1\}} \big(E - (\epsilon_1 u_1 + \cdots + \epsilon_k u_k) \big) \Big) > m(E)^{2^k} - \delta \Big\}$$

is syndetic in (the discrete) group $(\mathbb{R}^d)^k$ (meaning that finitely many shifts of R_δ cover $(\mathbb{R}^d)^k$).

In order to establish this result one needs a variant of Furstenberg's correspondence principle that holds for general, possibly uncountable, discrete amenable groups. (See Theorem 3.1 in Section 3.)

We conclude this section with an example that provides an answer to a question posed by the referee, which concerns the nilpotency assumption in Theorem 0.5. Let T and S be arbitrary invertible measure preserving transformations of a probability space $(X, \mathcal{B}, \mu)$ and let $A \in \mathcal{B}$, $\mu(A) > 0$. It is not hard to see that UC-liminf$_{m \in \mathbb{Z}}$ UC-lim$_{n \in \mathbb{Z}} \mu(A \cap T^n A \cap S^m A \cap T^n S^m A) \geq \mu(A)^4$, which implies that UC-limsup$_{(n,m) \in \mathbb{Z}^2} \mu(A \cap T^n A \cap S^m A \cap T^n S^m A) \geq \mu(A)^4$. It is therefore natural to inquire whether UC-liminf$_{(n,m) \in \mathbb{Z}^2} \mu(A \cap T^n A \cap S^m A \cap T^n S^m A) > 0$ if T and S do not generate a nilpotent group. The following example shows that the answer to this question is negative even when T and S generate a 2-step solvable group. (See [**BL2**] for a discussion of the similar phenomenon in the case of averages of the form $\frac{1}{N} \sum_{n=1}^{N} \mu(T^n A \cap S^n A)$.)

Let $\mathbb{T}$ be the one-dimensional torus $\mathbb{R}/\mathbb{Z}$ and let $X = \mathbb{T}^{\mathbb{Z}}$ equipped with the normalized Haar measure μ. Let S be "the right coordinate shift" of X, $(Sx)_i = x_{i-1}$, $i \in \mathbb{Z}$, and let T be the "coordinate-wise rotation" of X defined by $(Tx)_i = x_i + 1/i$ for $i > 0$, $(Tx)_i = x_i$ for $i \leq 0$. The transformations T and S generate a 2-step solvable group. Let $A = \{ x \in X : x_0 \in [0, 1/3] \}$. For any $x \in X$ and any $n, m \in \mathbb{N}$ one has $(S^{-m} T^n S^m x)_0 = x_0 + n/m$, which implies that $\mu(S^m A \cap$

---

[5] *A left-invariant mean* is a linear functional m on the space $BC(G)$ of bounded continuous real-valued functions on G such that $m(1) = 1$, $m(f) \geq 0$ if $f \geq 0$, and $m(f(tx)) = m(f(x))$ for all f and all $t \in G$.

$T^n S^m A) = \mu(A \cap S^{-m} T^n S^m A) = 0$ whenever $1/3 \le n/m \le 2/3$. Since "the angle" $\{(n,m) : 0 < m/3 \le n \le 2m/3\}$ in $\mathbb{Z}^2$ contains arbitrarily large squares, we have UC-liminf$_{(n,m) \in \mathbb{Z}^2} \mu(A \cap T^n A \cap S^m A \cap T^n S^m A) = 0$.

The structure of the rest of the paper is as follows. In Section 1 we state and prove the Fubini principle alluded to above. In Section 2 we prove Theorem 0.8 and obtain its corollaries. Section 3 is devoted to establishing a general form of Furstenberg's correspondence principle and utilizing it to derive some combinatorial applications.

1. The Fubini principle – double and repeated Cesàro limits

LEMMA 1.1. *Let G, H be amenable groups and let $(h,g) \mapsto v_{h,g}$, $(h,g) \in H \times G$, be a bounded continuous mapping from $H \times G$ to a Banach space V. Assume that* UC-lim$_{(h,g) \in H \times G} v_{h,g}$ *exists and for every $g \in G$,* UC-lim$_{h \in H} v_{h,g}$ *exists; then*

$$\text{UC-lim}_{g \in G} \text{UC-lim}_{h \in H} v_{h,g} = \text{UC-lim}_{(h,g) \in H \times G} v_{h,g}.$$

PROOF. Let $v = $ UC-lim$_{(h,g) \in H \times G} v_{h,g}$, and for every $g \in G$, let

$$v_g = \text{UC-lim}_{h \in H} v_{h,g}.$$

Let (Φ_α) be a Følner net in G and (Ψ_β) be a Følner net in H. Then $(\Psi_\beta \times \Phi_\alpha)$ is a Følner net in $H \times G$ (where by $(\beta_1, \alpha_1) < (\beta_2, \alpha_2)$ we mean $\beta_1 < \beta_2$ and $\alpha_1 < \alpha_2$). Let $\delta > 0$. Let τ, τ' be left Haar measures on G and H respectively, and let $\tau'' = \tau' \times \tau$. Find β_0, α_0 such that

$$\frac{1}{\tau''(\Psi_\beta \times \Phi_\alpha)} \int_{\Psi_\beta \times \Phi_\alpha} v_{h,g} \, d\tau''(h,g) \overset{\delta}{\approx} v$$

whenever $\beta > \beta_0$ and $\alpha > \alpha_0$. It then follows that for any $\alpha > \alpha_0$,

$$\frac{1}{\tau(\Phi_\alpha)} \int_{\Phi_\alpha} v_g \, d\tau(g) = \frac{1}{\tau(\Phi_\alpha)} \int_{\Phi_\alpha} \lim_\beta \frac{1}{\tau(\Psi_\beta)} \left(\int_{\Psi_\beta} v_{h,g} \, d\tau'(h) \right) d\tau(g)$$

$$= \lim_\beta \frac{1}{\tau''(\Psi_\beta \times \Phi_\alpha)} \int_{\Psi_\beta \times \Phi_\alpha} v_{h,g} \, d\tau''(h,g) \overset{\delta}{\approx} v.$$

$\square$

For σ-compact amenable groups one has a version of Lemma 1.1 that involves Følner sequences instead of Følner nets. For such a group G and a mapping $g \mapsto v_g$, $g \in G$, from G to a Banach space V, we have UC-lim$_{g \in G} v_g = v$ iff $\lim_{N \to \infty} \frac{1}{\tau(\Phi_n)} \int_{\Phi_n} v_g \, d\tau(g) = v$ for every Følner sequence (Φ_n) in G.

LEMMA 1.2. *Let G, H be σ-compact amenable groups and let $v_{h,g}$, $(h,g) \in H \times G$, be a continuous mapping from $H \times G$ to a Banach space V. Assume that* UC-lim$_{(h,g) \in H \times G} v_{h,g}$ *exists and for every $h \in H$,* UC-lim$_{g \in G} v_{h,g}$ *exists; then*

$$\text{UC-lim}_{h \in H} \text{UC-lim}_{g \in G} v_{h,g} = \text{UC-lim}_{(h,g) \in H \times G} v_{h,g}.$$

PROOF. Let $v = $ UC-lim$_{(h,g) \in H \times G} v_{h,g}$, and for every $g \in G$, let

$$v_g = \text{UC-lim}_{h \in H} x_{h,g}.$$

Let (Φ_n) be any Følner sequence in G and (Ψ_m) be a Følner sequence in H. Let τ, τ' be left Haar measures on G and H respectively, and let $\tau'' = \tau' \times \tau$. Choose an increasing sequence of integers (m_n) such that for every n,

$$\left\| \frac{1}{\tau'(\Psi_{m_n})} \int_{\Phi_{m_n}} v_{h,g}\, d\tau'(h) - v_g \right\| < \frac{1}{n}$$

for all $g \in \Phi_n$. Then for any n,

$$\left\| \frac{1}{\tau''(\Psi_{m_n} \times \Phi_n)} \int_{\Psi_{m_n} \times \Phi_n} v_{h,g}\, d\tau''(h,g) - \frac{1}{\tau(\Psi_n)} \int_{\Phi_n} v_g\, d\tau(g) \right\| < \frac{1}{n}.$$

$(\Psi_{m_n} \times \Phi_n)$ is a Følner sequence in $H \times G$, so $\frac{1}{\tau''(\Psi_{m_n} \times \Phi_n)} \int_{\Psi_{m_n} \times \Phi_n} v_{h,g}\, d\tau''(h,g) \longrightarrow v$ as $n \longrightarrow \infty$. Thus, $\frac{1}{\tau(\Phi_n)} \int_{\Phi_n} v_g\, d\tau(g) \longrightarrow v$. $\square$

To deduce, from the largeness of uniform Cesàro limits, the syndeticity of the "sets of large intersection" (part (b) of Theorem 0.6), we will need the following fact:

LEMMA 1.3. *Let G be a locally compact amenable group and let a_g, $g \in G$, be a mapping $G \longrightarrow \mathbb{R}$ with UC-$\lim_{g \in G} a_g = a$. Then for any $\delta > 0$ the set $R = \{g \in G : a_g > a - \delta\}$ is syndetic in G.*

PROOF. Choose a Følner net (Φ_α) in G. If R is not syndetic, then there is a net (g_α) in G such that for any α, $g_\alpha \Phi_\alpha \cap R = \emptyset$. But $(g_\alpha \Phi_\alpha)$ is also a Følner net in G, and for this net we have $\limsup_{\alpha \to \infty} \frac{1}{|g_\alpha \Phi_\alpha|} \sum_{g \in \Phi_\alpha} a_g \le a - \delta$, which contradicts UC-$\lim_{g \in G} a_g = a$. $\square$

2. Large cubic averages: proof of Theorem 0.6

PROOF OF THEOREM 0.6. Part (b) of the theorem follows from part (a) and Lemma 1.3, so we only have to prove (a). Notice that, putting $f_{j_1,\dots,j_{k-1},1} = 1$ and $f_{j_1,\dots,j_{k-1},j_k} = f_{j_1,\dots,j_{k-1}}$ for $j_k \ge 2$ for all $j_1,\dots,j_{k-1}$, we have from (i) that the limit

$$\underset{\substack{(g_1,\dots,g_{k-1}) \in G_1 \times \cdots \times G_{k-1}}}{\text{UC-}\lim} \int_X \prod_{\substack{1 \le j_1 \le r_1 \\ \vdots \\ 1 \le j_{k-1} \le r_{k-1}}} \Big(T_{1,j_1}(g_1) \cdots T_{k-1,j_{k-1}}(g_{k-1}) f_{j_1,\dots,j_{k-1}} \Big) d\mu$$

exists for any collection $f_{j_1,\dots,j_{k-1}}$ of functions from $L^\infty(X)$.

Let $f \in L^\infty(X)$, $f \ge 0$. For any $g_1 \in G_1,\dots,g_k \in G_k$, we can rewrite

$$\int_X \prod_{\substack{1 \le j_1 \le r_1 \\ \vdots \\ 1 \le j_k \le r_k}} \Big(T_{1,j_1}(g_1) \cdots T_{k-1,j_{k-1}}(g_{k-1}) T_{k,j_k}(g_k) f \Big) d\mu$$

$$= \int_X \prod_{\substack{1 \le j_1 \le r_1 \\ \vdots \\ 1 \le j_{k-1} \le r_{k-1}}} \Big(T_{1,j_1}(g_1) \cdots T_{k-1,j_{k-1}}(g_{k-1}) \Big(\prod_{j_k=1}^{r_k} \big(T_{k,j_k}(g_k) f \big) \Big) \Big) d\mu.$$

By assumption (i), the limit

$$\operatorname*{UC-lim}_{(g_1,\ldots,g_k)\in G_1\times\cdots\times G_k}\int_X \prod_{\substack{1\le j_1\le r_1 \\ \vdots \\ 1\le j_k\le r_k}} \Big(T_{1,j_1}(g_1)\cdots T_{k,j_k}(g_k)f\Big)d\mu$$

exists, also for any $g_k\in G_k$ the limit

$$\operatorname*{UC-lim}_{(g_1,\ldots,g_{k-1})\in G_1\times\cdots\times G_{k-1}}$$

$$\int_X \prod_{\substack{1\le j_1\le r_1 \\ \vdots \\ 1\le j_{k-1}\le r_{k-1}}} \Big(T_{1,j_1}(g_1)\cdots T_{k-1,j_{k-1}}(g_{k-1})\Big(\prod_{j_k=1}^{r_k}\big(T_{k,j_k}(g_k)f\big)\Big)\Big)d\mu$$

exists, and, by induction on k, is $\ge \big(\int_X \prod_{j_k=1}^{r_k}(T_{k,j_k}(g_k)f)\,d\mu\big)^{r_1\cdots r_{k-1}}$. Thus Lemma 1.1 applies and, combined with assumption (ii), implies

$$\operatorname*{UC-lim}_{(g_1,\ldots,g_k)\in G_1\times\cdots\times G_k}\int_X \prod_{\substack{1\le j_1\le r_1 \\ \vdots \\ 1\le j_k\le r_k}} \Big(T_{1,j_1}(g_1)\cdots T_{k,j_k}(g_k)f\Big)d\mu$$

$$=\operatorname*{UC-lim}_{(g_1,\ldots,g_{k-1})\in G_1\times\cdots\times G_{k-1}}$$

$$\int_X \prod_{\substack{1\le j_1\le r_1 \\ \vdots \\ 1\le j_{k-1}\le r_{k-1}}} \Big(T_{1,j_1}(g_1)\cdots T_{k-1,j_{k-1}}(g_{k-1})\Big(\prod_{j_k=1}^{r_k}\big(T_{k,j_k}(g_k)f\big)\Big)\Big)d\mu$$

$$\ge \operatorname*{UC-lim}_{g_k\in G_k}\Big(\int_X \prod_{j_k=1}^{r_k}\big(T_{k,j_k}(g_k)f\big)\,d\mu\Big)^{r_1\cdots r_{k-1}}$$

$$\ge \Big(\operatorname*{UC-lim}_{g_k\in G_k}\int_X \prod_{j_k=1}^{r_k}\big(T_{k,j_k}(g_k)f\big)\,d\mu\Big)^{r_1\cdots r_{k-1}} \ge \Big(\int_X f\,d\mu\Big)^{r_1\cdots r_k}.$$

$\square$

3. Furstenberg's correspondence principle for general amenable groups and its applications

When G is a countable amenable group, one can get combinatorial corollaries of Theorem 0.6 by invoking a version of Furstenberg's correspondence principle for amenable groups (see [**B3**], Theorem 6.4.17). For general amenable locally compact groups, a variant of correspondence principle was obtained in [**BF**]. This variant allows us to obtain combinatorial corollaries of multiple recurrence results for *continuous* actions and guarantees existence of combinatorial patterns only in properly dilated large sets in topological amenable groups (see the details in [**BF**], Section 1). Since some of the results obtained in Theorem 0.6 hold true for actions of discrete uncountable groups, it is natural to inquire whether there exists a properly general version of Furstenberg's correspondence principle which would guarantee the existence (and abundance) of "cubic" patterns in any large set of, say, $\mathbb{R}^d$, considered with the discrete topology. The goal of this section is to establish such a general principle and to derive some of its corollaries. To make the discussion precise, one has, of course, to define first what is meant by a "large" set.

If G is a countable amenable group, the standard way of defining the basic notion of largeness for a set $E \subseteq G$ is to declare E to be large if it has positive upper density $\bar{d}_{(\Phi_n)}(E) := \limsup_{n\to\infty} |E \cap \Phi_n|/|\Phi_n|$ with respect to some Følner sequence (Φ_n) in G. It is not hard to show that $\bar{d}_{(\Phi_n)}(E) > 0$ if and only if there exists an invariant mean on the space $B(G)$ of bounded real-valued functions on G such that $m(1_E) > 0$. If G is an uncountable discrete amenable group, one still has at ones disposal both approaches to defining the notion of largeness. However, a word of caution is in order here. Namely, in uncountable discrete groups one no longer has the luxury of working with Følner sequences and has to switch to Følner nets in order to define the notion of upper dentisty and establish its translation invariance (see [**HS**]). After that, in full analogy with the case of countable amenable groups, one can show that a set $E \subseteq G$ has positive upper density if and only if for some invariant mean m on $B(G)$ one has $m(E) > 0$. (When convenient, we will write $m(E)$ for $m(1_E)$.) We prefer to work with the invariant means from the outset. The following version of Furstenberg's correspondence principle has the more familiar correspondence principle for countable amenable groups ([**B3**]) as a special case.

THEOREM 3.1. *Let G be a discrete amenable group. Let m be a left-invariant mean on $B(G)$. Let $E \subseteq G$, $m(E) > 0$. Then there exists a probability measure preserving system $(X, \mathcal{B}, \mu, (T_g)_{g\in G})$, where X is a compact space, $\mathcal{B}$ is the Borel σ-algebra on X, and T is an action of G on X by homeomorphisms, and a set $A \in \mathcal{B}$ with $\mu(A) = m(E)$, such that for any $k \in \mathbb{N}$ and any $g_1, \ldots, g_k \in G$ one has*

$$(3.1) \qquad m\big(E \cap g_1^{-1}E \cap \ldots \cap g_k^{-1}E\big) = \mu\big(A \cap T_{g_1}^{-1}A \cap \ldots \cap T_{g_k}^{-1}A\big).$$

PROOF. The perspicacious reader has already guessed that we will set X to be βG, the Stone-Čech compactification of G. Let μ be the unique probability measure on the Borel σ-algebra of βG corresponding to m. The correspondence is implemented by the formula $m(f) = \int_{\beta G} \hat{f}\, d\mu$, where $f \in B(G)$ and $\hat{f}$ denotes the continuous extension of f to βG. For any set $E \subseteq G$, let $\overline{E}$ be the closure if $E \in \beta G$. Sets of the form $\overline{E}$, $E \subseteq G$, are closed and open and form the basis of open sets in βG. For any $g \in G$, the map $h \mapsto gh$, $h \in G$, has a unique continuous extension to βG, which we will denote by T_g. The maps T_g, $g \in G$, are μ-preserving homeomorphisms of βG. (The μ-invariance follows from the invariance of the mean m.) Now, for any $f_0, f_1, \ldots, f_k \in B(G)$ on has

$$m\left(\prod_{i=0}^k f_i\right) = \int_{\beta G} \prod_{i=0}^k \hat{f}_i\, d\mu.$$

Applying this to $f_0 = 1_E$ and $f_i = 1_{g_i^{-1}E}$, $i = 1, \ldots, k$, we get (3.1). $\square$

The following combinatorial result immediately follows from Theorem 3.1 and Theorem 0.8:

THEOREM 3.2. *Let G be a (discrete) nilpotent group and let m be an invariant mean on $B(G)$. If $E \subseteq G$ satisfies $m(E) > 0$, then, for any $k \in \mathbb{N}$ and $\delta > 0$, the* *set*

$$\left\{(g_1,\ldots,g_k) \in G^k : m\left(\bigcap_{\epsilon_1,\ldots,\epsilon_k \in \{0,1\}} (g_1^{\epsilon_1} \cdots g_k^{\epsilon_k})^{-1} E\right) > m(E)^{2^k} - \delta\right\}$$

is syndetic in G^k (that is, finitely many translates of this set cover G^k).

Theorem 0.10, in its turn, implies the following combinatorial fact:

THEOREM 3.3. *Let $V_1,\ldots,V_k$ be finite dimensional vectors spaces over a (discrete) field F, let $P_1,\ldots,P_k$ be polynomial mappings from V_i to a finite dimensional vector space W over F, let m be an invariant mean on $B(W)$, and let $E \subseteq W$ satisfies $m(E) > 0$. Then for any $\delta > 0$, the set*

$$\left\{(u_1,\ldots,u_k) \in V_1 \oplus \cdots \oplus V_k :\right.$$
$$\left. m\left(\bigcap_{\epsilon_1,\ldots,\epsilon_k \in \{0,1\}} \big(\epsilon_1 P_1(u_1) + \cdots + \epsilon_k P_k(u_k)\big) E\right) > m(E)^{2^k} - \delta\right\}$$

is syndetic in the group $V_1 \oplus \cdots \oplus V_k$.

REMARK. A more familiar form of Furstenberg's correspondence principle deals with large sets in countable groups (see, for example, [**B2**], [**B3**]), and in this case one can guarantee that the resulting measure preserving system $(X,\mathcal{B},\mu,(T_g)_{g\in G})$ is regular, meaning that X is a compact metric space and $\mathcal{B}$ is the Borel σ-algebra on X. The regularity of $(X,\mathcal{B},\mu,(T_g)_{g\in G})$ plays an instrumental role in ergodic proofs of Szemerédi's theorem and its extensions (see [**Fu1**], [**FuKa**], [**BL1**], [**BM**]). To quote from [**Fu2**], p.103: "For certain of the constructions to be carried out it will be necessary to choose between equivalent measure spaces, confining ones attention to *regular* spaces.". The goal of the next proposition is to show that when G is countable, one can replace the measure preserving system $(X,\mathcal{B},\mu,(T_g)_{g\in G})$, appearing in the proof of Theorem 3.1, by a regular one.

PROPOSITION 3.4. *Let $(T_g)_{g\in G}$ be a measure preserving action of a countable group G on a probability space $(X,\mathcal{B},\mu)$ and let $A \in \mathcal{B}$. Then there exists a regular probabilty measure preserving system $(\widetilde{X},\widetilde{\mathcal{B}},\tilde{\mu},(\widetilde{T}_g)_{g\in G})$ and a set $\widetilde{A} \in \mathcal{B}$ such that, for any $k \in \mathbb{N}$ and any $g_1,\ldots,g_k \in G$, one has*

$$\tilde{\mu}\big(\widetilde{A} \cap \widetilde{T}_{g_1}^{-1}\widetilde{A} \cap \cdots \cap \widetilde{T}_{g_k}^{-1}\widetilde{A}\big) = \mu\big(A \cap T_{g_1}^{-1}A \cap \cdots \cap T_{g_k}^{-1}A\big).$$

PROOF. Let $f = 1_A$ and let $\mathcal{A}$ be the closure in $L^\infty(X,\mathcal{B},\mu)$ of the algebra generated by the functions $T_g f$, $g \in G$, and their complex conjugates. $\mathcal{A}$ is a separable commutative C^*-algebra, and so, by Gelfand's theorem, there exists a compact metric space $\widetilde{X}$ such that $\mathcal{A}$ is isomorphic to $C(\widetilde{X})$. Let $\tilde{f} \in C(\widetilde{X})$ be the image of f under this isomorphism. Since $f = 1_A$ is an idempotent, $\tilde{f}$ is also an idempotent, and thus is of the form $\tilde{f} = 1_{\widetilde{A}}$ for some (clopen) set $\widetilde{A} \in \widetilde{\mathcal{B}}$, where $\widetilde{\mathcal{B}}$ is the Borel σ-algebra on $\widetilde{X}$. The measure μ, interpreted as a linear functional on $\mathcal{A}$, gives rise to a positive linear functional on $C(\widetilde{X})$, which, by Riesz's theorem, can be represented by a measure $\tilde{\mu}$ on $\widetilde{\mathcal{B}}$. Clearly, $\tilde{\mu}(\widetilde{A}) = \mu(A)$.

The isometric operators induced on $\mathcal{A}$ by T_g, $g \in G$, give rise to isometries of $C(\widetilde{X})$, which, by a classical theorem of Banach, determine (or, rather, are determined) by homomorphisms $\widetilde{T}_g \colon \widetilde{X} \longrightarrow \widetilde{X}$, which in our case are also $\tilde{\mu}$-preserving.

18 V. BERGELSON AND A. LEIBMAN

Let $g_0 = e$, $g_1, \ldots, g_k \in G$. It is clear that the functions $T_{g_i} f = 1_{T_{g_i}^{-1} A}$ and their products correspond to the functions $1_{\widetilde{T}_{g_i}^{-1} \widetilde{A}}$ and the products thereof; hence,

$$\tilde{\mu}\left(\widetilde{A} \cap \widetilde{T}_{g_1}^{-1} \widetilde{A} \cap \cdots \cap \widetilde{T}_{g_k}^{-1} \widetilde{A}\right) = \int_{\widetilde{X}} \prod_{i=0}^{k} 1_{\widetilde{T}_{g_i}^{-1} \widetilde{A}} d\tilde{\mu} = \int_{X} \prod_{i=0}^{k} 1_{T_{g_i}^{-1} A} d\mu$$

$$= \mu\left(A \cap T_{g_1}^{-1} A \cap \cdots \cap T_{g_k}^{-1} A\right).$$

$\square$

Acknowledgment. We thank Donald Robertson for corrections and helpful comments.

References

[B1] Vitaly Bergelson, *The multifarious Poincaré recurrence theorem*, Descriptive set theory and dynamical systems (Marseille-Luminy, 1996), London Math. Soc. Lecture Note Ser., vol. 277, Cambridge Univ. Press, Cambridge, 2000, pp. 31–57. MR1774423 (2001h:37006)

[B2] Vitaly Bergelson, *Ergodic Ramsey theory*, Logic and combinatorics (Arcata, Calif., 1985), Contemp. Math., vol. 65, Amer. Math. Soc., Providence, RI, 1987, pp. 63–87, DOI 10.1090/conm/065/891243. MR891243 (88h:05003)

[B3] Vitaly Bergelson, *Ergodic theory and Diophantine problems*, Topics in symbolic dynamics and applications (Temuco, 1997), London Math. Soc. Lecture Note Ser., vol. 279, Cambridge Univ. Press, Cambridge, 2000, pp. 167–205. MR1776759 (2001g:37005)

[BF] Vitaly Bergelson and Hillel Furstenberg, *WM groups and Ramsey theory*, Topology Appl. **156** (2009), no. 16, 2572–2580, DOI 10.1016/j.topol.2009.04.007. MR2561208 (2011b:05280)

[BL1] V. Bergelson and A. Leibman, *Polynomial extensions of van der Waerden's and Szemerédi's theorems*, J. Amer. Math. Soc. **9** (1996), no. 3, 725–753, DOI 10.1090/S0894-0347-96-00194-4. MR1325795 (96j:11013)

[BL2] V. Bergelson and A. Leibman, *Failure of the Roth theorem for solvable groups of exponential growth*, Ergodic Theory Dynam. Systems **24** (2004), no. 1, 45–53, DOI 10.1017/S0143385703000427. MR2041260 (2005e:37006)

[BLM] V. Bergelson, A. Leibman, and R. McCutcheon, *Polynomial Szemerédi theorems for countable modules over integral domains and finite fields*, J. Anal. Math. **95** (2005), 243–296, DOI 10.1007/BF02791504. MR2145566 (2006a:11028)

[BLMo] V. Bergelson, A. Leibman, and C. G. Moreira, *From discrete- to continuous-time ergodic theorems*, Ergodic Theory Dynam. Systems **32** (2012), no. 2, 383–426, DOI 10.1017/S0143385711000848. MR2901353

[BM] Vitaly Bergelson and Randall McCutcheon, *An ergodic IP polynomial Szemerédi theorem*, Mem. Amer. Math. Soc. **146** (2000), no. 695, viii+106, DOI 10.1090/memo/0695. MR1692634 (2000m:28018)

[FK] Nikos Frantzikinakis and Bryna Kra, *Polynomial averages converge to the product of integrals*, Israel J. Math. **148** (2005), 267–276, DOI 10.1007/BF02775439. Probability in mathematics. MR2191231 (2006i:37011)

[Fu1] Harry Furstenberg, *Ergodic behavior of diagonal measures and a theorem of Szemerédi on arithmetic progressions*, J. Analyse Math. **31** (1977), 204–256. MR0498471 (58 #16583)

[Fu2] H. Furstenberg, *Recurrence in ergodic theory and combinatorial number theory*, Princeton University Press, Princeton, N.J., 1981. M. B. Porter Lectures. MR603625 (82j:28010)

[FuKa] H. Furstenberg and Y. Katznelson, *An ergodic Szemerédi theorem for commuting transformations*, J. Analyse Math. **34** (1978), 275–291 (1979), DOI 10.1007/BF02790016. MR531279 (82c:28032)

[HS] Neil Hindman and Dona Strauss, *Density in arbitrary semigroups*, Semigroup Forum **73** (2006), no. 2, 273–300, DOI 10.1007/s00233-006-0622-5. MR2280825 (2009b:11019)

[HK1] Bernard Host and Bryna Kra, *Nonconventional ergodic averages and nilmanifolds*, Ann. of Math. (2) **161** (2005), no. 1, 397–488, DOI 10.4007/annals.2005.161.397. MR2150389 (2007b:37004)

[HK2] Bernard Host and Bryna Kra, *Averaging along cubes*, Modern dynamical systems and applications, Cambridge Univ. Press, Cambridge, 2004, pp. 123–144. MR2090768 (2005h:37004)

[Kh] A.Y. Khintchine, Eine Verschärfung des Poincaréschen "Wiederkehrsatzes" *Comp. Math.* **1** (1934), 177–179.

[La] Paul G. Larick, *Results in polynomial recurrence for actions of fields*, ProQuest LLC, Ann Arbor, MI, 1998. Thesis (Ph.D.)–The Ohio State University. MR2697530

[P] Alan L. T. Paterson, *Amenability*, Mathematical Surveys and Monographs, vol. 29, American Mathematical Society, Providence, RI, 1988. MR961261 (90e:43001)

[Po] H. Poincaré, Sur le problème des trois corps et les équations de la Dynamique, *Acta Mathematica* **13** (1890), 1–270.

[W] Miguel N. Walsh, *Norm convergence of nilpotent ergodic averages*, Ann. of Math. (2) **175** (2012), no. 3, 1667–1688, DOI 10.4007/annals.2012.175.3.15. MR2912715

[Z] P. Zorin-Kranich, Norm convergence of multiple ergodic averages on amenable groups, preprint. Available at arXiv:1111.7292.

DEPARTMENT OF MATHEMATICS, THE OHIO STATE UNIVERSITY, OHIO 43210
E-mail address: vitaly@math.ohio-state.edu

DEPARTMENT OF MATHEMATICS, THE OHIO STATE UNIVERSITY, OHIO 43210
E-mail address: leibman@math.ohio-state.edu

Contemporary Mathematics
Volume **631**, 2015
http://dx.doi.org/10.1090/conm/631/12593

Liouville property on G-spaces

C. R. E. Raja

Dedicated to Prof. S. G. Dani on his 65th birthday

ABSTRACT. Let G be a locally compact group and E be a G-space. An irreducible probability measure μ on G is said to have Liouville property on E if G-invariant functions on E are the only continuous bounded functions on E that satisfy the mean value property with respect to μ. We first prove that the random walk induced by μ on E is transient outside a closed set and on the closed set μ has Liouville provided G is metrizable and E is compact and metrizable. We mainly consider actions on vector spaces and projective spaces. We show that measures on $GL(V)$ that are supported inside a ball of radius less than 1 have Liouville property on V. We also prove that measures on $GL(\mathbb{R}^2)$ have Liouville property on the projective line. We next exhibit subgroups of $GL(V)$ so that irreducible measures on such subgroups have Liouville on the projective space $\mathbb{P}(V)$ of V. We also prove that irreducible measures on $SL(V)$ have Liouville property on $\mathbb{P}(\mathcal{SL}(V))$ where $\mathcal{SL}(V)$ is the Lie algebra of $SL(V)$.

1. Introduction

Let G be a locally compact group and μ be a regular Borel probability measure on G. Let G_μ be the closed subgroup generated by the support of μ. We consider a locally compact space E on which the group G acts by homeomorphisms. A bounded measurable function on E that satisfies the mean value property with respect to μ is called a μ-harmonic function and $H_\mu(E)$ is the space of all bounded μ-harmonic functions on E. In case of left or right action of G on itself, the space of harmonic functions was introduced by Furstenberg [5] and studied by others. In the context of left (or right) action of G on itself, an earlier work of Blackwell, Choquet and Deny on abelian groups showed that constant functions are the only continuous bounded μ-harmonic functions on abelian groups G for adapted μ (that is, $G_\mu = G$) - such a result is known as Choquet-Deny theorem or Liouville property (cf. [11] for recent developments in Choquet-Deny results on groups). Here, we consider Liouville property for group actions. In this situation we say that μ has Liouville property on a G-space E if G_μ-invariant functions are the only continuous functions in $H_\mu(E)$.

By considering the adjoint $\check{\mu}$ of μ, it may be seen that $\check{\mu}$ has Liouville on G implies μ has Liouville on any G-space E: recall $\check{\mu}$ is defined by $\int f(g) d\check{\mu}(g) =$

2010 *Mathematics Subject Classification.* Primary 54C40, 14E20; Secondary 46E25, 20C20.
Key words and phrases. Random walks, transience, harmonic functions, Liouville property, projective linear transformations.

$\int f(g^{-1})d\mu(g)$. However we provide examples in Section 7 to show that there are measures on $GL(V)$ that have Liouville property on V but neither the measure nor its adjoint has Liouville property on $GL(V)$.

It may be easily observed that there is a largest S_μ-invariant closed (possibly empty) subset L_μ of E such that any continuous function in $H_\mu(E)$ is S_μ-invariant when restricted to L_μ: S_μ is the closed semigroup generated by the support of μ.

Recently, R. Feres and E. Ronshausen [4] studied Liouville property for group actions and it was shown that if Γ is a countable group acting on the circle S^1 or $[0, 1]$, then any irreducible (that is, $S_\mu = G$) symmetric measure on Γ has Liouville property on S^1 or $[0, 1]$.

It is also shown in [4] that the random walk generated by irreducible μ is transient on $E \setminus L_\mu$ for actions of countable groups on compact spaces using boundaries of countable groups. We extend this result to a much more general class of actions using results about harmonic functions on (metrizable) groups (see Proposition 2.1).

We mainly consider actions on vector spaces and on the corresponding projective spaces. For a finite-dimensional vector space V over $\mathbb{R}$, and a linear transformation α of V, $||\alpha||$ denotes the operator norm of α. We first observe the following result which serves many interesting examples.

THEOREM 1.1. *Let V be a finite-dimensional vector space over $\mathbb{R}$ and G be a subgroup of $GL(V)$ such that $\{g \in G \mid ||g|| < 1\}$ is a non-empty open set. Suppose μ is an adapted probability measure on G such that $\mu(\{g \in G \mid ||g|| \leq a\}) = 1$ for some $0 < a < 1$. Then μ has Liouville property on V.*

For a G-space E, we say that G has Liouville property on E if any irreducible probability measure on G has Liouville on E.

We next consider projective linear actions on projective spaces. For a finite-dimensional vector space V, $\mathbb{P}(V)$ denotes the corresponding projective space. We first prove Liouville property for actions on the projective line $\mathbb{P}^1 = \mathbb{P}(\mathbb{R}^2)$: it may be recalled that S^1 and the projective line $\mathbb{P}^1 = \mathbb{P}(\mathbb{R}^2)$ are homeomorphic and [4] proved Liouville property for (symmetric measures on) countable group actions on S^1.

THEOREM 1.2. *Let G be a closed subgroup of $GL(\mathbb{R}^2)$. Then G has Liouville property on $\mathbb{P}^1$.*

We next look at locally compact subgroups of $GL(V)$.

THEOREM 1.3. *Let G be locally compact subgroup of $GL(V)$. Suppose G has an unipotent subgroup U such that U has only one linearly independent invariant vector in V. Then G has Liouville property on $\mathbb{P}(V)$.*

EXAMPLE 1.1. *The following Lie subgroups of $GL(V)$ have unipotent subgroups that have only one linearly independent invariant vector in V.*

(1) *G is the group of all invertible upper triangular matrices and U may be taken as the group of all upper triangular matrices with ones on the main diagonal.*

(2) $(2n + 1)$-*dimensional Heisenberg group and any of its extensions. For instance,* $SL_2(\mathbb{R}) \ltimes H_1$ *with*

$$H_1 = \left\{ \begin{pmatrix} 1 & a & x \\ 0 & 1 & b \\ 0 & 0 & 1 \end{pmatrix} \mid a, x, b \in \mathbb{R} \right\}$$

as the unipotent group.

As an illustration and application of the method involved in the proof of Theorem 1.3, we consider the conjugate action of $SL(V)$ on its Lie algebra $\mathcal{SL}(V)$. It may be noted that the Lie algebra $\mathcal{SL}(V)$ of $SL(V)$ is the space of all trace zero matrices in $GL(V)$ and the adjoint action is given by Ad $(g)(x) = gxg^{-1}$ for all $g \in SL(V)$ and $x \in \mathcal{SL}(V)$.

THEOREM 1.4. *Let V be a finite-dimensional vector space over $\mathbb{C}$. For the adjoint action of $SL(V)$ on the Lie algebra $\mathcal{SL}(V)$, $SL(V)$ has Liouville property on $\mathbb{P}(\mathcal{SL}(V))$.*

As a byproduct we obtain that actions such as distal, proximal and minimal have Liouville property (see Corollary 2.1). We make a few miscellaneous remarks: (1) on amenability of the group G and Liouville property on G-spaces, and (2) Liouville property for equicontinuous Markov operator associated to the random walk on E generated by μ.

2. Preliminaries

Let G be a locally compact group and μ be a (regular Borel) probability measure on G.

Let G_μ (resp. S_μ) denote the closed subgroup (resp. semigroup) generated by the support of μ. The measure μ is called adapted (resp. irreducible) if $G_\mu = G$ (resp. $S_\mu = G$).

A G-space E is a locally compact space with a continuous G-action.

A bounded (Borel measurable real-valued) function f on E is called μ-harmonic if $Pf(x) := \int f(gx)d\mu(g) = f(x)$ for all $x \in E$.

Let $H_\mu(E)$ denote the space of bounded μ-harmonic functions on E and $C_b(E)$ denote the space of bounded continuous functions on E.

We say that the measure μ has Liouville property (or Choquet-Deny) on E if any continuous bounded μ-harmonic function on E is G_μ-invariant. We say that (the action of) G has Liouville property on E if every irreducible probability measure on G has Liouville property on E.

For right-translation action of G on itself, we simply say that μ has Choquet-Deny if any continuous bounded right μ-harmonic function is constant on the cosets of G_μ. It is easy to verify that if μ has Choquet-Deny, then $\check{\mu}$ has Liouville property on any G-space: $\check{\mu}$ is given by $\check{\mu}(E) = \mu(E^{-1})$ where $E^{-1} = \{x^{-1} \mid x \in E\}$.

2.1. Recurrence and Transience. Let (X_n) (resp. (Z_n)) be the canonical left (resp. right) random walk on G defined by μ. Then for $x \in E$, $(X_n x)$ defines a random walk on E starting at x.

For a given $x \in E$, we consider the following recurrence property R_x and transience property T_x introduced in [8] for $(X_n x)$:

R_x: There exists a compact set $K \subset E$ such that a.s. $X_n x \in K$ infinitely often.

T_x: For any compact set $K \subset E$, a.e. ω, there exists $n(w) \in \mathbb{N}$ with $X_n(\omega)x \notin K$ for $n \geq n(\omega)$.

Several situations for the validity of R_x or T_x are given in [**8**].

For the action of G on E and a probability measure μ on G, we have a largest closed S_μ-invariant subset L of E such that any continuous $f \in H_\mu(E)$, f is S_μ-invariant on L. It is shown in [**4**] that the random walk is transient on $E \setminus L$ using Poisson boundary. Here we first extend this result to general actions using a well-known result about harmonic functions.

PROPOSITION 2.1. *Assume that G is metrizable and E is a compact metrizable G-space. Then there is a S_μ-invariant closed set L such that f is S_μ-invariant on L for any continuous $f \in H_\mu(E)$ and limit points of almost all paths $(X_n x)$ are in L for any $x \in E$. The action of G on $E \setminus L$ satisfies T_x for each $x \in E \setminus L$. In particular, if R_x is valid for all $x \in E$, then any bounded continuous μ-harmonic function on E is G_μ-invariant.*

The following lemma easily follows from a similar well-known result about μ-harmonic functions on groups. We include the proof for clarity and completeness. Recall that μ^k is the k-th convolution power of μ.

LEMMA 2.1. *For any $f \in H_\mu(E)$ and $x \in E$, we have for $k > 0$, μ^k-almost every $g \in G$, $\lim(f(gX_n x) - f(X_n x)) = 0$ a.e. ω.*

PROOF. Let $F(g) = f(g^{-1}x)$. Then F is a (right) $\check{\mu}$-harmonic function on G. Let (X_n) be the left random walk induced by μ. Then $(\tilde{X}_n = X_n^{-1})$ is the right random walk induced by $\check{\mu}$. It can easily be verified that $F(\tilde{X}_n)$ is a martingale, that is, $E(F(\tilde{X}_{n+1})|\mathcal{A}_n) = F(\tilde{X}_n)$ where $\mathcal{A}_n$ is the σ-algebra generated by $F(\tilde{X}_1), \cdots, F(\tilde{X}_n)$. Since $F(\tilde{X}_n)$ is bounded,

$$E((F(\tilde{X}_{n+1}) - F(\tilde{X}_n))^2)$$
$$= E(E((F(\tilde{X}_{n+1}) - F(\tilde{X}_n))^2|\mathcal{A}_n))$$
$$= E(E(F(\tilde{X}_{n+1})^2|\mathcal{A}_n)) - 2E(F(\tilde{X}_n)E(F(\tilde{X}_{n+1})|\mathcal{A}_n)) + E(F(\tilde{X}_n)^2)$$
$$= E(F(\tilde{X}_{n+1})^2) - E(F(\tilde{X}_n)^2)$$

for all n. This implies that $\sum_{i=0}^n E((F(\tilde{X}_{i+1}) - F(\tilde{X}_i))^2) = E(F(\tilde{X}_{n+1})^2) - E(F(\tilde{X}_0)^2)$. Since $E(F(\tilde{X}_n)^2)$ is uniformly bounded, by martingale convergence theorem $F(\tilde{X}_n)$ converges almost surely and in L^2, hence $E(F(\tilde{X}_{n+1})^2)$ converges. This implies that $\sum E((F(\tilde{X}_{n+1}) - F(\tilde{X}_n))^2) < \infty$ and hence $\sum E((F(\tilde{X}_n g^{-1}) - F(\tilde{X}_n))^2) < \infty$ for μ-almost every g. This implies that $\sum (F(\tilde{X}_n g^{-1}) - F(\tilde{X}_n))^2 < \infty$ a.e. ω. Thus, $F(\tilde{X}_n g^{-1}) - F(\tilde{X}_n) \to 0$ a.e. ω. Repeating the steps with μ^k instead of μ, we get the result. $\square$

PROOF OF PROPOSITION 2.1. Let L be the (largest) subset of E consisting of all $x \in E$ such that $f(x) = f(gx)$ for all $g \in S_\mu$ and all continuous $f \in H_\mu(E)$. Then L is a closed S_μ-invariant subset of E.

Suppose X_n is the left random walk on G defined by μ and $f \in H_\mu(E)$. Fix $x \in E$. Regularity of μ implies that G_μ is σ-compact, hence S_μ is σ-compact. Since G is metrizable, S_μ is a σ-compact metrizable space, hence S_μ is second countable. This implies by Lemma 2.1 that there is a countable dense set D_f in S_μ such that

$$(1) \qquad \lim(f(gX_n x) - f(X_n x)) = 0$$

for almost all ω and all $g \in D_f$: note that S_μ is the smallest closed set containing the support of μ^k for all $k > 0$. If y is a limit point of $(X_n x)$ for a path $(X_n x)$ which satisfies (1) for all $g \in D_f$ and $f \in H_\mu(E)$ is continuous, then $f(gy) = f(y)$ for all $g \in S_\mu$. Since E is compact and metrizable, $C_b(E) \cap H_\mu(E)$ is a separable metric space with uniform metric, hence limit points of almost all paths are in L.

Let $x \in E \setminus L$ and K be any compact subset of $E \setminus L$. If $X_n(\omega)x \in K$ infinitely often, then $X_n(\omega)x$ has a limit point $y \in K$. By the first part such ω form a null set. Thus, there is an almost finite random integer N_x such that $X_n(\omega)x \notin K$ for all $n \geq N_x(\omega)$. $\qquad\square$

2.2. A sufficient condition. We now provide a sufficient condition useful to prove the main results.

The following is a well-known fact concerning harmonic functions, which we state without proof (cf. Proposition 45, Chapter I of [**7**]).

LEMMA 2.2. *Let f be a continuous bounded μ-harmonic function on E. Then the sets $\{x \in E \mid f(x) = \inf_{a \in E} f(a)\}$ and $\{x \in E \mid f(x) = \sup_{a \in E} f(a)\}$ are S_μ-invariant closed sets (possibly empty).*

Liouville property for μ rests in showing that the two S_μ-invariant sets in the above Lemma 2.2 intersect (instead of E, one considers $\overline{G_\mu x}$). This motivates us to provide the following sufficient condition for Liouville property. It is a version of Corollary 2.2 of [**4**].

PROPOSITION 2.2. *If f is a continuous μ-harmonic function on E such that f has a minimum and maximum on E and S_μ-invariant sets overlap, then f is constant. In particular, if for any $x \in E$, $\overline{Gx}$ is compact and S_μ-invariant subsets of $\overline{Gx}$ overlap, then G has Liouville property on E.*

REMARK 2.1. *The condition that S_μ-invariant subsets of $\overline{Gx}$ overlap is not a necessary condition. Example 6.1 shows that there are actions having Liouville property but have orbits violating this condition. In section 5, we provide a class of measures/actions for which the above sufficient condition is necessary.*

PROOF. Let f be a continuous μ-harmonic function having maximum and minimum on E. Let $E_s = \{x \in E \mid f(x) = \sup_{y \in E} f(y)\}$ and $E_i = \{x \in E \mid f(x) = \inf_{y \in E} f(y)\}$. Then E_s and E_i are nonempty closed subsets in E. By Lemma 2.2, both E_s and E_i are S_μ-invariant closed subsets of E. If S_μ-invariant sets pairwise overlap, then there is a $x \in E_s \cap E_i$ and hence f is constant on E. Apply the first result to each orbit closure to obtain the second result. $\qquad\square$

As a consequence we get the Liouville property for proximal actions, minimal actions and distal actions on compact spaces. Recall that a semigroup S acting on a compact space E is called proximal (resp. distal) if for any two distinct points $x, y \in E$, the closure of $\{(gx, gy) \mid g \in S\}$ meets (resp. does not meet) the diagonal in $E \times E$.

COROLLARY 2.1. *Let G be a locally compact group acting on a compact space E and μ be a probability measure on G.*

    1 *If the action of S_μ on E is orbitwise proximal, that is, proximal on the closure of any orbit, then μ has Liouville property on E.*

    2 *If the orbit closures $\overline{S_\mu x}$ are minimal, then μ has Liouville property on E.*

3 *If the action of S_μ on E is distal, then μ has Liouville property on E.*

PROOF. We assume that $E = \overline{S_\mu a}$ for some $a \in E$.

If the action is orbitwise proximal, then S_μ is proximal on E. Since E is compact, $\overline{S_\mu x} \cap \overline{S_\mu y} \neq \emptyset$ for any $x, y \in E$. This implies that any two S_μ-invariant sets overlap. Now the result follows from Proposition 2.2.

If the orbit closures are minimal, then the result easily follows from Proposition 2.2.

If the action is distal, then the orbit closures are minimal, hence the result follows from 2.   □

3. Actions on vector spaces

We now look at actions on vector spaces. Given a sequence (μ_n) of probability measures on a locally compact space E, we say that $\mu_n \to \mu$ in the weak* topology to a probability measure μ on E if $\mu_n(f) = \int f d\mu_n \to \int f d\mu = \mu(f)$ for any continuous bounded function f on E. It may be noted that $\mu_n \to \mu$ in the weak* topology if and only if $\mu_n(f) \to \mu(f)$ for any continuous function f with compact support on E (cf. 1.1.9 of [**9**]).

PROOF OF THEOREM 1.1. Since $\mu(\{g \in G \mid ||g|| \leq a\}) = 1$ for some $0 < a < 1$, $\mu^n(\{g \in G \mid ||g|| \leq a^n\}) = 1$ for $n \geq 1$. Let $v \in V$. Then $||gv|| \leq a^n ||v||$ for all g in the support of μ^n. Let $\epsilon > 0$ and ψ be any continuous function with compact support on V. Then using uniform continuity of ψ, we get that $|\psi(gv) - \psi(0)| < \epsilon$ for all g in the support of μ^n, for large n. This implies that $\mu^n * \delta_v(\psi) = \int \psi(gv) d\mu^n(g) \to \psi(0) = \delta_0(\psi)$, hence $\mu^n * \delta_v \to \delta_0$ in the weak* topology.

Let f be a continuous bounded μ-harmonic function on V. Then $f(v) = \mu^n * \delta_v(f) \to f(0)$ for any $v \in V$, hence f is constant. Thus, μ has Liouville property on V.   □

4. Actions on projective spaces

We now consider actions on projective spaces. We first prove a useful result on unipotent actions on projective spaces. A split solvable algebraic group is a solvable algebraic group whose maximal torus is a split torus: e.g. unipotent algebraic groups, $(\mathbb{R}^*)^n \times \mathbb{R}^n$ where $\mathbb{R}^* = \mathbb{R} \setminus \{0\}$, more generally, group of all upper triangular matrices.

LEMMA 4.1. *Let V be a finite-dimensional vector space over $\mathbb{R}$ and G be a closed subgroup of $GL(V)$. If G is amenable, then action of G on any G-minimal subset of $\mathbb{P}(V)$ factors through a compact group. If the algebraic closure of G is a split solvable algebraic group, then any G-minimal subset of $\mathbb{P}(V)$ consists of a G-fixed point.*

PROOF. Let E be a G-minimal subset of $\mathbb{P}(V)$. If G is amenable, then E supports a G-invariant probability measure ρ. Since E is G-minimal, support of ρ is E. Let $I_\rho = \{g \in GL(V) \mid g(x) = x \text{ for all } x \in E\}$ and $\mathcal{I}_\rho = \{g \in GL(V) \mid g(\rho) = \rho\}$. Then I_ρ and $\mathcal{I}_\rho$ are algebraic groups and $\mathcal{I}_\rho / I_\rho$ is a compact algebraic group (cf. [**3**] and [**6**]). Since ρ is G-invariant, $G \subset \mathcal{I}_\rho$. If the algebraic closure of G is a split solvable algebraic group, then $G \subset I_\rho$.   □

We next consider actions on the projective line.

PROOF OF THEOREM 1.2. Let $\phi\colon \mathbb{R}^2 \setminus \{0\} \to \mathbb{P}^1$ be the canonical projection. Suppose $\mathbb{R}^2$ is not G-irreducible. Then there is a nonzero vector $v \in \mathbb{R}^2$ such that for each $g \in G$, $gv = t_g v$ for some $t_g \in \mathbb{R}$. If all $g \in G$ is diagonalizable over $\mathbb{R}$, then G is abelian, hence G has Liouville on $\mathbb{P}^1$. So, assume that there is a $g_0 \in G$ that is not diagonalizable over $\mathbb{R}$. Thus, v is the only eigenvector of $g_0 \in G$. Let G_0 be the group generated by g_0. Then $\phi(v)$ is the only G_0-invariant vector and the algebraic closure of G_0 is a split solvable algebraic group. By Lemma 4.1, any G_0-minimal subset of $\mathbb{P}^1$ consists of a G_0-fixed point. Since $\phi(v)$ is the only G_0-fixed point, we get that any G-invariant closed subset of $\mathbb{P}^1$ contains $\phi(v)$. Now the result follows from Proposition 2.2.

We now assume that $\mathbb{R}^2$ is G-irreducible. Suppose G is not a relatively compact subset of projective linear transformations of $\mathbb{P}^1$, that is, $PGL(\mathbb{R}^2)$. Then there is a sequence (g_n) in G such that (g_n) has no convergent subsequence in $PGL(\mathbb{R}^2)$. Passing to a subsequence, we may assume that $\frac{g_n}{\|g_n\|} \to h$ where h is a linear transformation on $\mathbb{R}^2$. Since $\|h\| = 1$, $h \neq 0$. Since (g_n) has no convergent subsequence in $PGL(\mathbb{R}^2)$, h is a rank-one transformation. Let w be a nonzero vector in the image of h. Then $g_n(\phi(x)) \to \phi(w)$ for all x not in the kernel of h. If x is a nonzero vector in the kernel of h, then since $\mathbb{R}^2$ is G-irreducible, there is a $g \in G$ such that gx is not in the kernel of h. Now $\frac{g_n g}{\|g_n g\|} \to \frac{hg}{\|hg\|}$. Since $h(gx) \neq 0$ we get that $g_n g(\phi(x)) \to \phi(hg(x)) = \phi(w)$. Thus, $\phi(w) \in \overline{G\phi(x)}$ for any non-zero $x \in \mathbb{R}^2$. Now the result follows from Proposition 2.2. $\qquad\square$

We next consider actions of locally compact subgroups of $GL(V)$.

PROOF OF THEOREM 1.3. Let U be an unipotent subgroup of G that has only one linearly independent invariant vector $v \in V$. Suppose $x \in \mathbb{P}(V)$ is U-invariant. Let $\phi\colon V \setminus \{0\} \to \mathbb{P}(V)$ be the canonical projection and $w \in V \setminus \{0\}$ be such that $\phi(w) = x$. Then for each $g \in U$ there is a $t_g \in \mathbb{R}$ such that $g(w) = t_g w$. Since $g \in U$ are unipotent, $g(w) = w$. Since U has only one linearly independent invariant vector $v \in V$, $w = tv$ for some $t \in \mathbb{R}$, that is, $\phi(v) = \phi(w)$.

Let E be a G-invariant closed subset of $\mathbb{P}(V)$. Then E is U-invariant and contains a U-minimal subset M. By Lemma 4.1, M consists of a fixed point of U, hence $M = \{\phi(v)\}$. Thus, any G-invariant closed subset contains $\phi(v)$. This implies by Proposition 2.2, that any continuous bounded μ-harmonic function is constant. $\qquad\square$

We now look at the conjugate action of $SL(V)$.

PROOF OF THEOREM 1.4. Let $e_{1,n}$ be a matrix whose (i,j)-th entry is nonzero only for $i = 1$ and $j = n$ and $\phi\colon \mathcal{SL}(V) \setminus \{0\} \to \mathbb{P}(\mathcal{SL}(V))$ be the canonical projection. We now prove that every $SL(V)$-invariant subset of $\mathbb{P}(\mathcal{SL}(V))$ contains $\phi(e_{1,n})$.

Choose subspaces $\{0\} = W_0 \subset W_1 \subset \cdots \subset W_d = V$ such that $\dim(W_i) = i$. Take

$$U = \cap_{1 \le k < d} \{g \in SL(V) \mid g(W_k) = W_k \text{ and } g(v) - v \in W_k \text{ for all } v \in W_{k+1}\}.$$

Then U is the unipotent algebraic group of all upper triangular unipotent matrices in $SL(V)$ (with respect to a basis from the subspace W_k). It is easy to notice that W_{i+1}/W_i is the subspace of all U-invariant vectors in W_d/W_i for all $0 \le i < d$ and it can also easily be seen that the center of U is $\{\exp(te_{1,n}) \mid t \in \mathbb{R}\}$.

Let $v \in \mathcal{SL}(V)$ be such that v is U-invariant, that is, $gv = vg$ for all $g \in U$. Let $v = v_s + v_n$ be the Jordan-Chevalley decomposition of v into a semisimple element v_s and a nilpotent element v_n such that $v_s v_n = v_n v_s$ and there are polynomials P and Q with $v_s = P(v)$ and $v_n = Q(v)$ (cf. Proposition 4.2 of [10]). Then since $gv = vg$ for all $g \in U$, $gv_s = v_s g$ and $gv_n = v_n g$ for all $g \in U$. Since $v_n u = uv_n$ for all $u \in U$, the subspace of U-invariant vectors in V is also v_n-invariant, hence W_1 is invariant under v_n. Since v_n is nilpotent and $\dim(W_1) = 1$, $v_n(W_1) = \{0\} = W_0$. We now claim by induction that $v_n(W_k) \subset W_{k-1}$ for all $k \geq 1$. Suppose $v_n(W_i) \subset W_{i-1}$ for some $1 \leq i < d$. Since $v_n u = uv_n$ for all $u \in U$, by considering the U-invariant subspace W_{i+1}/W_i of W_d/W_i, we conclude that $v_n(W_{i+1}/W_i) \subset W_{i+1}/W_i$. Since v_n is nilpotent and $\dim(W_{i+1}/W_i) = 1$, we get that $v_n(W_{i+1}) \subset W_i$. This implies that $\exp(v_n) \in U$. Since v_n commutes with every element of U, $\exp(v_n)$ is in the center of U. Thus, $v_n = te_{1,n}$ for some $t \in \mathbb{C}$. Since $uv_s = v_s u$ for all $u \in U$, it can easily be seen that v_s has only one eigenvalue and hence $v_s = tI$ for some $t \in \mathbb{C}$. Since v has trace zero, $v_s = 0$. Thus, $v = v_n = te_{1,n}$. This shows that U has only one linearly independent invariant vector $e_{1,n}$ in $\mathcal{SL}(V)$. Now the result may be proved as in Theorem 1.3. $\square$

5. Miscellaneous remarks

5.1. Amenability and Liouville property. A well-known conjecture due to Furstenberg [5] states that a locally compact σ-compact group G is amenable if and only if G admits adapted probability measures having Liouville property. Several proofs of this conjecture are available (cf. [12], [13] and [15]). In this case one could prove a similar result for Liouville property of group actions using the above conjecture: recall that a probability measure μ on a locally compact group G is called spread-out if a convolution power of μ, say μ^k, is not singular with respect to the Haar measure on G.

PROPOSITION 5.1. *Let G be a locally compact σ-compact group and μ be an adapted spread-out probability measure on G. Then the following are equivalent:*

(1) *$\check{\mu}$ has Choquet-Deny;*
(2) *μ has Liouville property on all G-spaces;*
(3) *μ has Liouville property on all compact G-spaces;*
(4) *μ has Liouville property on all compact affine G-spaces.*

In particular, G is amenable if and only if G has an adapted spread-out probability measure μ such that μ has Liouville property on all compact G-spaces.

PROOF. It is sufficient to prove that $(4) \Rightarrow (1)$. Suppose μ has Liouville property on all compact affine G-spaces. Let $C_b^r(G)$ be the Banach space of all right-uniformly continuous bounded functions on G. Let E be the unit ball in the dual $C_b^r(G)^*$ of $C_b^r(G)$. Then E is compact in the weak* topology. G acts on $C_b^r(G)$ by $gf(x) = f(xg)$ for all $x, g \in G$ and $f \in C_b^r(G)$. The action is continuous and acts by isometries. By duality each $g \in G$ defines a continuous map g^* on E. Now, $g \mapsto g^{-1*}$ defines an action of G on E.

Let f be a continuous bounded $\check{\mu}$-harmonic function on G, that is,

$$\int f(xg^{-1})d\mu(g) = f(x), \ x \in G.$$

As μ is spread-out, $\check{\mu}$ is also spread-out, hence f is right uniformly continuous.

Define the function ϕ on E by $\phi(\sigma) = < \sigma, f >$ for $\sigma \in E$. Then ϕ is a continuous bounded function on E. Now,

$$\int \phi(g^{-1*}\sigma)d\mu(g) = \int < g^{-1*}\sigma, f > d\mu(g) \quad \begin{aligned} &= \int < \sigma, g^{-1}f > d\mu(g) \\ &= < \sigma, \int g^{-1}fd\mu(g) > \\ &= < \sigma, f > = \phi(\sigma). \end{aligned}$$

Thus, ϕ is a μ-harmonic function on E. By assumption ϕ is constant on G-orbits, that is, $< x^{-1*}\sigma, f > = < \sigma, f >$ for all $x \in G$ and $\sigma \in E$. For any $x \in G$, since $\delta_x \in E$ we have $f(x) = \phi(\delta_x) = \phi(x^{-1}\delta_e) = \phi(\delta_e) = f(e)$. Thus, f is constant. $\square$

5.2. Equicontinuous Markov operator. If E is a G-space and μ is a probability measure on G, then we define a Markov operator P on $C_b(E)$ by

$$Pf(x) = \int f(gx)d\mu(g), \ f \in C_b(E), \ x \in E.$$

We say that the Markov operator P defined by μ on E is equicontinuous if to each $f \in C_b(E)$, there exists a subsequence (k_n) of integers and $F \in C_b(E)$ such that $\{\frac{1}{k_n}\sum_{i=1}^{k_n}P^if\} \to F$ pointwise.

REMARK 5.1.

(1) *If the closed semigroup (or equivalently the closed subgroup) generated by the support of μ is compact, then P is equicontinuous on any G-space E which may be seen as follows. Since the closed semigroup generated by the support of μ is compact, $\frac{1}{n}\sum_{k=1}^{n}\mu^k$ converges to a probability measure λ in the weak*-topology, that is, $\frac{1}{n}\sum_{k=1}^{n}\mu^k(f) \to \lambda(f)$ for all continuous bounded functions f on G (cf. [**14**]). Thus, if E is a G-space, then for $x \in E$, $\frac{1}{n}\sum_{k=1}^{n}\mu^k * \delta_x \to \lambda * \delta_x$ in the weak* topology on E, hence $\frac{1}{n}\sum_{k=1}^{n}P^kf(x) \to \lambda * \delta_x(f)$ for all continuous bounded function f on E. This proves that P is equicontinuous on E.*

(2) *If V is a finite-dimensional vector space and μ is a measure on $GL(V)$ such that S_μ is strongly irreducible on V, that is, no finite union of proper subspaces is invariant under S_μ, then by Proposition 3.1 of [**2**] we get that P defined by μ on $\mathbb{P}(V)$ is equicontinuous.*

We now characterize equicontinuous P that has Liouville property.

THEOREM 5.1. *Let E be a compact G-space and μ be an adapted probability measure on G such that the corresponding Markov operator P on E is equicontinuous. Then the following are equivalent:*

(1) *λ has Liouville property on E for all adapted probability measures λ on G;*

(2) *μ has Liouville property on E;*

(3) *for any $x \in E$, $\overline{S_\mu x}$ has a unique minimal subset;*

(4) *for any $x \in E$, S_μ-invariant subsets of $\overline{S_\mu x}$ overlap.*

PROOF. $(1) \Rightarrow (2)$ is evident and $(4) \Rightarrow (1)$ follows from Proposition 2.2. It only remains to show that $(2) \Rightarrow (3) \Rightarrow (4)$.

If there is a $x \in E$ such that $\overline{S_\mu x}$ contains disjoint nonempty closed S_μ-minimal sets E_1 and E_2. Let f be a continuous function on E such that $f(E_1) = \{1\}$ and $f(E_2) = \{2\}$. Since P is equicontinuous, there exists (k_n) such that $\frac{1}{k_n}\sum_{i=1}^{k_n}P^if$ converges to a continuous function F on E. It can easily be verified that $PF = F$

but $F(E_1) = \{1\}$ and $F(E_2) = \{2\}$. Since $E_1, E_2 \subset \overline{S_\mu x}$, F is not S_μ-invariant but μ-harmonic. This proves that $(2) \Rightarrow (3)$.

Let X and Y be closed S_μ-invariant subsets of $\overline{S_\mu x}$ for $x \in E$. Then X and Y contain S_μ-minimal subsets. This proves that $(3) \Rightarrow (4)$.   □

6. Examples

EXAMPLE 6.1. *Consider the following linear action of $\mathbb{R}^2$ on $\mathbb{R}^4$ given by*

$$(t, s) \mapsto \begin{pmatrix} 1 & 0 & 0 & 0 \\ 0 & 1 & 0 & 0 \\ 0 & 0 & e^{t-s} & 0 \\ 0 & 0 & 0 & e^{s-t} \end{pmatrix}$$

for $s, t \in \mathbb{R}$. Then the orbit of $v = (1,1,1,0)$ is $\{(1,1,e^{t-s},0) \mid t, s \in \mathbb{R}\}$. Let $\mathbb{P}^3(\mathbb{R})$ be the corresponding projective space and $\pi \colon \mathbb{R}^4 \setminus \{0\} \to \mathbb{P}^3(\mathbb{R})$ be the canonical projection. We now claim that the closure of the orbit of $\pi(v)$ has two disjoint invariant sets.

Choose sequences (t_n) and (s_n) in $\mathbb{R}$ such that $t_n - s_n \to -\infty$. Then orbit closure of v contains $(1,1,0,0)$.

Choose sequences (t_n) and (s_n) in $\mathbb{R}$ such that $t_n - s_n \to \infty$. Then

$$e^{s_n - t_n}(1, 1, e^{t_n - s_n}, 0) \to (0, 0, 1, 0).$$

This implies that $\pi(0,0,1,0)$ is in the closure of the orbit of $\pi(v)$.

It can easily be seen that $(1,1,0,0)$ is an invariant vector in $\mathbb{R}^4$ and $\pi(0,0,1,0)$ is an invariant point in $\mathbb{P}^3(\mathbb{R})$. Thus, the closure of the orbit of $\pi(v)$ has two disjoint invariant sets $\{\pi(1,1,0,0)\}$ and $\{\pi(0,0,1,0)\}$.

But since $\mathbb{R}^2$ is abelian, $\mathbb{R}^2$ has Liouville property and hence any action of $\mathbb{R}^2$ also has Liouville property.

The following example produces a measure on the $ax + b$-group that does not have Choquet-Deny but has Liouville for its action on $\mathbb{R}^2$.

EXAMPLE 6.2. *Let $G = \{ \begin{pmatrix} t^2 & a \\ 0 & t \end{pmatrix} \mid t > 0, \ a \in \mathbb{R} \}$. Then G is a solvable group and G is the $ax + b$-group. Any measure μ on G supported on $\{ \begin{pmatrix} t^2 & a \\ 0 & t \end{pmatrix} \mid 0 < t < 1/5, \ |a| < 1/5 \}$ satisfies the condition in Theorem 1.1. Hence μ has Liouville on $\mathbb{R}^2$. But μ itself is not Liouville, that is, there are non-constant continuous bounded μ-harmonic functions on G - this could be seen from section 5.1.2 of [1].*

The next example provides measures μ on $GL(V)$ that has Liouville property on V but neither μ nor $\check{\mu}$ has Choquet-Deny.

EXAMPLE 6.3. *Assume that V has dimension at least two. Let $0 < a < 1$ and μ be a probability measure on $GL(V)$ supported on $\{g \in GL(V) \mid \|g\| \leq a\}$. Then by Theorem 1.1, μ has Liouville on V. If G_μ is nonamenable, then neither μ or $\check{\mu}$ can have Liouville property (or Choquet-Deny). It may be noted that G_μ is nonamenable if the support of μ is $\{g \in GL(V) \mid \|g\| \leq a\}$.*

ACKNOWLEDGEMENT . *I thank Prof. P-E. Caprace for noticing a mistake in the previous version of Proposition 5.1 and for suggesting $C_b^r(G)$.*

References

[1] M. Babillot, *An introduction to Poisson boundaries of Lie groups*, Probability measures on groups: recent directions and trends, Tata Inst. Fund. Res., Mumbai, 2006, pp. 1–90. MR2213476 (2007b:60015)

[2] Y. Benoist and J-F. Quint, Random walks on projective spaces, Preprint.

[3] S. G. Dani, *On ergodic quasi-invariant measures of group automorphism*, Israel J. Math. **43** (1982), no. 1, 62–74, DOI 10.1007/BF02761685. MR728879 (85d:22017)

[4] Renato Feres and Emily Ronshausen, *Harmonic functions over group actions*, Geometry, rigidity, and group actions, Chicago Lectures in Math., Univ. Chicago Press, Chicago, IL, 2011, pp. 59–71. MR2807829 (2012e:60011)

[5] Harry Furstenberg, *Boundary theory and stochastic processes on homogeneous spaces*, Harmonic analysis on homogeneous spaces (Proc. Sympos. Pure Math., Vol. XXVI, Williams Coll., Williamstown, Mass., 1972), Amer. Math. Soc., Providence, R.I., 1973, pp. 193–229. MR0352328 (50 #4815)

[6] Harry Furstenberg, *A note on Borel's density theorem*, Proc. Amer. Math. Soc. **55** (1976), no. 1, 209–212. MR0422497 (54 #10484)

[7] Yves Guivarc'h, Michael Keane, and Bernard Roynette, *Marches aléatoires sur les groupes de Lie* (French), Lecture Notes in Mathematics, Vol. 624, Springer-Verlag, Berlin-New York, 1977. MR0517359 (58 #24454)

[8] Y. Guivarc'h and C. R. E. Raja, *Recurrence and ergodicity of random walks on linear groups and on homogeneous spaces*, Ergodic Theory Dynam. Systems **32** (2012), no. 4, 1313–1349, DOI 10.1017/S0143385711000149. MR2955316

[9] Herbert Heyer, *Probability measures on locally compact groups*, Springer-Verlag, Berlin-New York, 1977. Ergebnisse der Mathematik und ihrer Grenzgebiete, Band 94. MR0501241 (58 #18648)

[10] James E. Humphreys, *Introduction to Lie algebras and representation theory*, Springer-Verlag, New York-Berlin, 1972. Graduate Texts in Mathematics, Vol. 9. MR0323842 (48 #2197)

[11] Wojciech Jaworski and C. Robinson Edward Raja, *The Choquet-Deny theorem and distal properties of totally disconnected locally compact groups of polynomial growth*, New York J. Math. **13** (2007), 159–174. MR2336237 (2008h:60020)

[12] V. A. Kaĭmanovich and A. M. Vershik, *Random walks on discrete groups: boundary and entropy*, Ann. Probab. **11** (1983), no. 3, 457–490. MR704539 (85d:60024)

[13] Joseph Rosenblatt, *Ergodic and mixing random walks on locally compact groups*, Math. Ann. **257** (1981), no. 1, 31–42, DOI 10.1007/BF01450653. MR630645 (83f:43002)

[14] M. Rosenblatt, *Limits of convolution sequences of measures on a compact topological semigroup*, J. Math. Mech. **9** (1960), 293–305. MR0118773 (22 #9544)

[15] G. A. Willis, *Probability measures on groups and some related ideals in group algebras*, J. Funct. Anal. **92** (1990), no. 1, 202–263, DOI 10.1016/0022-1236(90)90075-V. MR1064694 (91i:43003)

STATISTICS AND MATHEMATICS UNIT, 8TH MILE MYSORE ROAD, INDIAN STATISTICAL INSTITUTE, BANGALORE 560059

E-mail address: creraja@isibang.ac.in

Contemporary Mathematics
Volume **631**, 2015
http://dx.doi.org/10.1090/conm/631/12594

A new connection between metric theory of Diophantine approximations and distribution of algebraic numbers

V. Bernik and F. Götze

ABSTRACT. Presented in this paper is a more detailed version of the talk given at the international conference "Recent trends in ergodic theory and Dynamical systems" (University of Vadodara, India, December 26–29, 2012).

In 1998 D. Kleinbock and G. Margulis have proved V. Sprindzuk's hypothesis on extremality of almost all points of nondegenerate manifolds. Their proof was based on the results of S. Dani and G. Margulis in the theory of dynamical systems.

In this article, metric theorems on solvability of systems of inequalities involving polynomials and their derivatives are used to obtain results on the distribution of algebraic numbers, as well as discriminants and resultants of integer polynomials.

1. Introduction

1.1. Approximation of real numbers by rational numbers. Diophantine approximation is a field of mathematics where real, complex and p-adic numbers are investigated with respect to their approximations by rational or algebraic numbers. Approximation by rational numbers is a classical branch of Diophantine approximation which includes famous results by Dirichlet [**D**], Liouville [**Li**], Thue [**T**] and Roth [**R**]. As another example, presented below is a fundamental metric result of Khinchine on the approximation of real numbers by rational numbers—a research direction where the first result was obtained by Borel [**Bor**]. Throughout the paper, μA denotes the Lebesgue measure of a measurable set $A \subset \mathbb{R}$, and $\dim B$ denotes Hausdorff dimension of $B \subset \mathbb{R}$. For a given function $\Psi : \mathbb{N} \to \mathbb{R}_{>0}$, we define $\mathcal{L}(\Psi)$ to be the set of $x \in \mathbb{R}$ for which the inequality

$$(1.1) \qquad \left| x - \frac{p}{q} \right| < \frac{\Psi(q)}{q}$$

has infinitely many solutions $(p, q) \in \mathbb{Z} \times \mathbb{N}$. We begin by recalling two classical results in metric theory of Diophantine approximation.

2010 *Mathematics Subject Classification.* Primary 11K60, 11J83.

Key words and phrases. Diophantine approximation, algebraic numbers, dynamical systems, Lebesgue measure.

The first author was supported by CRC 701, University of Bielefeld.

The second author was supported by CRC 701, University of Bielefeld.

THEOREM 1.1 (Khintchine, [**K**]). *Let $q\Psi(q)$ be a monotone function and I be an interval in $\mathbb{R}$. Then*

$$\mu(\mathcal{L}(\Psi) \cap I) = \begin{cases} 0 & \text{if } \sum_{q=1}^{\infty} \Psi(q) < \infty, \\ \mu I & \text{if } \sum_{q=1}^{\infty} \Psi(q) = \infty \end{cases}$$

THEOREM 1.2 (Jarnik-Besicovitch, [**J**, **Bes**]). *Let $v > 1$, and for $q \in \mathbb{N}$ let $\Psi_v(q) = q^{-v}$. Then*

$$\dim \mathcal{L}(\Psi_v) = \frac{2}{v+1}.$$

Although several unsolved problems remain in the theory of approximation by rational numbers [**BDV**, **H**], the authors regard this theory as nearly complete.

1.2. Approximation of real numbers by algebraic numbers. On the other hand, approximation of real and complex numbers by algebraic numbers presents numerous unsolved problems [**Bu**]. The main interest in this type of approximation is related to studies of fundamental number sets, e.g., the set of algebraic numbers or the set of algebraic integers. Let

$$P_n(x) = a_n x^n + \cdots + a_1 x + a_0$$

be an integer polynomial of degree $\deg P_n = n$ if $a_n \neq 0$ and let its height be defined as $H = H(P_n) = \max_{0 \leq j \leq n} |a_j|$. Positive constants depending only on n are denoted by $c(n)$. If necessary, these constants are numbered: $c_j = c_j(n)$, $j = 1, 2, \ldots$. Throughout, $I = [a, b] \subset \mathbb{R}$ is a fixed interval. Using the Dirichlet box principle, it is easy to show that for $\Psi(H) = H^{-w}$, $w \leq n$, the inequality

$$(1.2) \qquad\qquad |P_n(x)| < c(n)\Psi(H)$$

has infinitely many solutions for all $x \in I$. Denote by $\mathcal{L}_n(\Psi)$ the set of $x \in \mathbb{R}$ for which the inequality (1.2) has infinitely many solutions in polynomials $P_n \in \mathbb{Z}[x]$. For $\Psi(H) = H^{-w}$, $w > n$, Sprindzuk [**Sp69**] proved that $\mu\mathcal{L}_n(H^{-w}) = 0$. Moreover, the following analogue of Khinchine's theorem holds:

$$(1.3) \qquad \mu(\mathcal{L}_n(\Psi) \cap I) = \begin{cases} 0 & \text{if } \sum_{H=1}^{\infty} H^{n-1}\Psi(H) < \infty, \\ \mu I & \text{if } \sum_{H=1}^{\infty} H^{n-1}\Psi(H) = \infty. \end{cases}$$

The first inequality in (1.3) was proved in [**B89**] and the second in [**Ber99**]. The proofs in [**Ber99**, **B89**, **Sp69**] are essentially derived from a series of statements on distributions of algebraic numbers, their discriminants, and resultants.

Many years ago, the first author discussed the problem of using metric theorems to study the distribution of algebraic numbers with Sprindzuk, however, the actual development of this idea has started in the beginning of this century in Bielefeld. The main difficulty in proving the second statement (1.3) is to show that the real algebraic numbers form an optimal regular system. The notion of regular systems was introduced by A. Baker and Schmidt in [**BS**]. Regular systems are a powerful tool for the generalization of the Jarnik-Besicovitch theorem to polynomials in (1.2) with $\Psi(H) = H^{-w}$, $w > n$, as well proving analogues of Khinchine's theorem in the case of divergence [**BS**, **Ber99**, **B83**, **BD**, **BBD08**, **DD**].

1.3. Approximation on non-degenerate curves. The result $\mu\mathcal{L}_2(H^{-w}) = 0$, $w > 2$, was obtained by Kubilius [**Ku**]. It was generalized by Schmidt in 1964 [**S**]. Let $f_1, \ldots, f_n$ be C^{n+1} real-valued functions defined on an interval I with a Wronskian

$$W(x) = \begin{vmatrix} f_1'(x) & \ldots & f_n'(x) \\ \ldots & \ldots & \ldots \\ f_1^{(n)}(x) & \ldots & f_n^{(n)}(x) \end{vmatrix} \neq 0$$

for almost all $x \in I$. A curve $G = (f_1, \ldots, f_n)$ satisfying this condition on the Wronskian is called non-degenerate. Denote by $\mathcal{M}_n(\Psi)$ the set of $x \in I$ such that the inequality

$$|F(x)| < \Psi(H)$$

holds for infinitely many functions

$$F(x) = a_n f_n(x) + \ldots + a_1 f_1(x) + a_0, \quad H = \max_{0 \leq j \leq n} |a_j|, \quad a_j \in \mathbb{Z}, \ 0 \leq j \leq n.$$

Schmidt [**S**] proved that $\mu\mathcal{M}_2(\Psi) = 0$ for $\Psi(H) = H^{-v}$ and $v > 2$. Baker [**B**] showed that $\dim \mathcal{M}_2(H^{-v}) = \frac{3}{v+1}$, $v > 2$. The result $\mu\mathcal{M}_3(H^{-v}) = 0$, $v > 3$, was proved in [**BB**]. Sprindzuk [**Sp69, Sp79**] conjectured that $\mu\mathcal{M}_n(H^{-v}) = 0$ for $v > n$. To solve this conjecture, a new method was developed by Kleinbock and Margulis [**KM**]; it is based on the results of Margulis and Dani in ergodic theory and the theory of dynamic systems [**D85, D86, DM, Mar**]. In particular, Kleinbock and Margulis [**KM**] showed that for $v_1 + v_2 > n - 1$ the system of Diophantine inequalities

$$|F(x)| < H^{-v_1}, \quad |F'(x)| < H^{-v_2}$$

has infinitely many solutions only for x lying in a certain set of measure zero. Earlier, results of such type were proved only for v_1 sufficiently larger than v_2. Currently, analogues of Khintchine's theorem for non-degenerate curves and surfaces have been proved in [**Ber02, BBKM, BKM**] and (for the case of simultaneous approximations) in [**Bes, BBD10**].

2. Effective theorems in metric theory of Diophantine approximations of dependent variables

In this section, we consider some results of a different type and discuss methods to obtain them. Let $\delta_0 > 0$ and $c_0 > \delta_0$ be some real numbers and $m \in \mathbb{Z}$, $1 \leq m < n$. For a positive integer Q, define the following sets of polynomials:

$$\mathcal{P}_n(Q) := \{P \in \mathbb{Z}[x], \ \deg P = n, \ H(P) \leq Q\},$$
$$\mathcal{P}_n'(Q) := \{P \in \mathcal{P}_n(Q), \ |a_n| > c_1 H(P)\},$$
$$\mathcal{P}_{\leq n}(Q) := \{P \in \mathbb{Z}[x], \ \deg P \leq n, \ H(P) \leq Q\}.$$

Denote by $\mathcal{B}_1$ the set of $x \in I$ such that for some real numbers $v_0, v_1, \ldots, v_m$, $v_j \geq -1$, $0 \leq j \leq m$, $\sum_{i=0}^m v_i = n - m$, the system of inequalities

(2.1)
$$\begin{cases} \delta_0 Q^{-v_0} & < \ |P(x)| & < \ c_0 Q^{-v_0}, \\ \delta_0 Q^{-v_j} & < \ |P^{(j)}(x)| & < \ c_0 Q^{-v_j}, & 1 \leq j \leq m, \\ \delta_0 Q & < \ |P^{(j)}(x)| & < \ c_0 Q, & m+1 \leq j \leq n \end{cases}$$

has a solution in irreducible polynomials $P \in \mathcal{P}'_n(Q)$. In addition to (2.1), let the sequence $d_1, \ldots, d_{m+1}$, given by

$$(2.2) \qquad d_j = v_{j-1} - v_j, \quad 1 \leq j \leq m, \quad d_{m+1} = v_m + 1,$$

be non-increasing. Furthermore, order the roots of P with respect to their distance to the point x:

$$(2.3) \qquad |x - \alpha_1| \leq |x - \alpha_2| \leq \ldots \leq |x - \alpha_n|.$$

The problem of finding conditions on the parameters $v_0, v_1, \ldots, v_m$ such that the solution set $\mathcal{B}_1$ of the system of inequalities (2.1) would satisfy $\mu\mathcal{B}_1 > d|I|$, $0 < d_0 < d < 1$, has been investigated for a long time for $m = 1$, and the results obtained have had numerous applications, see [**BBG10a, BBG10b, BBG10c, BGK08a, BGK08b**].

In the case of arbitrary order of derivatives m, the system (2.1) was studied in [**Ber12**]. The following result has turned out to be of great importance for certain applications.

THEOREM 2.1. *There exist positive constants δ_0 and c_0 such that the systems (2.1) and (2.2) is satisfied for all $x \in \mathcal{B}_1$ where $\mu\mathcal{B}_1 > 3/4|I|$.*

In this section we prove that Theorem 2.1 can be used to obtain some information on the distribution of the roots $\alpha_1, \alpha_2, \ldots, \alpha_{m+1}$ in a neighborhood of x, where x satisfies (2.1)–(2.3).

THEOREM 2.2. *If at the point x for some polynomial P the systems (2.1) and (2.2) hold then there exist the roots $\alpha_1, \alpha_2, \ldots, \alpha_{m+1}$ of the polynomial P satisfy the following inequalities*

$$(2.4) \qquad \begin{array}{rcl} |x - \alpha_j| & < & c_2 Q^{-v_{j-1} + v_j}, \quad 1 \leq j \leq m, \\ |x - \alpha_{m+1}| & < & c_3 Q^{-v_m - 1}. \end{array}$$

We are going to make repeated use of the following lemma from [**Sp69**].

LEMMA 2.3. *Let α_1 be the root of the polynomial P lying closest to x. Then*

$$|x - \alpha_1| < 2^{n-1} |P(x)| |P'(\alpha_1)|^{-1}.$$

Now we are going to prove Theorem 2.2.

PROOF. We proceed by induction. For $j = 1$ the statement follows from (2.3) and Lemma 2.3. Now let us show that (2.4) holds for $j = 2$. Let $s = 2(n - 1)$. First, if the inequality

$$|x - \alpha_2| < s|x - \alpha_1| < snc_0 \delta_0^{-1} Q^{-v_0 + v_1}$$

holds, then the statement of the theorem follows from the fact that the sequence $d_1, \ldots, d_{m+1}$ is non-increasing. Second, if

$$|x - \alpha_2| \geq s|x - \alpha_1|,$$

then from (2.3) we obtain

$$(2.5) \qquad \frac{1}{2}\mathcal{M}_2 \leq |P'(x)| \leq n\mathcal{M}_2,$$

where $\mathcal{M}_2 = |a_n||x - \alpha_2| \ldots |x - \alpha_n|$. By (2.3) and (2.5),

$$|x - \alpha_2||P''(x)| < \binom{n}{2}\mathcal{M}_2 < 2\binom{n}{2}|P'(x)|.$$

The last inequality implies (2.4) for $j = 2$.

For an arbitrary j, let us again consider two cases. If the inequality

$$|x - \alpha_j| \leq 2\binom{n-1}{j-1}|x - \alpha_{j-1}|$$

holds, then (2.4) is true. In the opposite case,

$$|x - \alpha_j| > 2\binom{n-1}{j-1}|x - \alpha_{j-1}|,$$

we have

(2.6)
$$\frac{1}{2}\mathcal{M}_j < |P^{(j)}(x)| < \binom{n}{j}\mathcal{M}_j,$$

where $\mathcal{M}_j = |a_n||x - \alpha_j|\dots|x - \alpha_n|$. From (2.3) and (2.6) it follows that

$$|x - \alpha_j||P^{(j)}(x)| < \binom{n}{j}\mathcal{M}_j < 2\binom{n}{j}|P^{(j-1)}(x)|,$$

which implies $|x - \alpha_j| < 2c_0\delta_0^{-1}\binom{n}{j}Q^{-v_{j-1}+v_j}$. □

3. Distribution of conjugate algebraic numbers

The question *"How close to each other can two conjugate algebraic numbers of degree n be?"* crops up in a variety of number-theoretic problems, as well as several applied problems. Over the previous 50 years a number of upper and lower bounds have been proved for this minimal distance. However, the exact answers are known only for the degrees 2 and 3.

To formalize the discussion, let us introduce several number sets. Throughout the paper we discuss algebraic numbers in $\mathbb{C}$—the set of complex numbers. Let $n \geq 2$. Recall that complex algebraic numbers are called *conjugate* (over $\mathbb{Q}$) if they are roots of the same irreducible (over $\mathbb{Q}$) polynomial with integer coefficients. Define k_n (respectively k_n^*) to be the infimum of k such that the inequality

$$|\alpha_1 - \alpha_2| > H(\alpha_1)^{-k}$$

holds for arbitrary conjugate algebraic numbers (respectively algebraic integers) $\alpha_1 \neq \alpha_2$ of degree n with a sufficiently large height $H(\alpha_1)$. Here and elsewhere $H(\alpha)$ denotes the height of an algebraic number α, which is the absolute height of the minimal polynomial of α over $\mathbb{Z}$. Clearly, $k_n^* \leq k_n$ for all n.

In 1964 Mahler [M] proved the upper bound $k_n \leq n-1$, which is apparently the best estimate to date. It is an easy exercise to show that $k_2 = 1$ (see, e.g., [BM10]). Furthermore, Evertse [E] and Schönhage [Sc] proved, independently, that $k_3 = 2$. In the case of algebraic integers we have $k_2^* = 0$ and $k_3^* \geq 3/2$. The last result was proved by Bugeaud and Mignotte [BM10], who have also shown that the equality $k_3^* = 3/2$ is equivalent to Hall's conjecture on the difference between integers x^3 and y^2. The latter is known to be a special case of the *abc*-conjecture of Masser and Oesterle, see [BM10] for further details and references.

For $n > 3$, the estimates for k_n are less satisfactory. Bugeaud and Mignotte [BM04, BM10] showed that

$$\begin{array}{llll} k_n \geq n/2 & \text{when} & n \geq 4 & \text{is even,} \\ k_n^* \geq (n-1)/2 & \text{when} & n \geq 4 & \text{is even,} \\ k_n \geq (n+2)/4 & \text{when} & n \geq 5 & \text{is odd,} \\ k_n^* \geq (n+2)/4 & \text{when} & n \geq 5 & \text{is odd.} \end{array}$$

The above results were obtained by presenting explicit families of irreducible polynomials of degree n with their roots lying sufficiently close. Bugeaud and Mignotte [**BM10**] point out that "at present there is no general theory for constructing integer polynomials of degree at least four with two roots close to each other." We make an attempt to address this issue. One particular consequence of our results is the following theorem that improves the lower bounds of Bugeaud and Mignotte in the apparently more difficult case of odd n:

THEOREM 3.1. *For any $n \geq 2$ the lower bound* $\min\{k_n, k_{n+1}^*\} \geq (n+1)/3$ *holds.*

In 2011 Bugeaud and Dujella [**BD11**] proved that

$$k_n \geq \frac{n}{2} + \frac{n-2}{4(n-1)}$$

for any integer $n \geq 4$. Recently, Bugeaud and Dujella [**BD13**] shown that there exist separable integer polynomials P of $\deg P = n$ such that $P(\alpha_i) = P(\alpha_j) = 0$ and

$$|\alpha_i - \alpha_j| < c(n)H(P)^{-\frac{2n-1}{3}}$$

for any integer $n \geq 4$.

Let us show how Theorem 3.1 can be obtained from Theorem 2.2. Moreover, we are going to find a lower bound for the number of polynomials such that their roots satisfy Theorem 3.1. Here, we are only going to prove the result obtained for k_n. The lower bound for k_{n+1}^* was proved in [**BBG10a**].

Set $m = 1$ in Theorem 2.2. Then at any point $x_1 \in \mathcal{B}_1$, with $\mu \mathcal{B}_1 > \frac{3}{4}|I|$, we have

$$(3.1) \qquad \begin{cases} \delta_0 Q^{-v_0} < |P(x_1)| < c_0 Q^{-v_0}, & v_0 + v_1 = n-1, \\ \delta_0 Q^{-v_1} < |P'(x_1)| < c_0 Q^{-v_1}, \\ \delta_0 Q < |P''(x_1)| < c_0 Q. \end{cases}$$

By Theorem 2.2, for $v_1 = \frac{n-2}{3}$ and $v_0 = \frac{2n-1}{3}$, there exist two roots α_1 and α_2 of the polynomial P lying in the neighborhood

$$(3.2) \qquad |x - x_1| < c_4 Q^{-(n+1)/3}$$

of x_1. Moreover, the remaining roots $\alpha_3, \dots, \alpha_n$ of P satisfy the inequality

$$(3.3) \qquad |\alpha_1 - \alpha_j| > c_5, \quad 3 \leq j \leq n,$$

since if $|\alpha_1 - \alpha_3| < c_5' \delta_0$ then the third inequality in (3.1) does not hold. By the Mean Value Theorem, we have

$$(3.4) \qquad P'(x_1) = P'(\alpha_1) + \tfrac{1}{2}P''(\eta_1)(x_1 - \alpha_1),$$

where η_1 is between x_1 and α_1. From Lemma 2.3 we obtain

$$|x_1 - \alpha_1| < nc_0 \delta_0^{-1} Q^{-v_0+v_1} = nc_0 \delta_0^{-1} Q^{-\frac{n+1}{3}}.$$

Thus, we have $|\tfrac{1}{2}P''(\eta_1)(x_1 - \alpha_1)| < c_6 Q^{-\frac{n-2}{3}}$. If we assume that the inequality $|P'(\alpha_1)| > (c_0 + c_6)Q^{-\frac{n-2}{3}}$ holds, then we obtain a contradiction to (3.4). Therefore, we have

$$(3.5) \qquad |P'(\alpha_1)| = |a_n(\alpha_1 - \alpha_2) \dots (\alpha_1 - \alpha_n)| < c_7 Q^{-\frac{n-2}{3}}.$$

Using (3.5) and the fact that $|a_n| > c_1 Q$, we obtain

$$(3.6) \qquad |\alpha_1 - \alpha_2| < c_8 Q^{-\frac{n+1}{3}}.$$

The inequality (3.6) shows that, starting with the point x_1, we have constructed a polynomial $P_1 := P$ with two roots α_1, α_2 lying close to each other. The polynomial P_1 can satisfy (2.1) and (2.4) only for points $x \in \mathcal{B}_1(P_1) \subset I$, such that $\mu \mathcal{B}_1(P_1) < c_9 Q^{-(n+1)/3}$. Choosing another point $x_2 \in \mathcal{B}_2 \subset \mathcal{B}_1 \setminus \mathcal{B}_1(P_1)$, we can use it to construct a polynomial $P_2 \neq P_1$ such that a pair of its roots again satisfies (3.6). From the estimate $\mu \mathcal{B}_1 > \frac{3}{4}|I|$, repeating this procedure allows us to construct at least $c_{10} Q^{(n+1)/3}$ polynomials P.

4. Distribution of resultants of integral polynomials

In this section we discuss the distribution of the resultants $R(P_1, P_2)$ of polynomials P_1 and P_2 from $\mathcal{P}_n(Q)$. It is well known that

$$(4.1) \qquad R(P_1, P_2) = a_n^n(P_1) a_n^n(P_2) \prod_{1 \le i \le n, 1 \le j \le n} (\alpha_i - \beta_j),$$

where α_i, $1 \le i \le n$, and β_j, $1 \le j \le n$, denote respectively the roots of P_1 and P_2; $a_n(P_1)$ and $a_n(P_2)$ are the leading coefficients of the respective polynomials. The resultant $R(P_1, P_2)$ equals zero if and only if the polynomials P_1 and P_2 have a common root. Since the resultant can be represented as the determinant of the Sylvester matrix of the coefficients of P_1 and P_2, it follows that $R(P_1, P_2)$ is integer. Furthermore,

$$(4.2) \qquad |R(P_1, P_2)| < c_{11} Q^{2n}$$

for $P_1, P_2 \in \mathcal{P}_n(Q)$. Let us prove the following theorem on the distribution of resultants.

THEOREM 4.1. *Let $m \in \mathbb{Z}$, $0 \le m < n$. There are at least $c_{12} Q^{\frac{2(n+1)}{(m+1)(m+2)}}$ pairs of different primitive irreducible polynomials (P_1, P_2) from $\mathcal{P}_n(Q)$ such that*

$$(4.3) \qquad 1 \le |R(P_1, P_2)| < c_{13} Q^{\frac{2(n-m-1)}{m+2}}.$$

Note that the first inequality in (4.3) is obvious since P_1 and P_2 are primitive and irreducible, and therefore do not have common roots.

There are a few interesting corollaries. For $m = 0$ we have at least $c_{14} Q^{n+1}$ pairs (P_1, P_2) that satisfy $|R(P_1, P_2)| < c_{15} Q^{n-1}$. For $m = n - 1$ we have at least $c_{16} Q^{2/n}$ pairs (P_1, P_2) that satisfy $|R(P_1, P_2)| < c_{17}$.

Now we are going to prove Theorem 4.1.

PROOF. Let the constants v_j, $0 \le j \le m$, be chosen so that

$$(4.4) \qquad v_0 = (m+1)v_m + m, \quad v_0 = (j+1)v_j - j v_{j+1}, \; 1 \le j \le m-1.$$

By the first equation of (4.4), we have

$$(4.5) \qquad v_m = \frac{v_0 - m}{m+1}.$$

Solving (4.4) in v_j for $j = 1, 2, \ldots, m-1$ yields

$$(4.6) \qquad v_{m-1} = \frac{2v_0 - m + 1}{m+1}, \quad v_k = \frac{(m-k+1)v_0 - k}{m+1}, \; 1 \le k \le m-2.$$

By (4.5) and (4.6), we obtain

$$(4.7) \qquad v_{j-1} - v_j = v_m + 1 = \frac{v_0 + 1}{m + 1}, \quad 0 \le j \le m.$$

In view of the condition

$$v_0 + v_1 + \ldots + v_m = n - m,$$

(4.5) and (4.6) together imply

$$v_0 = \frac{2n - m}{m + 2}.$$

Thus, the roots $\alpha_1, \alpha_2, \ldots, \alpha_m$ of the polynomial P_1 such that $P_1(x_1)$ satisfying (3.1) lie in the interval with the centre at the point x_1:

$$(4.8) \qquad |x - x_1| < c_{18} Q^{-\frac{v_0 + 1}{m + 1}}.$$

For the same reason, the roots $\beta_1, \beta_2, \ldots, \beta_m$ of the polynomial P_2 lie in the same interval.

Now we can estimate the product

$$\begin{aligned} T(m) &= \prod_{1 \le i \le m+1, \, 1 \le j \le m+1} |\alpha_i - \beta_j| < c'_{18} Q^{-\frac{v_0 + 1}{m + 1}(m+1)^2} \\ &= c'_{18} Q^{-\left(\frac{2n - m}{m + 2} + 1\right)(m+1)} = c'_{18} Q^{-\frac{2(n+1)(m+1)}{m + 2}} \end{aligned}$$

from above by the value $c'_{18} Q^{-\frac{2(n+1)(m+1)}{m + 2}}$. As a result, we obtain the following estimate for $R(P_1, P_2)$:

$$(4.9) \qquad |R(P_1, P_2)| < c_{19} Q^{2n - \frac{2(n+1)(m+1)}{m + 2}} = c_{19} Q^{\frac{2(n - m - 1)}{m + 2}}.$$

For an arbitrary point $x_1 \in \mathcal{B}_1$, we have constructed two polynomials with the resultant satisfying (4.8) (see theorem 3.1). The polynomials P_1 and P_2 can form new pairs with other polynomials no more than $(n - 1)^2$ times. The measure of the union G_{J_1} of intervals (4.8) over all such points is less than $c_{20} Q^{-\frac{2(n+1)}{(m+1)(m+2)}}$. Now choose a different point $x_2 \in \mathcal{B}_1 \setminus G_{J_1}$ and construct a new pair of polynomials with the resultant satisfying (4.9).

Repeating this procedure until we can no longer find a suitable point in $\mathcal{B}_1$ yields at least $c_{21} Q^{\frac{2(n+1)}{(m+1)(m+2)}}$ pairs of polynomials satisfying the above conditions. $\square$

5. Distribution of discriminants of integral polynomials

For a natural number Q and a real number v, $0 \le v \le n-1$, define the following sets of polynomials (see section 2):

$$\mathcal{P}_n(Q, v) := \{P \in \mathcal{P}_n(Q), \, 1 \le |D(P)| < Q^{2n-2-2v}\},$$

$$\mathcal{P}_{\le n}(Q, v) := \{P \in \mathcal{P}_{\le n}(Q), \, 1 \le |D(P)| < Q^{2n-2-2v}\},$$

where $D(P) = a_n^{2n-2} \prod_{1 \le i < j \le n} (\alpha_i - \alpha_j)^2$ is the discriminant of the polynomial P, and $\alpha_1, \alpha_2, \ldots, \alpha_n$ are the roots of this polynomial.

We are interested in finding the cardinality of the sets $\mathcal{P}_n(Q, v)$ and $\mathcal{P}_{\le n}(Q, v)$. Currently, this problem is solved for $n = 2, 3$. It was proved in [**Ko**] that

$$c_{22} Q^{4 - \frac{5}{3}v} < \#\mathcal{P}_3(Q, v) < c_{23} Q^{4 - \frac{5}{3}v}$$

for $0 \le v < \frac{3}{5}$. For $0 \le v < \frac{1}{2}$, it is shown in [**BGK08b**] that

$$(5.1) \qquad \#\mathcal{P}_n(Q, v) > c_{24} Q^{2n-2-2v}.$$

Let us demonstrate how the estimate (5.1) can be improved by applying theorem 2.1 and theorem 2.2.

THEOREM 5.1. *The following estimates hold:*

$$(5.2) \qquad \#\mathcal{P}_n(Q,v) > c_{25} Q^{n+1-\frac{n+2}{n}v}, \ 0 \le v \le n-1,$$

$$(5.3) \qquad \#\mathcal{P}_{\le n}(Q,v) > c_{26} \max_{2 \le k \le n} Q^{k+1-\frac{k+2}{k}v}, \ 0 \le v \le k-1.$$

REMARK 5.2. If $v = 0$, then the maximum value in the right-hand side of (5.3) is obtained for $k = n$. If $v = k-1$, then the right-hand side of (5.3) equals $Q^{2/k}$, and the maximum value is obtained for $k = 2$.

Now we are going to prove Theorem 5.1.

PROOF. The proof is based on the application of theorems 2.1 and 2.2. Let $\gamma = \frac{2}{n(n-1)}$, $m = n-1$, $v_j = -1 + (n-j)\gamma$, $1 \le j \le n-1$, $v_0 = n - v$. Then the inequalities (2.2) hold, and the system (2.4) has the form

$$|x - \alpha_1| < c_{27} Q^{-n-1+\frac{n+2}{n}v}, \quad |x - \alpha_j| < c_{28} Q^{-\frac{2v}{n(n-1)}}, \quad 2 \le j \le n.$$

Using the inequality $|\alpha_i - \alpha_j| \le |x - \alpha_i| + |x - \alpha_j|$ and (2.4), we have

$$|\alpha_i - \alpha_j| < c_{29} Q^{-\frac{2v}{n(n-1)}},$$

for all $1 \le i < j \le n$. Therefore, $\prod_{1 \le i < j \le n} (\alpha_i - \alpha_j)^2 < c_{30} Q^{-2v}$, and for an irreducible polynomial P we obtain

$$(5.4) \qquad 1 \le |D(P)| < c_{31} Q^{2n-2-2v}.$$

Thus, at the point $x_1 = x \in \mathcal{B}_1$, $\mu \mathcal{B}_1 > 3/4|I|$, we can construct a polynomial P_1 with the discriminant $D(P_1)$ satisfying (5.4). The polynomial P_1 satisfies (5.4) in the neighborhood of x_1, which is defined by the first inequality in (2.6). Since the number of roots of P_1 does not exceed n, taking a sum over all such intervals of the form $|x - \alpha_1| < c_{32} Q^{-n-1+(n+2)v/n}$ while keeping the polynomial P_1 fixed allows us to estimate the measure of these intervals from above as $c_{33} Q^{-n-1+(n+2)v/n}$. Let t be the number of constructed polynomials P in $x \in \mathcal{B}_1$ that satisfy (5.4). By (2.4), we obtain $t c_{33} Q^{-n-1+(n+2)v/n} > \frac{3}{4}|I|$, which implies

$$(5.5) \qquad t > c_{34} Q^{n+1-\frac{n+2}{n}v}.$$

Thus, the inequality (5.2) is proved.

Now we are going to prove the inequality (5.3). Similarly to (5.5), for every k, $2 \le k \le n$, we obtain

$$\#\mathcal{P}_k(Q,v) > c_{35} Q^{k+1-\frac{k+2}{k}v}, \quad 0 \le v \le k-1.$$

It is easy to see that for every k, $2 \le k \le n$, there exists a nontrivial interval $I \subset [0, k-1]$ such that $f(k) = k + 1 - \frac{k+2}{k}v > f(l)$, $\forall l \in I, l \ne k$. Therefore, the inequality (5.3) holds. $\qquad \square$

6. Algebraic numbers in small intervals

Let us consider the question of whether an algebraic number α of degree $\deg \alpha = n$ and height $H(\alpha) \leq Q$ exists in a small interval $I \in \mathbb{R}$, $|I| = c_{36}Q^{-\mu}$, $0 \leq \mu \leq 1$.

THEOREM 6.1. *For any $n \geq 1$, there exists an interval I_1, $|I_1| = \frac{1}{2}Q^{-1}$, that does not contain an algebraic number α of degree $\deg \alpha = n$ and height $H(\alpha) \leq Q$.*

PROOF. Let an algebraic number α with $\deg \alpha = n$ and $H(\alpha) \leq Q$ be a root of a polynomial $P_n = \sum_{i=0}^{n} a_i x^i \in \mathcal{P}_n(Q)$. If we assume that $a_0 = 0$ then $P(\alpha) = \alpha(a_n \alpha^{n-1} + \ldots + a_1)$ and $a_n \alpha^{n-1} + \ldots + a_1 = 0$, contradicting the fact that $\deg \alpha = n$. For $a_0 \neq 0$ and $|\alpha| < Q^{-1}/2$ we have

$$1 \leq |a_0| = \left|(-a_n \alpha^{n-1} - a_{n-1}\alpha^{n-2} - a_2\alpha - a_1)\alpha\right| < 2Q \cdot Q^{-1}\tfrac{1}{2} = 1,$$

which is contradictory. Therefore, we can take $I_1 = (0, \frac{1}{2}Q^{-1})$.   □

THEOREM 6.2. *There exist positive constants c_{37} and c_{38} such that for any $\epsilon > 0$ every interval I_2, $|I_2| > c_{37}Q^{-1+\epsilon}$, contains at least $c_{38}Q^4|I_2|$ algebraic numbers α with $\deg \alpha \leq 3$ and $H(\alpha) \leq Q$.*

The statement of the theorem is trivial for $\epsilon \geq 1$. For $0 < \epsilon < 1$, a detailed proof of the theorem is given in [**BGK12**], and is based on the following fact.

Let $\mathcal{L}_3 = \mathcal{L}_3(Q, \delta_1)$ denote the set of $x \in I$ such that a system of inequalities

$$(6.1) \qquad \begin{cases} |P(x)| & < \quad Q^{-3}, \\ |P'(x)| & < \quad \delta_1 Q \end{cases}$$

has a solution in polynomials $P \in \mathcal{P}_3(Q)$.

THEOREM 6.3. *There exists a constant $\delta_1 > 0$ such that*

$$(6.2) \qquad \mu\mathcal{L}_3 < \tfrac{1}{4}|I|.$$

From inequality (6.2), for a set $\mathcal{B}_2 = I \setminus \mathcal{L}_3$ we have $\mu\mathcal{B}_2 \geq \frac{3}{4}|I|$. Taking a point $x \subset \mathcal{B}_2$, there exists a polynomial $P(x)$ such that the first inequality in (6.1) holds (this easily follows, e.g., from Minkowski's theorem on linear forms). Then we must have $|P'(x)| \geq \delta_1 Q$. Thus, there is an algebraic number α lying in a small neighborhood of $x \in \mathcal{B}_2$. Such algebraic numbers form a regular system, see [**BS**, **Ber99**].

Acknowledgements. The authors are grateful to the anonymous referee for the very useful comments on this paper.

References

[BS]   A. Baker and Wolfgang M. Schmidt, *Diophantine approximation and Hausdorff dimension*, Proc. London Math. Soc. (3) **21** (1970), 1–11. MR0271033 (42 #5916)

[B]   R. C. Baker, *Sprindzuk's theorem and Hausdorff dimension*, Mathematika **23** (1976), no. 2, 184–197. MR0432558 (55 #5546)

[BB]   V. Beresnevich and V. Bernik, *On a metrical theorem of W. Schmidt*, Acta Arith. **75** (1996), no. 3, 219–233. MR1387861 (97b:11095)

[Ber99] V. Beresnevich, *On approximation of real numbers by real algebraic numbers*, Acta Arith. **90** (1999), no. 2, 97–112. MR1709049 (2000f:11087)

[Ber02] V. Beresnevich, *A Groshev type theorem for convergence on manifolds*, Acta Math. Hungar. **94** (2002), no. 1-2, 99–130, DOI 10.1023/A:1015662722298. MR1905790 (2003d:11109)

[BBKM]   V. V. Beresnevich, V. I. Bernik, D. Y. Kleinbock, and G. A. Margulis, *Metric Diophantine approximation: the Khintchine-Groshev theorem for nondegenerate manifolds*, Mosc. Math. J. **2** (2002), no. 2, 203–225. Dedicated to Yuri I. Manin on the occasion of his 65th birthday. MR1944505 (2004b:11107)

[BDV]   Victor Beresnevich, Detta Dickinson, and Sanju Velani, *Measure theoretic laws for lim sup sets*, Mem. Amer. Math. Soc. **179** (2006), no. 846, x+91, DOI 10.1090/memo/0846. MR2184760 (2007d:11086)

[BBG10a] Victor Beresnevich, Vasili Bernik, and Friedrich Götze, *The distribution of close conjugate algebraic numbers*, Compos. Math. **146** (2010), no. 5, 1165–1179, DOI 10.1112/S0010437X10004860. MR2684299 (2011i:11107)

[BBG10b] V. V. Beresnevich, V. I. Bernik, and F. Götze, *On the distribution of the values of the resultants of integral polynomials* (Russian, with English and Russian summaries), Dokl. Nats. Akad. Nauk Belarusi **54** (2010), no. 5, 21–23, 125. MR2809829

[BBG10c] V. V. Beresnevich, V. I. Bernik, and F. Götze, *Simultaneous approximations of zero by an integral polynomial, its derivative, and small values of discriminants* (Russian, with English and Russian summaries), Dokl. Nats. Akad. Nauk Belarusi **54** (2010), no. 2, 26–28, 125. MR2808979

[Ber12]   Victor Beresnevich, *Rational points near manifolds and metric Diophantine approximation*, Ann. of Math. (2) **175** (2012), no. 1, 187–235, DOI 10.4007/annals.2012.175.1.5. MR2874641

[B83]   V. I. Bernik, *Application of the Hausdorff dimension in the theory of Diophantine approximations* (Russian), Acta Arith. **42** (1983), no. 3, 219–253. MR729734 (85k:11035)

[B89]   V. I. Bernik, *The exact order of approximating zero by values of integral polynomials* (Russian), Acta Arith. **53** (1989), no. 1, 17–28. MR1045454 (91d:11079)

[BD]   V.I. Bernik and M.M. Dodson, *Metric Diophantine approximation on manifolds*, vol. 137, Cambridge Tracts in Mathematics, CUP, 1999.

[BKM]   V. Bernik, D. Kleinbock, and G. A. Margulis, *Khintchine-type theorems on manifolds: the convergence case for standard and multiplicative versions*, Internat. Math. Res. Notices **9** (2001), 453–486, DOI 10.1155/S1073792801000241. MR1829381 (2002g:11102)

[BGK08a] V. Bernik, F. Götze, and O. Kukso, *On the divisibility of the discriminant of an integral polynomial by prime powers*, Lith. Math. J. **48** (2008), no. 4, 380–396, DOI 10.1007/s10986-008-9025-5. MR2470800 (2009i:11093)

[BGK08b] Vasili Bernik, Friedrich Götze, and Olga Kukso, *Lower bounds for the number of integral polynomials with given order of discriminants*, Acta Arith. **133** (2008), no. 4, 375–390, DOI 10.4064/aa133-4-6. MR2457267 (2009h:11119)

[BBD08]   V. Bernik, N. Budarina, and D. Dickinson, *A divergent Khintchine theorem in the real, complex, and p-adic fields*, Lith. Math. J. **48** (2008), no. 2, 158–173, DOI 10.1007/s10986-008-9005-9. MR2425109 (2009g:11083)

[BBD10]   Natalia Budarina, Detta Dickinson, and Vasili Bernik, *Simultaneous Diophantine approximation in the real, complex and p-adic fields*, Math. Proc. Cambridge Philos. Soc. **149** (2010), no. 2, 193–216, DOI 10.1017/S0305004110000162. MR2670212 (2011i:11108)

[BGK12]   Vasilii Bernik, Friedrich Götze, and Olga Kukso, *Regular systems of real algebraic numbers of third degree in small intervals*, Analytic and probabilistic methods in number theory, TEV, Vilnius, 2012, pp. 61–68. MR3025458

[Bes]   A. S. Besicovitch, *Sets of Fractional Dimensions (IV): On Rational Approximation to Real Numbers*, J. London Math. Soc. **S1-9**, no. 2, 126, DOI 10.1112/jlms/s1-9.2.126. MR1574327

[Bor]   E. Borel, *Les probabilités dénombrables et leurs applications arithmétiques*, Rend. Circ. Math. Palermo, **27** (1909), 247–271.

[Bu]   Yann Bugeaud, *Approximation by algebraic integers and Hausdorff dimension*, J. London Math. Soc. (2) **65** (2002), no. 3, 547–559, DOI 10.1112/S0024610702003137. MR1895732 (2003d:11110)

[BM04]   Yann Bugeaud and Maurice Mignotte, *On the distance between roots of integer polynomials*, Proc. Edinb. Math. Soc. (2) **47** (2004), no. 3, 553–556, DOI 10.1017/S0013091503000257. MR2096618 (2005f:11037)

[BM10]   Yann Bugeaud and Maurice Mignotte, *Polynomial root separation*, Int. J. Number
         Theory **6** (2010), no. 3, 587–602, DOI 10.1142/S1793042110003083.   MR2652896
         (2011d:11054)

[BD11]   Yann Bugeaud and Andrej Dujella, *Root separation for irreducible integer polynomi-*
         als, Bull. Lond. Math. Soc. **43** (2011), no. 6, 1239–1244, DOI 10.1112/blms/bdr085.
         MR2861545 (2012k:11033)

[BD13]   Yann Bugeaud and Andrej Dujella, *Root separation for irreducible integer polynomi-*
         als, Bull. Lond. Math. Soc. **43** (2011), no. 6, 1239–1244, DOI 10.1112/blms/bdr085.
         MR2861545 (2012k:11033)

[D85]   S. G. Dani, *Divergent trajectories of flows on homogeneous spaces and Diophantine ap-*
         proximation, J. Reine Angew. Math. **359** (1985), 55–89, DOI 10.1515/crll.1985.359.55.
         MR794799 (87g:58110a)

[D86]   S. G. Dani, *On orbits of unipotent flows on homogeneous spaces. II*, Ergodic Theory
         Dynam. Systems **6** (1986), no. 2, 167–182. MR857195 (88e:58052)

[DM]   S. G. Dani and G. A. Margulis, *Limit distributions of orbits of unipotent flows and*
         values of quadratic forms, I. M. Gel′fand Seminar, Adv. Soviet Math., vol. 16, Amer.
         Math. Soc., Providence, RI, 1993, pp. 91–137. MR1237827 (95b:22024)

[DD]   H. Dickinson and M. M. Dodson, *Extremal manifolds and Hausdorff dimension*, Duke
         Math. J. **101** (2000), no. 2, 271–281, DOI 10.1215/S0012-7094-00-10126-3. MR1738177
         (2001f:11116)

[D]   L.G. P. Dirichlet, *Verallgemeinerung eines Satzes aus der Lehre von den Ket-*
         tenbr euchen nebst einige Anwendungen auf die Theorie der Zahlen, S.-B. Preuss.
         Akad. Wiss., 1842, 93–95.

[E]   Jan-Hendrik Evertse, *Distances between the conjugates of an algebraic number*, Publ.
         Math. Debrecen **65** (2004), no. 3-4, 323–340. MR2107951 (2005h:11148)

[H]   Glyn Harman, *Metric number theory*, London Mathematical Society Monographs.
         New Series, vol. 18, The Clarendon Press, Oxford University Press, New York, 1998.
         MR1672558 (99k:11112)

[J]   V. Jarnik, *Diophantischen Approximationen und Hausdorffsches Mass*, Mat. Sb., **36**
         (1929), 371–382.

[K]   A. Khintchine, *Einige Sätze über Kettenbrüche, mit Anwendungen auf die Theorie der*
         Diophantischen Approximationen (German), Math. Ann. **92** (1924), no. 1-2, 115–125,
         DOI 10.1007/BF01448437. MR1512207

[KM]   D. Y. Kleinbock and G. A. Margulis, *Flows on homogeneous spaces and Diophan-*
         tine approximation on manifolds, Ann. of Math. (2) **148** (1998), no. 1, 339–360, DOI
         10.2307/120997. MR1652916 (99j:11083)

[Ko]   D. V. Koleda, *An upper bound for the number of integral polynomials of third de-*
         gree with a given bound for discriminants (Russian, with English and Russian sum-
         maries), Vestsī Nats. Akad. Navuk Belarusī Ser. Fīz.-Mat. Navuk **3** (2010), 10–16, 124.
         MR2809006 (2012a:11038)

[Ku]   I. Kubilyus, *On the application of I. M. Vinogradov's method to the solution of a*
         problem of the metric theory of numbers (Russian), Doklady Akad. Nauk SSSR (N.S.)
         67 (1949), 783–786. MR0030984 (11,82j)

[Li]   J. Liouville, *Remarques relatives à des classes très étendues de quantitès dont la valeur*
         n'est ni rationnelle, ni même réductible à des irrationnelles algébriques, C. R. Acad.
         Sci. Paris, **18** (1844), 883–885.

[M]   K. Mahler, *An inequality for the discriminant of a polynomial*, Michigan Math. J. **11**
         (1964), 257–262. MR0166188 (29 #3465)

[Mar]   G. A. Margulis, *On the action of unipotent groups in the space of lattices*, Lie groups
         and their representations (Proc. Summer School, Bolyai, János Math. Soc., Budapest,
         1971), Halsted, New York, 1975, pp. 365–370. MR0470140 (57 #9907)

[R]   K. F. Roth, *Rational approximations to algebraic numbers*, Mathematika **2** (1955),
         1–20; corrigendum, 168. MR0072182 (17,242d)

[S]   Wolfgang M. Schmidt, *Metrische Sätze über simultane Approximation abhängiger*
         Grössen (German), Monatsh. Math. **68** (1964), 154–166. MR0171753 (30 #1980)

[Sc]   Arnold Schönhage, *Polynomial root separation examples*, J. Symbolic Comput. **41**
         (2006), no. 10, 1080–1090, DOI 10.1016/j.jsc.2006.06.003. MR2262084 (2007m:12001)

[Sp69]   V. G. Sprindžuk, *Mahler's problem in metric number theory*, Translated from the Russian by B. Volkmann. Translations of Mathematical Monographs, Vol. 25, American Mathematical Society, Providence, R.I., 1969. MR0245527 (39 #6833)

[Sp79]   Vladimir G. Sprindžuk, *Metric theory of Diophantine approximations*, V. H. Winston & Sons, Washington, D.C.; A Halsted Press Book, John Wiley & Sons, New York-Toronto, Ont.-London, 1979. Translated from the Russian and edited by Richard A. Silverman; With a foreword by Donald J. Newman; Scripta Series in Mathematics. MR548467 (80k:10048)

[T]   A. Thue, *Über Annäherungswerte algebraischer Zahlen*, J. reine angew. Math., **135** (1909), 284–305.

INSTITUTE OF MATHEMATICS, ACADEMY OF SCIENCES OF BELARUS, SURGANOVA 11, MINSK 220072, BELARUS
E-mail address: bernik@im.bas-net.by

UNIVERSITY OF BIELEFELD, BIELEFELD 33501, GERMANY
E-mail address: goetze@math.uni-bielefeld.de

Contemporary Mathematics
Volume **631**, 2015
http://dx.doi.org/10.1090/conm/631/12595

The gap distribution of slopes on the golden L

Jayadev S. Athreya, Jon Chaika, and Samuel Lelièvre

ABSTRACT. We give an explicit formula for the limiting gap distribution of slopes of saddle connections on the golden L, or any translation surface in its $SL(2, \mathbb{R})$-orbit, in particular the double pentagon. This is the first explicit computation of the distribution of gaps for a flat surface that is not a torus cover.

1. Introduction

1.1. The golden L. The golden L is a *translation surface* obtained from an L-shaped polygon (with length ratios equal to the golden ratio $\varphi = \frac{1+\sqrt{5}}{2}$) by gluing opposite sides by horizontal and vertical translations (see Figure 1). It has genus two and a single cone-type singularity of angle 6π resulting from the identification of all vertices of the L-shaped polygon to a single point after the side gluings.

In this paper, we describe an explicit computation of the distribution of gaps between slopes of *saddle connections* on the golden L, where a saddle connection is a straight line trajectory starting and ending at the cone point of the golden L. This can be viewed as a geometric generalization of the gap distribution for Farey fractions [1].

1.2. Translation structure. The side identifications of the golden L are by translations, which are holomorphic, so it inherits a Riemann surface structure, as well as a holomorphic one-form (from the form dz in the plane, which is preserved by translations). This one-form ω on this Riemann surface has a single zero, of order two, at the cone-point of the golden L.

We identify the golden L to the one-form ω_{L}, the underlying Riemann surface being implied. The golden L and the closely related double pentagon have been popular objects of study and a testing ground for various properties of translation surfaces [8, 9].

J.S.A. partially supported by NSF grant DMS 1069153, and NSF grants DMS 1107452, 1107263, 1107367 "RNMS: GEometric structures And Representation varieties" (the GEAR Network).

J.C. partially supported by NSF postdoctoral fellowship DMS 1004372.

S.L. partially supported by ANR projet blanc GeoDyM.

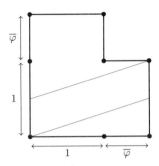

FIGURE 1. Left: The golden L has sides glued pairwise by horizontal and vertical translations. All vertices, marked by dots, become a single point after side gluings. A saddle connection crossing a pair of identified sides is shown.

1.3. Saddle connections and holonomy. Associated to each oriented saddle connection γ is a *holonomy vector*, which we will call a saddle connection vector,

$$\mathbf{v}_\gamma = \int_\gamma \omega \in \mathbb{C},$$

which records how far and in what direction γ travels. The set of saddle connection vectors,

$$\Lambda_\omega = \{\, \mathbf{v}_\gamma : \gamma \text{ an oriented saddle connection on the golden L} \,\} \subset \mathbb{C},$$

is a discrete subset of the plane with *quadratic asymptotics* [**18**]. Veech [**23**] showed that

$$|\Lambda_\omega \cap B(0, R)| \simeq cR^2,$$

where $a(R) \simeq b(R)$ indicates that the ratio of $a(R)$ and $b(R)$ goes to 1 as $R \to \infty$, and where $c = \frac{10}{3\pi^2}(3\varphi + 1)$. Here, $\frac{10}{3\pi}$ is the inverse of the volume $\frac{3\pi}{10}$ of $\mathbb{H}^2/\Gamma$, where $\Gamma = \triangle(2, 5, \infty)$ is the Hecke $(2, 5, \infty)$ triangle group, which is the *Veech group* of the golden L, see §2.3. See Appendix C for details on this computation.

1.4. Slopes and uniform distribution. The object of this paper is to study the distribution of the set of *slopes* of Λ_ω. Since the set Λ_ω is symmetric about the coordinate axes as well as about the first and second diagonals, it is enough to study slopes of vectors in the first quadrant below the first diagonal,

$$\mathbb{S} = \left\{ \text{slope}(z) = \frac{\text{Im}(z)}{\text{Re}(z)} : z \in \Lambda_\omega,\ 0 \leq \text{Im}(z) \leq \text{Re}(z) \right\} \subset [0, 1].$$

We view $\mathbb{S}$ as the union of the nested sets

$$\mathbb{S}_R = \left\{ \text{slope}(z) = \frac{\text{Im}(z)}{\text{Re}(z)} : z \in \Lambda_\omega,\ 0 \leq \text{Im}(z) \leq \text{Re}(z) \leq R \right\}.$$

In [**23**], Veech shows that not only the cardinality $N(R) = |\mathbb{S}_R|$ grows quadratically (as discussed above), but also the sets $\mathbb{S}_R$ become *equidistributed* in $[0, 1]$ with respect to Lebesgue measure. That is, the uniform probability measure on the finite set $\mathbb{S}_R$ weak-∗-converges to the Lebesgue probability measure on $[0, 1]$:

$$\frac{1}{N(R)} \sum_{s \in \mathbb{S}_R} \delta_s \xrightarrow{\text{w*}} \lambda$$

(where δ_s denotes the Dirac mass at s). This result can be interpreted as saying that to the first order, the directions of saddle connections on the golden L appear randomly.

1.5. Gap distributions. A finer assessment of randomness arises from the *gap distribution* of the slopes. For $R \geq 1$ (so $\mathbb{S}_R$ is nonempty), index the elements of $\mathbb{S}_R$ in increasing order:

$$0 = s_R^{(0)} < s_R^{(1)} < s_R^{(2)} < \ldots < s_R^{(N(R)-1)} = 1$$

and consider the set of scaled differences or *gaps* (scaled by R^2 since $N(R)$ grows in R^2):

$$\mathbb{G}_R = \left\{ R^2 \left(s_R^{(i+1)} - s_R^{(i)} \right) : 0 \leq i < N(R) - 1 \right\}.$$

We are interested in the limiting behavior of the probability measure supported on $\mathbb{G}_R$, in particular, for $0 \leq a < b \leq +\infty$, the existence and evaluation of

$$(1.1) \qquad\qquad \lim_{R \to \infty} \frac{|\mathbb{G}_R \cap (a,b)|}{N(R)}.$$

If the slopes were 'truly random', obtained from sampling a sequence of independent identically distributed random variables following the uniform law on $[0,1]$, it is a simple exercise in probability theory to show that associated gap distribution would be *exponential*, that is, the above limit would be $e^{-a} - e^{-b}$.

Our main result is the existence and computation of the gap distribution (1.1).

THEOREM 1.1. *There is a limiting probability distribution function $f : [0, \infty) \to [0, \infty)$, with*

$$\lim_{R \to \infty} \frac{|\mathbb{G}_R \cap (a,b)|}{N(R)} = \int_a^b f(x)dx.$$

This function f is continuous, piecewise real-analytic, with seven points of non-differentiability, and the real-analytic pieces have explicit expressions involving usual functions.

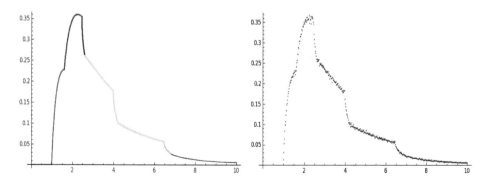

FIGURE 2. The limiting and empirical distributions for gaps of saddle connection slopes on the golden L. The empirical distribution is for saddle connection vectors of slope at most 1 and horizontal component less than 10^4.

REMARK 1. The formulas for the real-analytic pieces of the probability distribution function f are given in Appendix A.

REMARK 2. The distribution has no support at 0, in fact $f(x) = 0$ for $0 \leq x \leq 1$. The tail of the distribution is *quadratic*, that is, for $t \gg 0$,

$$\int_t^\infty f(x)dx \sim t^{-2},$$

where $\sim$ indicates that, for large enough t, the ratio is bounded between two positive constants. In particular, as is clear from Figure 2, the distribution is *not* exponential.

1.6. Typical surfaces. In [2], the first two authors considered gap distributions for *typical* translation surfaces, that is, surfaces in a set of full measure for the Masur-Veech measure (or, in fact, any ergodic $SL(2, \mathbb{R})$-invariant measure) on a connected component of a stratum of the moduli space Ω_g of genus $g \geq 2$ translation surfaces.

It was shown that the limiting distribution exists, and is the same for almost every surface, and, as above, the tail is quadratic. In contrast to our setting, the distribution does have support at 0 (for generic surfaces for the Masur-Veech measure). For *lattice surfaces*, of which the golden L is an example, the distribution does *not* have support at 0, in fact, there are *no* small gaps. This is essentially equivalent to the *no small triangles* condition of Smillie-Weiss [21]. However, the only explicit computations in [2] were for branched covers of tori, and relied on previous work of Marklof-Strömbergsson [16] on the space of affine lattices. To the best of our knowledge, the current paper gives the first explicit computation of a gap distribution of saddle connections for a surface which is not a branched cover of a torus.

1.7. Strategy of proof. We follow the work of [3] where the first author and Y. Cheung, inspired by the work of Boca-Cobeli-Zaharescu [5] (also see [6] for a survey), gave an ergodic-theoretic proof of Hall's Theorem [13] on the gap distribution of Farey fractions, using the horocycle flow on the modular surface. The key idea is to construct a Poincaré section for the horocycle flow on an appropriate moduli space, and to compute the distribution of the return time function with respect to an appropriate measure. This strategy can be used in many different situations, see [1] for a description of some of them.

Acknowledgments We would like to thank C. Uyanik and G. Work for a careful reading of the initial version of this paper, and the anonymous referee for their thoughtful comments. We would also like to thank the Mathematisches Forschungsinstitut Oberwolfach (MFO), the organizers of the workshop "Billiards, Flat Surfaces, and Dynamics on Moduli Spaces", May 2011, as well as the organizers of the conference "Geometrie Ergodique", June 2011. J.C. and S.L. would like to thank the University of Illinois for its hospitality.

2. Strata, $SL(2, \mathbb{R})$-action, and Veech groups

In this section we relax the notations ω, Λ_ω from their use in the previous section.

2.1. Strata of translation surfaces. A (compact, genus g) *translation surface* is given by a holomorphic one-form ω on a compact genus g Riemann surface. Leaving the Riemann surface implicit, we simply refer to ω as the translation surface. More geometrically, a translation surface is given by a union of polygons $P_1 \cup \cdots \cup P_n \subset \mathbb{C}$, and gluings of parallel sides by translations, such that each side is glued to exactly one other, and the total angle at each vertex class is an integer multiple of 2π. Since translations are holomorphic, and preserve dz, we obtain a complex structure and a holomorphic differential on the glued up surface. The zeros of the differential are at the identified vertices with total angle greater than 2π. A zero of order k is a point with total angle $2\pi(k+1)$. The sum of the angle excess is $2\pi(2g-2)$, where g is the genus of the glued up surface. Equivalently the orders of the zeros sum up to $2g - 2$.

Thus, the space of genus g translation surfaces can be stratified by integer partitions of $2g - 2$. If $\underline{k} = (k_1, \ldots, k_s)$ is a partition of $2g - 2$, we denote by $\mathcal{H}(\underline{k})$ the moduli space of translation surfaces ω such that the multiplicities of the zeros are given by $k_1, \ldots, k_s$. The golden L is a genus 2 surface with one zero of order 2 (total angle 6π), that is, $\omega \in \mathcal{H}(2)$.

2.2. Action of $\mathrm{SL}(2, \mathbb{R})$. The group $\mathrm{SL}(2, \mathbb{R})$ acts on each stratum $\mathcal{H}(\underline{k})$ via its action by linear maps on the plane: given a surface ω glued from polygons P_1, $\ldots$, P_n in the plane, define $g \cdot \omega$ as $gP_1 \cup \cdots \cup gP_n$, with the same side gluings for $g \cdot \omega$ as for ω.

2.3. Veech groups and lattice surfaces. For a partition $\underline{k}$ of $2g - 2$, with $g \geq 2$, and a typical surface $\omega \in \mathcal{H}(\underline{k})$, the stabilizer Γ of ω in $\mathrm{SL}(2, \mathbb{R})$, known as the *Veech group* of ω, is trivial. However, for a *dense* subset, the Veech group is a *lattice* in $\mathrm{SL}(2, \mathbb{R})$. These surfaces are known as *lattice surfaces* or *Veech surfaces*. While these lattices are never co-compact, Smillie, in an unpublished work (see Wright-Gekhtman [26] for a nice exposition) showed that the $\mathrm{SL}(2, \mathbb{R})$-orbit of a translation surface ω is a *closed* subset of $\mathcal{H}(\underline{k})$ if and only if ω is a lattice surface, and in this setting it can be identified with the quotient $\mathrm{SL}(2, \mathbb{R})/\Gamma$. Thus, a general principle is the following:

Principle. *If a problem about the geometry of a translation surface can be framed in terms of its $\mathrm{SL}(2, \mathbb{R})$ orbit, and ω is a lattice surface, then the problem can be reduced to studying the dynamics of the $\mathrm{SL}(2, \mathbb{R})$ action on the homogeneous space $\mathrm{SL}(2, \mathbb{R})/\Gamma$.*

2.4. Veech group orbits and saddle connections. Recall that a *saddle connection* γ on a translation surface ω is a geodesic in the flat metric determined by ω on the underlying Riemann surface, with both endpoints at zeros of ω and no zero of ω in its interior. The *holonomy vector* of γ is given by

$$\mathbf{v}_\gamma = \int_\gamma \omega \in \mathbb{C}.$$

The set of holonomy vectors

$$\Lambda_\omega = \{\mathbf{v}_\gamma : \gamma \text{ is a saddle connection on } \omega\} \subset \mathbb{C}$$

is a discrete subset of the plane, which varies equivariantly under the $\mathrm{SL}(2, \mathbb{R})$-action, that is $\Lambda_{g\omega} = g\Lambda_\omega$. When ω is a lattice surface, the set Λ_ω is a finite union of orbits of the Veech group Γ acting linearly on $\mathbb{R}^2 \backslash \{0\}$ (see, for example [14]). That is, there are vectors $\mathbf{v}_1, \ldots, \mathbf{v}_k \in \mathbb{R}^2 \backslash \{0\}$ such that

$$\Lambda_\omega = \bigcup_{i=1}^{k} \Gamma \mathbf{v}_i.$$

In the setting of the golden L, we have two vectors $\mathbf{v}_1 = \frac{\sqrt{5}-1}{2}$ and $\mathbf{v}_2 = 1$, viewed as complex numbers, and $\Lambda_\omega = \Gamma \mathbf{v}_1 \cup \Gamma \mathbf{v}_2$, where $\Gamma = \Gamma(\omega) = \triangle(2, 5, \infty)$. We fix the notation $\Gamma = \triangle(2, 5, \infty)$ in the rest of this paper. Since we are interested in the *slopes* of vectors, and $\mathbf{v}_1$ and $\mathbf{v}_2$ are collinear, we work with the orbit $\Gamma \mathbf{v}_2$, and we write $\Lambda = \Gamma \mathbf{v}_2 \subset \Lambda_\omega$.

3. Horocycle flows and first return maps

In the rest of the paper, ω denotes the golden L, Γ denotes its Veech group, the $(2, 5, \infty)$ triangle group, and Λ denotes the subset $\Gamma \mathbf{v}_2$ of Λ_ω described at the end of section 2.

3.1. Horocycle flow and slope gaps. The key tool in the proof of Theorem 1.1 is the construction of a first return map for the horocycle flow on the $\mathrm{SL}(2, \mathbb{R})$-orbit of ω, which, as discussed above, can be identified with the homogeneous space $\mathrm{SL}(2, \mathbb{R})/\Gamma$. For our purposes, the horocycle flow is given by the (left) action of the subgroup:

$$\left\{ h_s = \begin{bmatrix} 1 & 0 \\ -s & 1 \end{bmatrix} : s \in \mathbb{R} \right\}.$$

We also define, for $a, b \in \mathbb{R}$ with $a \neq 0$, the matrix

$$g_{a,b} = \begin{bmatrix} a & b \\ 0 & a^{-1} \end{bmatrix}.$$

Slope gaps. Note that if $z \in \mathbb{C}$, and $\mathrm{slope}(z) = \mathrm{Im}(z)/\mathrm{Re}(z)$ is the slope, we have

$$\mathrm{slope}(h_s z) = \mathrm{slope}(z) - s,$$

since $\mathrm{Re}(h_s z) = \mathrm{Re}(z)$, $\mathrm{Im}(h_s z) = \mathrm{Im}(z) - s\,\mathrm{Re}(z)$. Thus, the action of h_s preserves *differences* in slopes. Thus to understand the slope gaps of Λ (or more generally of $g\Lambda$ for $g \in \mathrm{SL}(2, \mathbb{R})$), it is useful to consider the orbit of Λ under h_s.

3.2. A Poincaré section. Following [1, 3], we build a cross-section to the flow h_s on the space $X = \mathrm{SL}(2, \mathbb{R})/\Gamma$. Let $\Omega_X \subset X$ be given by

$$(3.1) \qquad \Omega_X = \{ g\Gamma : g\Lambda \cap (0, 1] \neq \emptyset \},$$

where we view $(0, 1]$ as the horizontal segment $\{0 < x \leq 1\} \subset \mathbb{C}$. That is, Ω_X consists of surfaces in the $\mathrm{SL}(2, \mathbb{R})$ orbit of ω which contain a short (length ≤ 1) horizontal vector.

THEOREM 3.1. *The subset $\Omega_X \subset X$ is a Poincaré section to the horocycle flow h_s on X. Moreover, the map $(a, b) \mapsto g_{a,b}\Gamma$ establishes a bijection*

$$\Omega = \{ (a, b) \in \mathbb{R}^2 : 0 < a \leq 1, 1 - a\varphi < b \leq 1 \} \longrightarrow \Omega_X.$$

In these coordinates, the return map $T : \Omega \to \Omega$ is a measure-preserving bijection, piecewise linear with countably many pieces. The return time function $R : \Omega \to \mathbb{R}^+$ defined by

$$R(a, b) = \min\{ s > 0 : h_s g_{a,b}\Gamma \in \Omega_X \}$$

is a piecewise rational function with three pieces, and is uniformly bounded below by 1. We call the map T the golden-L BCZ map.

3.3. Connection to slope gaps. The connection between Theorem 3.1 and slope gap distributions can be seen as follows. For $g \in \mathrm{SL}(2, \mathbb{R})$, consider the set of positive slopes $\mathbb{S}_1^g = \{s_1 < s_2 < \ldots < s_N < \ldots\}$ of elements of $g\Lambda \cap U_1$, where U_1 is the vertical strip

$$U_1 = (0, 1] \times (0, +\infty) = \{z \in \mathbb{C} : \mathrm{Im}(z) > 0, \mathrm{Re}(z) \in (0, 1]\}.$$

These slopes are the positive times s when the orbit $\{h_s g\Gamma\}_{s>0}$ intersects Ω_X. That is,

$$h_{s_i} g\Gamma \in \Omega_X, \quad i = 1, 2, \ldots$$

Let $(a, b) \in \Omega$ be such that $h_{s_1} g\Gamma = g_{a,b}\Gamma$. Then we have, for $i \in \mathbb{N}$, $R(T^i(a, b)) = s_{i+2} - s_{i+1}$. That is, the set of gaps $\mathbb{G}_g^N = \{s_{i+1} - s_i : i = 1, 2, 3, \ldots N\}$ is given by the roof function R evaluated along the orbit of the return map T up till time $N - 1$. For $t > 0$, the proportion of gaps of size at most t can be expressed as a *Birkhoff sum* of the indicator function χ_t of the set $R^{-1}([t, \infty)) \subset \Omega$, via

$$\frac{1}{N} |\mathbb{G}_g^N \cap [t, \infty)| = \frac{1}{N} \sum_{i=0}^{N-1} \chi_t(T^i(a, b)).$$

Thus, the dynamics and ergodic theory of the return map (and in particular the distribution of the roof function along orbits) are crucial to understanding the gap distributions of slopes.

3.4. Proof of Theorem 3.1. Let $a_t = g_{e^{t/2}, 0}$ and $u_s = g_{1,s}$. The action of a_t is known as the *geodesic* flow and the action of u_s is the (opposite) horocycle flow. For $x = g\Gamma \in X$, let $\ell(x) = \min\{|g\mathbf{v}| : \mathbf{v} \in \Lambda\}$ denote the length of the shortest nonzero vector in $g\Lambda$. For any compact subset $K \subset X$, there is an $\epsilon > 0$ such that for any $x \in K$, $\ell(x) > \epsilon$. Given $x = g\Gamma \in X$, the following are equivalent (see, e.g., [**14**]):

- u_s-**periodicity**: there exists $s_0 \neq 0$ so that $u_{s_0} x = x$.
- a_{-t}-**divergence**: as $t \to +\infty$, $a_{-t} x \to \infty$ in X. In particular, $\ell(a_{-t}x) \to 0$.
- x has a **horizontal saddle connection**, that is, there exists $\mathbf{v} \in g\Lambda \cap \mathbb{R}$.
- **upper triangular form**: we can write $x = g_{a,b}\Gamma$, where $a = \min\{|\mathbf{v}| : \mathbf{v} \in g\Lambda \cap \mathbb{R}\}$ is the length of the shortest horizontal vector in $g\Lambda$.

Similarly, h_s-periodicity is equivalent to a_t-divergence and having a vertical saddle connection.

Recall that Ω_X is the set of surfaces in the $\mathrm{SL}(2, \mathbb{R})$-orbit of the golden L with a length ≤ 1 horizontal long saddle connection vector. By the above, these can be expressed as $g_{a,b'}\Gamma \in X$, with $0 < a \leq 1$, and $b' \in \mathbb{R}$. Since the parabolic element $u_{-\varphi}$ is in Γ, we can apply it k times on the right to $g_{a,b'}$ to obtain $g_{a,b'-k\varphi a}$, where $b = b' - k\varphi a \in (1 - \varphi a, 1]$. This condition determines b uniquely (and given any starting b', it determines k). Conversely, any surface of the form $g_{a,b}\omega$ with $(a, b) \in \Omega$ has the saddle connection vector $a \in \mathbb{C}$, and thus clearly has a horizontal vector of length at most 1. Thus, we have shown that Ω_X and Ω are in bijection.

To calculate the return time and the return map, we need to understand the vector in $g_{a,b}\Lambda$ of smallest positive slope in the vertical strip U_1. In Figure 3, we

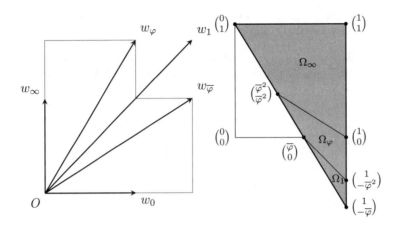

FIGURE 3. Left: Vectors w_j in Λ (the orbit $\Gamma\mathbf{v}_2$) of slopes $j = 0$, $\overline{\varphi}$, 1, φ, ∞.
Right: The partition of Ω into Ω_1, Ω_φ, Ω_∞.

show a selection of vectors in Λ. Particularly relevant to our discussion are the vectors $w_1 = (\varphi, \varphi)$, $w_\varphi = (1, \varphi)$, $w_\infty = (0, 1)$.

Consider the partition of Ω into the three subdomains Ω_1, Ω_φ, Ω_∞ defined by:

$$\Omega_1 = \big\{\, (a,b) \in \Omega : \ \overline{\varphi} \le a \le 1, \ 1 - a\varphi < b \le \overline{\varphi} - a \,\big\}$$

$$\Omega_\varphi = \big\{\, (a,b) \in \Omega : \ \overline{\varphi}^2 \le a \le 1, \ \overline{\varphi} - a < b \le \overline{\varphi} - a\overline{\varphi} \,\big\}$$

$$\Omega_\infty = \big\{\, (a,b) \in \Omega : \ 0 < a \le 1, \ \overline{\varphi} - a\overline{\varphi} < b \le 1 \,\big\}$$

These subdomains are illustrated in Figure 3.

A direct calculation, left as an exercise to the reader, shows that for each j in $\{1, \varphi, \infty\}$, for any (a,b) in Ω_j, the vector $g_{a,b}w_j$ is the one with smallest slope in $g_{a,b}\Lambda \cap U_1$.

In each zone, the return time is given by the slope of the corresponding vectors. Thus,

- if $(a,b) \in \Omega_1$, $g_{a,b}w_1 = \varphi((a+b) + a^{-1}i)$, and $R(a,b) = \frac{a^{-1}}{a+b} = \frac{1}{a(a+b)}$;

- if $(a,b) \in \Omega_\varphi$, $g_{a,b}w_\varphi = (a+b\varphi) + a^{-1}\varphi i$, and $R(a,b) = \frac{a^{-1}\varphi}{a+b\varphi} = \frac{1}{a(a\overline{\varphi}+b)}$;

- if $(a,b) \in \Omega_1$, $g_{a,b}w_\infty = b + a^{-1}i$, and $R(a,b) = \frac{a^{-1}}{b} = \frac{1}{ab}$.

To compute the return map, Note that $h_s g_{a,b} = \begin{bmatrix} a & b \\ -sa & -sb + a^{-1} \end{bmatrix}$ is then

respectively $\begin{bmatrix} a & b \\ \frac{-1}{a+b} & \frac{1}{a+b} \end{bmatrix}$, $\begin{bmatrix} a & b \\ \frac{-1}{a\overline{\varphi}+b} & \frac{\overline{\varphi}}{a\overline{\varphi}+b} \end{bmatrix}$, $\begin{bmatrix} a & b \\ \frac{-1}{b} & 0 \end{bmatrix}$.

Computing the canonical representative $g_{T(a,b)}$ in the class $h_s g_{a,b}\Gamma$ yields:

- in Ω_1: $\begin{bmatrix} a & b \\ \frac{-1}{a+b} & \frac{1}{a+b} \end{bmatrix} \begin{bmatrix} \varphi & 1 \\ \varphi & \varphi \end{bmatrix} u_\varphi^{k_1} = \begin{bmatrix} (a+b)\varphi & a+b\varphi + k_1\varphi^2(a+b) \\ 0 & \frac{\overline{\varphi}}{a+b} \end{bmatrix}$, $k_1 = -\left\lfloor \frac{a+b\varphi-1}{\varphi^2(a+b)} \right\rfloor$,

- in Ω_φ: $\begin{bmatrix} a & b \\ \frac{-1}{a\overline{\varphi}+b} & \frac{\overline{\varphi}}{a\overline{\varphi}+b} \end{bmatrix} \begin{bmatrix} 1 & 0 \\ \varphi & 1 \end{bmatrix} u_\varphi^{k_\varphi hi} = \begin{bmatrix} a+b\varphi & b + k_\varphi\varphi(a+b\varphi) \\ 0 & \frac{\overline{\varphi}}{a\overline{\varphi}+b} \end{bmatrix}$, $k_\varphi = -\left\lfloor \frac{b-1}{\varphi(a+b\varphi)} \right\rfloor$,

- in Ω_∞: $\begin{bmatrix} a & b \\ \frac{-1}{b} & 0 \end{bmatrix} \begin{bmatrix} 0 & -1 \\ 1 & 0 \end{bmatrix} u_\varphi^{k_\infty} = \begin{bmatrix} b & -a + k_\infty \varphi b \\ 0 & \frac{1}{b} \end{bmatrix}$, $k_\infty = -\left\lfloor \frac{-a-1}{\varphi b} \right\rfloor$.

Thus, we have the following formulas for $T(a,b)$:

- in Ω_1: $T(a,b) = \big((a+b)\varphi, a+b\varphi + k_1\varphi^2(a+b)\big)$, $k_1 = -\left\lfloor \frac{a+b\varphi-1}{\varphi^2(a+b)} \right\rfloor$,

- in Ω_φ: $T(a,b) = (a+b\varphi, b+k_\varphi\varphi(a+b\varphi))$, $k_\varphi = -\left\lfloor \frac{b-1}{\varphi(a+b\varphi)} \right\rfloor$,

- in Ω_∞: $T(a,b) = (b, -a+k_\infty\varphi b)$, $k_\infty = -\left\lfloor \frac{-a-1}{\varphi b} \right\rfloor$.

This completes the proof of Theorem 3.1. $\qquad\square$

4. Ergodicity and equidistribution

4.1. Ergodic theory for T. The construction of the map T as a first return map for horocycle flow on $X = \mathrm{SL}(2,\mathbb{R})/\Gamma$ allows us to classify the ergodic invariant measures for T and that long periodic orbits of T equidistribute, as consequences of the corresponding results for h_s acting on $\mathrm{SL}(2,\mathbb{R})/\Gamma$, due to Sarnak [20] and Dani-Smillie [7].

THEOREM 4.1. *The Lebesgue probability measure m given by $dm = \frac{2}{\varphi}dadb$ is the unique ergodic invariant probability measure for T not supported on a periodic orbit. In particular it is the unique ergodic absolutely continuous invariant measure (acim). For every (a,b) not periodic under T and any function $f \in L^1(\Omega, dm)$, we have that*

$$\lim_{N\to\infty} \frac{1}{N} \sum_{i=0}^{N-1} f(T^i(a,b)) = \int_\Omega f\, dm$$

Moreover, if $\{(a_r, b_r)\}_{r=1}^\infty$ is a sequence of periodic points with periods $N(r) \to \infty$ as $r \to \infty$, we have, for any bounded function f on Ω,

$$\lim_{r\to\infty} \frac{1}{N(r)} \sum_{i=0}^{N(r)-1} f(T^i(a_r, b_r)) = \int_\Omega f\, dm.$$

PROOF. By standard theory of first return maps, if (a,b) is not periodic with respect to T, then $g_{a,b}\Gamma$ is not h_s-periodic. By the results of Dani-Smillie [7, Theorem 1], non-periodic orbits of h_s equidistribute (i.e., are Birkhoff regular) with respect to Haar measure μ on $\mathrm{SL}(2,\mathbb{R})/\Gamma$, which can be described as (a multiple of) $dadbds$ when we realize $\mathrm{SL}(2,\mathbb{R})/\Gamma$ as a suspension space over Ω (see also Appendix B for detailed volume computations). Thus, the corresponding non-periodic orbit of T on Ω must equidistribute with respect to Lebesgue measure on Ω. Similarly, by Sarnak [20, Theorem 1], long periodic horocycles on $\mathrm{SL}(2,\mathbb{R})/\Gamma$ equidistribute, and thus, long periodic orbits of T on Ω must equidistribute. Since the roof function R is bounded below by 1, the length of the discrete period $N(r) \to \infty$ implies that the length of the continuous period of $g_{a_r,b_r}\Gamma$ must also go to infinity (in fact, it is at least $N(r)$). $\qquad\square$

4.2. Proof of main theorems. Before proving Theorem 1.1, we state and prove a more general result:

THEOREM 4.2. *Let $x = g\Gamma \in X$ be such that x is not h_s-periodic. Let*

$$\mathbb{S}_g = \{0 \le s_1 < s_2 < \ldots < s_N < \ldots\}$$

be the set of slopes of elements of $g\Lambda$ in the vertical strip U_1. Let

$$\mathbb{G}_x^N = \{s_{i+1} - s_i : i = 1, 2, \ldots, N\}$$

denote the associated gap set. Then for any $t \geq 0$,

$$\lim_{N \to \infty} \frac{1}{N} \left| \mathbb{G}_x^N \cap [t, \infty) \right| = m(\{\, (a, b) \in \Omega : R(a, b) \geq t \,\}).$$

If $x = g\Gamma$ is h_s-periodic, define $x_r = a_{-r}g\Gamma = a_{-r}x$. Then there is a $P(r)$ so that for any $N \geq P(r)$, $\mathbb{G}_{x_r}^N = \mathbb{G}_{x_r}^{P(r)}$. We then have

$$\lim_{r \to \infty} \frac{1}{P(r)} \left| \mathbb{G}_{x_r}^{P(r)} \cap [t, \infty) \right| = m(\{\, (a, b) \in \Omega : R(a, b) \geq t \,\}).$$

PROOF. As observed in §3.3, the proportion $\frac{1}{N}|\mathbb{G}_x^N \cap [t, \infty)|$ of the first N slope gaps of size at most t in the strip U_1 is a Birkhoff sum of the indicator function of the super-level set $\{(a, b) \in \Omega : R(a, b) \geq t\}$. The first statement then follows from the first statement of Theorem 4.1. For the second statement, note that if x is periodic under h_s with period s_0, $a_{-r}x$ is periodic with period $e^r s_0$, by the conjugation relation $a_r h_s a_{-r} = h_{se^{-r}}$. $P(r)$ is the corresponding period for the map T. The second statement of the theorem then follows from the second statement of Theorem 4.1. □

4.3. Proof of Theorem 1.1.

LEMMA 4.3. *Let x be the golden L and $x_R = a_{-2 \log R} x$. We have*

$$\frac{|\mathbb{G}_R(\Lambda) \cap [t, \infty)|}{N(R)} = \frac{1}{N(R)} |\mathbb{G}_{x_R}^{N(R)} \cap [t, \infty)|.$$

PROOF. The golden L is h_s-periodic, with period φ, since the matrix $\begin{bmatrix} 1 & 0 \\ -\varphi & 1 \end{bmatrix} \in$ Γ. We are interested in the slopes of saddle connections with horizontal component at most R, and slope at most 1. The map T does not see any of the saddle connections slopes for vectors of length more than 1. However, renormalizing by the matrix $a_{-2 \log R} = \begin{bmatrix} 1/R & 0 \\ 0 & R \end{bmatrix}$, we can consider the point $x_R = a_{-2 \log R}\Gamma \in X$. Note that this matrix scales slopes and thus differences of slopes by R^2. Each $\gamma \in \mathbb{S}_R$ corresponds with a saddle connection $a_{-2 \log R}\gamma$ on x_R which is in Ω. The point x_R has period $N(R)$ under T. □

PROOF OF THEOREM 1.1. By Theorem 4.2 we have

$$\frac{1}{N(R)} \left| \mathbb{G}_{x_R}^{N(R)} \cap (t, \infty) \right| \to m(\{(a, b) \in \Omega : R(a, b) \geq t\}).$$

So by the previous lemma

$$\frac{1}{N(R)} \left| \mathbb{G}_R(\Lambda) \cap [t, \infty) \right| \to m(\{(a, b) \in \Omega : R(a, b) \geq t\}).$$

□

4.4. Spacings and statistics. The equidistribution of periodic points also yields significant further information on higher-order spacings and statistics for the gap distribution. We record one representative result on h-spacings, and refer the reader to [**3**, §1.5] for further results of this type in the setting of the torus, whose proofs can be easily modified to this setting.

THEOREM 4.4. *Let h be a positive integer, and let $t_1, t_2, \ldots, t_h > 0$. Then the h-spacing distribution*

$$\frac{1}{N(R)}\Big|\Big\{ 1 \le i \le N(R) : R^2(s_R^{(i+k)} - s_R^{(i+k-1)}) > t_k, \ 1 \le k \le h \Big\}\Big|$$

converges, as $R \to \infty$, to

$$m(\{(a,b) \in \Omega : R(T^j(a,b)) \ge t_j, 0 \le j < h\})$$

PROOF. Apply the periodic case of Theorem 4.2 to the indicator function of the set

$$\{(a,b) \in \Omega : R(T^j(a,b)) \ge t_j, 0 \le j < h\}.$$

$\square$

5. Further questions

Our method to explicitly compute the gap distribution in this setting leads us to several natural questions.

5.1. Real analyticity.

QUESTION. *Is the gap distribution for a generic (with respect to Masur-Veech measure on a stratum $\mathcal{H}(\underline{k})$ surface real analytic?*

This distribution was shown to exist in [**2**], and in [**1**], a possible method of computing it (and, in fact, the generic distribution for any h_s-invariant measure) by constructing a first return map for the action of h_s on the stratum $\mathcal{H}(\underline{k})$ was suggested. Piecewise real analyticity of the generic distribution for any h_s-invariant measure follows from the proof in [**1**], however, explicitly computing this return map (or indeed the return time, which is the only requirement for the gap distribution) seems difficult. Our computation here gives an explicit proof that the distribution for the golden L is piecewise real-analytic, on eight pieces, and in [**22**], the case of the regular octagon is also explicitly computed. For the torus, the distribution was computed by [**13**], and was real-analytic on 3 pieces. This leads to:

QUESTION. *How does the number of pieces in the gap distribution for lattice surfaces vary? Is there a bound in any fixed stratum?*

The number of pieces is some measure of the 'complexity' of the Veech surface.

5.2. Support at 0. In [**2**] it was shown that lattice surfaces have no small (normalized) gaps, and that in contrast, that the gap distribution for generic surfaces has support at 0. This leaves open the question:

QUESTION. *Is there a translation surface whose gap distribution has no support at 0 but does have small gaps. That is, for all $\epsilon > 0$, for all $R \gg 0$, there exists a gap of slopes of saddle connections of length at most R less than ϵ/R^2, but there is an $\epsilon_0 > 0$ so that the proportion of gaps of size at most ϵ_0/R^2 goes to 0 as $R \to \infty$.*

A possible set of candidates for such a surface might be *completely periodic* surfaces which are not lattice surfaces.

A. Explicit formulas for the probability distribution function

NOTATION. • We use a bar to denote the inverse, so $\overline{\alpha}$ and $\overline{\varphi}$ denote α^{-1} and φ^{-1}.
• We denote by ath the inverse hyperbolic tangent function: $\mathrm{ath}(x) = (1/2)\ln((1+x)/(1-x))$.
• We denote by r the function $x \mapsto r(x) = \sqrt{1-4x}$.

The cumulative distribution function for the gaps in slopes is the function F which to α associates the probability $F(\alpha)$ that a gap is less than α. The probability distribution function f is the derivative of F.

We can see F as the sum of three partial cdfs F_s for the zones Ω_s, for each $s \in \{1, \varphi, \infty\}$, where $F_s(\alpha) = \mathrm{area}(\Omega_s \cap (R_s(x,y) < \alpha))$, giving the formulas $F_\infty(\alpha) = \mathrm{area}(\Omega_\infty \cap (xy > \overline{\alpha}))$, $F_\varphi(\alpha) = \mathrm{area}(\Omega_\varphi \cap (x(\overline{\varphi}x + y) > \overline{\alpha}))$, and $F_1(\alpha) = \mathrm{area}(\Omega_1 \cap (x(x+y) > \overline{\alpha}))$.

The different configurations, as α varies, of the intersection of the hyperbolas $R_s(x,y) = \alpha$ with the domains Ω_s determine different evaluations of these formulas; which by differentiating give the partial pdfs:

• Respectively for: $0 < \alpha < 1$, $1 < \alpha < 4\varphi$, $4\varphi < \alpha < \varphi^4$, $\varphi^4 < \alpha$,
  $F_\infty(\alpha)$ equals: 0, $1 - \overline{\alpha}(1 + \ln\alpha)$, $1 - \overline{\alpha}(1 + \ln\alpha - 4\,\mathrm{ath}\,r(\varphi\overline{\alpha})) - \overline{\varphi}r(\varphi\overline{\alpha})$,
  $\qquad\qquad 3\overline{\varphi}/2 - \overline{\alpha}(1 + 2\ln(\overline{\varphi}\alpha/2) + 2\ln(1 - r(\varphi\overline{\alpha}))) + (\overline{\varphi}/2)r(\varphi\overline{\alpha})$;
  $f_\infty(\alpha)$ equals: 0, $\overline{\alpha}^2 \ln\alpha$, $\overline{\alpha}^2(\ln\alpha - 4\,\mathrm{ath}\,r(\varphi\overline{\alpha}))$,
  $\qquad\qquad \overline{\alpha}^2(2\ln(\overline{\varphi}\alpha/2) + 2\ln(1 - r(\varphi\overline{\alpha})))$.

• Respectively for: $0 < \alpha < \varphi$, $\varphi < \alpha < 4$, $4 < \alpha < \varphi^3$, $\varphi^3 < \alpha$,
  $F_\varphi(\alpha)$ equals: 0, $\overline{\varphi} - \overline{\alpha}(1 + \ln\overline{\varphi}\alpha)$, $\overline{\varphi} - \overline{\alpha}(1 + \ln\overline{\varphi}\alpha - 4\,\mathrm{ath}\,r(\overline{\alpha})) - r(\overline{\alpha})$, $\overline{\varphi}^4$;
  $f_\varphi(\alpha)$ equals: 0, $\overline{\alpha}^2 \ln\overline{\varphi}\alpha$, $\overline{\alpha}^2(\ln\overline{\varphi}\alpha - 4\,\mathrm{ath}\,r(\overline{\alpha}))$, 0.

• Respectively for: $0 < \alpha < \varphi$, $\varphi < \alpha < 4\overline{\varphi}$, $4\overline{\varphi} < \alpha < \varphi^2$, $\varphi^3 < \alpha$,
  $F_1(\alpha)$ equals: 0, $\overline{\varphi} - \overline{\alpha}(1 + \ln\overline{\varphi}\alpha)$, $\overline{\varphi} - \overline{\alpha}(1 + \ln\overline{\varphi}\alpha - 2\,\mathrm{ath}\,r(\overline{\varphi}\alpha)) - (\varphi/2)r(\overline{\varphi}\alpha)$,
  $\qquad\qquad \overline{\varphi}^5/2$;
  $f_1(\alpha)$ equals: 0, $\overline{\alpha}^2 \ln\overline{\varphi}\alpha$, $\overline{\alpha}^2(\ln\overline{\varphi}\alpha - 2\,\mathrm{ath}\,r(\overline{\varphi}\alpha))$, 0.

B. Volume computations

Define a measurable partition of $X = \mathrm{SL}(2, \mathbb{R})/\Gamma$ into X_1, X_φ, X_∞ where each X_j is the part of X spanned by Ω_j under the flow (h_s), until its first return to Ω. The complement of the union of the X_j's is the union of periodic orbits for the flow (h_s), which has measure zero. The partial volumes $V_j = \mathrm{vol}(X_j) = \int_{\Omega_j} R$ are obtained by integrating the return time function $R(a,b)$ over the domains Ω_j.

- Integrating R over Ω_1,

$$V_1 = \int_{a=\overline{\varphi}}^{1} \int_{b=1-a\varphi}^{\overline{\varphi}-a} \frac{1}{a(a+b)} \, da \, db = \int_{\overline{\varphi}}^{1} \left[\ln(a+b)\right]_{1-a\varphi}^{\overline{\varphi}-a} \frac{da}{a}$$

$$= \int_{\overline{\varphi}}^{1} \left(\ln(\overline{\varphi}) - \ln(1-a\overline{\varphi})\right) \frac{da}{a}$$

$$= -(\ln\overline{\varphi})^2 - \int_{\overline{\varphi}^2}^{\overline{\varphi}} \ln(1-t) \frac{dt}{t} = -(\ln\varphi)^2 + \mathrm{Li}_2(\overline{\varphi}) - \mathrm{Li}_2(\overline{\varphi}^2),$$

where the last step uses $\ln\overline{\varphi} = -\ln\varphi$ and the definition $\mathrm{Li}_2(t) = \int_t^0 \ln(1-t)\,dt/t$.

- Integrating R over Ω_φ,

$$V_\varphi = \int_{a=\overline{\varphi}^2}^{\overline{\varphi}} \int_{b=1-a\varphi}^{\overline{\varphi}-a\overline{\varphi}} \frac{1}{a(a\overline{\varphi}+b)} \, da \, db + \int_{a=\overline{\varphi}}^{1} \int_{b=\overline{\varphi}-a}^{\overline{\varphi}-a\overline{\varphi}} \frac{1}{a(a\overline{\varphi}+b)} \, da \, db$$

$$= \int_{\overline{\varphi}^2}^{\overline{\varphi}} \left[\ln(a\overline{\varphi}+b)\right]_{1-a\varphi}^{\overline{\varphi}-a\overline{\varphi}} \frac{da}{a} + \int_{a=\overline{\varphi}}^{1} \left[\ln(a\overline{\varphi}+b)\right]_{\overline{\varphi}-a}^{\overline{\varphi}-a\overline{\varphi}} \frac{da}{a}$$

$$= \int_{\overline{\varphi}^2}^{\overline{\varphi}} \left(\ln(\overline{\varphi}) - \ln(1-a)\right) \frac{da}{a} + \int_{\overline{\varphi}}^{1} \left(\ln(\overline{\varphi}) - \ln(\overline{\varphi}-a\overline{\varphi}^2)\right) \frac{da}{a}$$

$$= (\ln\overline{\varphi})(\ln\overline{\varphi} - \ln\overline{\varphi}^2) - \int_{\overline{\varphi}^2}^{\overline{\varphi}} \ln(1-t) \frac{dt}{t} - \int_{\overline{\varphi}^2}^{\overline{\varphi}} \ln(1-t) \frac{dt}{t}$$

$$= -(\ln\overline{\varphi})^2 - 2\int_{\overline{\varphi}^2}^{\overline{\varphi}} \ln(1-t) \frac{dt}{t} = -(\ln\varphi)^2 - 2\,\mathrm{Li}_2(\overline{\varphi}^2) + 2\,\mathrm{Li}_2(\overline{\varphi}).$$

- Integrating R over Ω_∞,

$$V_\infty = \int_{a=0}^{\overline{\varphi}^2} \int_{b=1-a\varphi}^{1} \frac{1}{ab} \, da \, db + \int_{a=\overline{\varphi}^2}^{1} \int_{b=\overline{\varphi}-a\overline{\varphi}}^{1} \frac{1}{ab} \, da \, db$$

$$= \int_0^{\overline{\varphi}^2} \left[\ln b\right]_{1-a\varphi}^{1} \frac{da}{a} + \int_{\overline{\varphi}^2}^{1} \left[\ln b\right]_{\overline{\varphi}-a\overline{\varphi}}^{1} \frac{da}{a}$$

$$= -\int_0^{\overline{\varphi}} \ln(1-t) \frac{dt}{t} + (\ln\overline{\varphi})(\ln\overline{\varphi}^2) - \int_{\overline{\varphi}^2}^{1} \ln(1-a) \frac{da}{a}$$

$$= 2(\ln\varphi)^2 + \mathrm{Li}_2(\overline{\varphi}) - \int_{\overline{\varphi}^2}^{0} \ln(1-t) \frac{dt}{t} - \int_0^{1} \ln(1-t) \frac{dt}{t}$$

$$= 2(\ln\varphi)^2 + \mathrm{Li}_2(\overline{\varphi}) - \mathrm{Li}_2(\overline{\varphi}^2) + \mathrm{Li}_2(1).$$

These three partial volumes add up to

$$V = \mathrm{Li}_2(1) + 4(\mathrm{Li}_2(\overline{\varphi}) - \mathrm{Li}_2(\overline{\varphi}^2)),$$

which, since $\mathrm{Li}_2(1) = \pi^2/6$ and $\mathrm{Li}_2(\overline{\varphi}) - \mathrm{Li}_2(\overline{\varphi}^2) = \pi^2/30$, can be expressed as

$$V = \pi^2/6 + 4\pi^2/30 = 9\pi^2/30 = 3\pi^2/10.$$

This is exactly the classically known volume of $X = \mathrm{SL}(2,\mathbb{R})/\Gamma$, as should be expected.

C. Quadratic Asymptotics

In this appendix we derive the asymptotic formula

$$\lim_{R \to \infty} \frac{N(R)}{R^2} = \frac{10}{3\pi}(3\varphi + 1),$$

where $N(R) = |\Lambda_\omega \cap B(0, R)|$ is the cardinality of the set of saddle connection holonomy vectors of length at most R. Recall that if $\mathbf{v}_1 = \frac{\sqrt{5}-1}{2}$ and $\mathbf{v}_2 = 1$ (viewed as elements of $\mathbb{C} \cong \mathbb{R}^2$), then $\Lambda_\omega = \Gamma\mathbf{v}_1 \cup \Gamma\mathbf{v}_2$. Veech [23, §16] showed that for any non-uniform lattice $\Gamma \subset \mathrm{SL}(2, \mathbb{R})$, and any vector $\mathbf{v}$ stabilized by a maximal parabolic subgroup $\Lambda \subset \Gamma$, we have ([23, (16.2)]):

$$\lim_{R \to \infty} \frac{|g\Gamma\mathbf{v} \cap B(0, R)|}{R^2} = c(\Gamma, \mathbf{v}),$$

where $c(\Gamma, \mathbf{v})$ is given as follows. Let $g_0 \in SL(2, \mathbb{R})$ be such that

$$g_0^{-1}\Lambda g_0 = \Lambda_0,$$

with

$$\Lambda_0 = \left\{ \begin{bmatrix} 1 & n \\ 0 & 1 \end{bmatrix} : n \in \mathbb{Z} \right\}.$$

Then setting $\mathbf{v}_0 = 1$, we have $g_0\mathbf{v}_0 = t\mathbf{v}$, and we have

$$c(\Gamma, \mathbf{v}) = t^2 \mathrm{vol}(\mathbb{H}^2/\Gamma)^{-1}.$$

In fact, there is a typographical error in [23, (16.2)], an extra factor of π in the right-hand side, as can be seen by computing the example $\Gamma = SL(2, \mathbb{Z}), \mathbf{v} = 1$.

Veech's formula and the decomposition of Λ_ω into two Veech group orbits allows us to reduce our problem to simply computing $c(\Gamma, \mathbf{v}_2)$, since

$$\lim_{R \to \infty} \frac{N(R)}{R^2} = c(\Gamma, \mathbf{v}_1) + c(\Gamma, \mathbf{v}_2),$$

and $\mathbf{v}_2 = \varphi\mathbf{v}_1$ implies that $c(\Gamma, \mathbf{v}_2) = \varphi^2 c(\Gamma, \mathbf{v}_1)$.

In our setting, $\mathbf{v}_2 = \mathbf{v}_0$, $\Lambda = \left\{ \begin{bmatrix} 1 & n\varphi \\ 0 & 1 \end{bmatrix} : n \in \mathbb{Z} \right\}$, $g_0 = \begin{bmatrix} \varphi^{1/2} & 0 \\ 0 & \varphi^{-1/2} \end{bmatrix}$, so

$$g_0\mathbf{v}_0 = \varphi^{1/2} = \varphi^{1/2}\mathbf{v}_2,$$

yielding $t = \varphi^{1/2}$. Thus,

$$c(\Gamma, \mathbf{v}_2) = t^2 \mathrm{vol}(\mathbb{H}^2/\Gamma)^{-1} = \varphi \frac{10}{3\pi},$$

and thus

$$\lim_{R \to \infty} \frac{N(R)}{R^2} = (1 + \varphi^2)c(\Gamma, \mathbf{v}_2) = (2 + \varphi)\varphi\frac{10}{3\pi} = (2\varphi + \varphi^2)\frac{10}{3\pi} = (3\varphi + 1)\frac{10}{3\pi},$$

where we repeatedly use $\varphi^2 = \varphi + 1$.

Finally, if we normalize instead our golden L to have area 1, we have to multiply our constant by our current area $(2\varphi - 1)$, yielding

$$(3\varphi + 1)\frac{10}{3\pi}(2\varphi - 1) = \frac{10}{3\pi}(6\varphi^2 - \varphi - 1) = \frac{10}{3\pi}(5\varphi^2)$$

Thus, we obtain (now normalizing our count by area of $B(0, R)$, πR^2), the extraordinary formula:

$$\lim_{R\to\infty} \frac{N^{(1)}(R)}{\pi R^2} = c$$

where $N^{(1)}(R)$ denotes the number of saddle connection holonomy vectors of length at most R on the unit-area golden L,

$$c = \frac{50}{3\pi^2}\varphi^2 = \frac{25}{3\pi^2} \times 2 \times \varphi^2.$$

This is a coefficient that involves the *Siegel-Veech* constant $\frac{25}{3}$ (see [**4**, Theorem 1.5]); the volume $\frac{3\pi}{10}$ of $\mathbb{H}^2/\Gamma$, an orientation factor of 2, and a normalization factor of φ^2.

References

[1] J. S. Athreya, *Gap Distributions and Homogeneous Dynamics*, to appear, Proceedings of the ICM Satellite conference on Geometry, Topology, and Dynamics in Negative Curvature.

[2] J. S. Athreya and J. Chaika, *The distribution of gaps for saddle connection directions*, Geom. Funct. Anal. **22** (2012), no. 6, 1491–1516, DOI 10.1007/s00039-012-0164-9. MR3000496

[3] J. S. Athreya and Y. Cheung, *A Poincaré section for horocycle flow on the space of lattices*, International Math Research Notices, 2013.

[4] Matt Bainbridge, *Euler characteristics of Teichmüller curves in genus two*, Geom. Topol. **11** (2007), 1887–2073, DOI 10.2140/gt.2007.11.1887. MR2350471 (2009c:32025)

[5] Florin P. Boca, Cristian Cobeli, and Alexandru Zaharescu, *A conjecture of R. R. Hall on Farey points*, J. Reine Angew. Math. **535** (2001), 207–236, DOI 10.1515/crll.2001.049. MR1837099 (2002f:11127)

[6] Florin P. Boca and Alexandru Zaharescu, *Farey fractions and two-dimensional tori*, Noncommutative geometry and number theory, Aspects Math., E37, Vieweg, Wiesbaden, 2006, pp. 57–77, DOI 10.1007/978-3-8348-0352-8.3. MR2327299 (2008f:11023)

[7] S. G. Dani and John Smillie, *Uniform distribution of horocycle orbits for Fuchsian groups*, Duke Math. J. **51** (1984), no. 1, 185–194, DOI 10.1215/S0012-7094-84-05110-X. MR744294 (85f:58093)

[8] Diana Davis, *Cutting sequences, regular polygons, and the Veech group*, Geom. Dedicata **162** (2013), 231–261, DOI 10.1007/s10711-012-9724-2. MR3009542

[9] Diana Davis, Dmitry Fuchs, and Serge Tabachnikov, *Periodic trajectories in the regular pentagon* (English, with English and Russian summaries), Mosc. Math. J. **11** (2011), no. 3, 439–461, 629. MR2894424 (2012k:37074)

[10] Noam D. Elkies and Curtis T. McMullen, *Gaps in $\sqrt{n}$ mod 1 and ergodic theory*, Duke Math. J. **123** (2004), no. 1, 95–139, DOI 10.1215/S0012-7094-04-12314-0. MR2060024 (2005f:11143)

[11] Alex Eskin and Howard Masur, *Asymptotic formulas on flat surfaces*, Ergodic Theory Dynam. Systems **21** (2001), no. 2, 443–478, DOI 10.1017/S0143385701001225. MR1827113 (2002g:37049)

[12] Alex Eskin, Howard Masur, and Anton Zorich, *Moduli spaces of abelian differentials: the principal boundary, counting problems, and the Siegel-Veech constants*, Publ. Math. Inst. Hautes Études Sci. **97** (2003), 61–179, DOI 10.1007/s10240-003-0015-1. MR2010740 (2005b:32029)

[13] R. R. Hall, *A note on Farey series*, J. London Math. Soc. (2) **2** (1970), 139–148. MR0253978 (40 #7191)

[14] Pascal Hubert and Thomas A. Schmidt, *An introduction to Veech surfaces*, Handbook of dynamical systems. Vol. 1B, Elsevier B. V., Amsterdam, 2006, pp. 501–526, DOI 10.1016/S1874-575X(06)80031-7. MR2186246 (2006i:37099)

[15] Pascal Hubert and Thomas A. Schmidt, *Diophantine approximation on Veech surfaces* (English, with English and French summaries), Bull. Soc. Math. France **140** (2012), no. 4, 551–568 (2013). MR3059850

[16] Jens Marklof and Andreas Strömbergsson, *The distribution of free path lengths in the periodic Lorentz gas and related lattice point problems*, Ann. of Math. (2) **172** (2010), no. 3, 1949–2033, DOI 10.4007/annals.2010.172.1949. MR2726104 (2012b:37103)

[17] Howard Masur, *Interval exchange transformations and measured foliations*, Ann. of Math. (2) **115** (1982), no. 1, 169–200, DOI 10.2307/1971341. MR644018 (83e:28012)

[18] Howard Masur, *The growth rate of trajectories of a quadratic differential*, Ergodic Theory Dynam. Systems **10** (1990), no. 1, 151–176, DOI 10.1017/S0143385700005459. MR1053805 (91d:30042)

[19] Howard Masur and John Smillie, *Hausdorff dimension of sets of nonergodic measured foliations*, Ann. of Math. (2) **134** (1991), no. 3, 455–543, DOI 10.2307/2944356. MR1135877 (92j:58081)

[20] Peter Sarnak, *Asymptotic behavior of periodic orbits of the horocycle flow and Eisenstein series*, Comm. Pure Appl. Math. **34** (1981), no. 6, 719–739, DOI 10.1002/cpa.3160340602. MR634284 (83m:58060)

[21] John Smillie and Barak Weiss, *Characterizations of lattice surfaces*, Invent. Math. **180** (2010), no. 3, 535–557, DOI 10.1007/s00222-010-0236-0. MR2609249 (2012c:37072)

[22] C. Uyanik and G. Work, *The distribution of gaps for saddle connections on the octagon*, in preparation.

[23] William A. Veech, *Siegel measures*, Ann. of Math. (2) **148** (1998), no. 3, 895–944, DOI 10.2307/121033. MR1670061 (2000k:37028)

[24] William A. Veech, *Geometric realizations of hyperelliptic curves*, Algorithms, fractals, and dynamics (Okayama/Kyoto, 1992), Plenum, New York, 1995, pp. 217–226. MR1402493 (98f:14022)

[25] Hermann Weyl, *Über die Gleichverteilung von Zahlen mod. Eins* (German), Math. Ann. **77** (1916), no. 3, 313–352, DOI 10.1007/BF01475864. MR1511862

[26] A. Wright and I. Gekhtman, *Smillie's Theorem on closed $SL(2, \mathbb{R})$-orbits for quadratic differentials*, preprint, http://math.uchicago.edu/~amwright/Smillie.pdf

DEPARTMENT OF MATHEMATICS, UNIVERSITY OF ILLINOIS URBANA-CHAMPAIGN, 1409 W. GREEN STREET, URBANA, ILLINOIS 61801
E-mail address: jathreya@illinois.edu

DEPARTMENT OF MATHEMATICS, UNIVERSITY OF UTAH, 155 SOUTH 1400 EAST, JWB 233, SALT LAKE CITY, UTAH 84112-0090
E-mail address: chaika@math.utah.edu

LABORATOIRE DE MATHÉMATIQUE D'ORSAY, UMR 8628 CNRS / UNIVERSITÉ PARIS-SUD, BÂTIMENT 425, 91405 ORSAY CEDEX, FRANCE
E-mail address: samuel.lelievre@gmail.com

Contemporary Mathematics
Volume **631**, 2015
http://dx.doi.org/10.1090/conm/631/12596

Uniformly recurrent subgroups

Eli Glasner and Benjamin Weiss

ABSTRACT. We define the notion of uniformly recurrent subgroup, URS in short, which is a topological analog of the notion of invariant random subgroup (IRS), introduced in Abert, Glasner, and Virag (2014). Our main results are as follows. (i) It was shown in Weiss (2012) that for an arbitrary countable infinite group G, any free ergodic probability measure preserving G-system admits a minimal model. In contrast we show here, using URS's, that for the lamplighter group there is an ergodic measure preserving action which does not admit a minimal model. (ii) For an arbitrary countable group G, every URS can be realized as the stability system of some topologically transitive G-system.

CONTENTS

Introduction

Let G be a locally compact second countable topological group. A *G-dynamical system* is a pair (X, G) where X is a compact metric space and G acts on X by homeomorphisms. Given a compact dynamical system (X, G), for $x \in X$ let $G_x = \{g \in G : gx = x\}$ be the *stability group at* x.

Let $\mathcal{S} = \mathcal{S}(G)$ be the compact metrizable space of all subgroups of G equipped with the Fell (or Chabauty) topology. Recall that given a Hausdorff topological

2010 *Mathematics Subject Classification.* Primary 37B05, 37A15, 20E05, 20E15, 57S20.
Key words and phrases. Invariant minimal subgroups, URS, IRS, stability group, stability system, essentially free action, free group.
This research was supported by a grant of Israel Science Foundation (ISF 668/13).

space X, a basis for the *Fell topology* on the hyperspace 2^X, comprising the closed subsets of X, is given by the collection of sets $\{\mathcal{U}(U_1, \ldots, U_n; C)\}$, where

$$\mathcal{U}(U_1, \ldots, U_n; C) = \{A \in 2^X : \forall\, 1 \le j \le n,\ A \cap U_j \ne \emptyset\ \&\ A \cap C = \emptyset\}.$$

Here $\{U_1, \ldots, U_n\}$ ranges over finite collections of open subsets of X and C runs over the compact subsets of X. The Fell topology is always compact and it is Hausdorff iff X is locally compact (see e.g. [6]). We let G act on $\mathcal{S}(G)$ by conjugation. This action makes $(\mathcal{S}(G), G)$ a G-dynamical system. In order to avoid confusion we denote this action by $(g, H) \mapsto g \cdot H$ $(g \in G, H \in \mathcal{S}(G))$. Thus for a subgroup $H < G$ we have $g \cdot H = H^g = gHg^{-1}$.

Perhaps the first systematic study of the space $\mathcal{S}(G)$ is to be found in Auslander and Moore's memoir [4]. It then played a central role in the seminal work of Stuck and Zimmer [25]. More recently the notion of IRS (invariant random subgroup) was introduced in the work of M. Abert, Y. Glasner and B. Virag [2]. Formally this object is just a G-invariant probability measure on $\mathcal{S}(G)$. This latter work served as a catalyst and lead to a renewed vigorous interest in the study of IRS's (see, among others, [1], [3], [7], [8], [9], and [26]). A brief historical discussion of the subject can be found in [2].

Pursuing the well studied analogies between ergodic theory and topological dynamics (see [21]) we propose to introduce a topological dynamical analogue of the notion of an IRS.

0.1. DEFINITION. A minimal subsystem of $\mathcal{S}(G)$ is called a *uniformly recurrent subgroup*, URS in short. (Recall that according to Furstenberg a point x in a compact dynamical system (X, G) is *uniformly recurrent* (i.e. for every neighborhood U of x the set $N(x, U) = \{g \in G : gx \in U\}$ is syndetic) if and only if the orbit closure $\mathrm{cls}\,\{gx : g \in G\}$ is a minimal set.) A topologically transitive subsystem of $\mathcal{S}(G)$ is called a *topologically transitive subgroup*, TTS in short.

For later use we also define a notion of nonsingular random subgroup.

0.2. DEFINITION. Recall that a *nonsingular action of G* is a measurable action of G on a standard Lebesgue probability space $(X, \mathcal{B}, \mu)$, where the action preserves the measure class of μ (i.e. $\mu(A) = 0 \iff \mu(gA) = 0$ for every $A \in \mathcal{B}$ and every $g \in G$). We will call a nonsingular measure on $\mathcal{S}(G)$ a *nonsingular random subgroup*, NSRS in short.

In Section 1 we define and study the stability system which is associated to a dynamical system and then consider some examples of groups possessing only trivial URS's. In Section 2 we examine homogeneity properties of URS's. In Section 3 we show how a great variety of URS's can arise for lamplighter groups. In Section 4 we obtain some applications of URS's to ergodic theory. In Section 5 the richness of the space $\mathcal{S}(\mathbb{F}_2)$ is demonstrated. Finally, in Section 6 we consider the question of realization of URS's as stability systems.

We thank N. Avni, P.-E. Caprace, Y. Glasner and S. Mozes for their helpful advice.

1. Stability systems

It is easy to check that, whenever (X, G) is a dynamical system, the map $\phi : X \to \mathcal{S}(G)$, $x \mapsto G_x$ is upper-semi-continuous; i.e. $x_i \to x$ implies $\limsup \phi(x_i) \subset \phi(x)$. In fact, if $x_i \to x$ and $g_i \to g$ (when G is discrete the latter just means that eventually $g_i = g$) are convergent sequences in X and G respectively, with $g_i \in G_{x_i}$, then $x_i = g_i x_i \to gx$, hence $gx = x$ so that $\limsup G_{x_i} \subset G_x$. Recall that whenever $\psi : X \to 2^Z$ is an upper-semi-continuous map, where X is compact metric and Z is locally compact and second countable, there exists a dense G_δ subset $X_0 \subset X$ where ψ is continuous at each point $x_0 \in X_0$ (see e.g. [**11**, page 95, Theorem 1]).

1.1. DEFINITION. Let $\pi : (X, G) \to (Y, G)$ be a homomorphism of G-systems; i.e. π is a continuous, surjective map and $\pi(gx) = g\pi(x)$ for every $x \in X$ and $g \in G$. We say that π is an *almost one-to-one extension* if there is a dense G_δ subset $X_0 \subset X$ such that $\pi^{-1}(\pi(x)) = \{x\}$ for every $x \in X_0$.

1.2. PROPOSITION. *Let (X, G) be a compact system. Denote by $\phi : X \to \mathcal{S}(G)$ the upper-semi-continuous map $x \mapsto G_x$ and let $X_0 \subset X$ denote the dense G_δ subset of continuity points of ϕ. Construct the diagram*

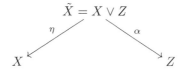

where

$$Z = \mathrm{cls}\,\{G_x : x \in X_0\} \subset \mathcal{S}(G),$$
$$\tilde{X} = \mathrm{cls}\,\{(x, G_x) : x \in X_0\} \subset X \times Z,$$

and η and α are the restrictions to $\tilde{X}$ of the projection maps. We have:

(1) *The map η is an almost one-to-one extension.*

If moreover (X, G) is minimal then

(2) *Z and $\tilde{X}$ are minimal systems.*

(3) *Z is the unique minimal subset of the set $\mathrm{cls}\,\{G_x : x \in X\} \subset \mathcal{S}(G)$ and $\tilde{X}$ is the unique minimal subset of the set $\mathrm{cls}\,\{(x, G_x) : x \in X\} \subset X \times \mathcal{S}(G)$.*

PROOF. (1) Let $\tilde{X}_0 = \{(x, G_x) : x \in X_0\}$. It is easy to see that the fact that $x \in X_0$ implies that the fiber $\eta^{-1}(\eta((x, G_x)))$ is the singleton $\{(x, G_x)\}$.

(2) Fix a point $x_0 \in X_0$. The minimality of (X, G) implies that the orbit of the point (x_0, G_{x_0}) is dense in $\tilde{X}$. On the other hand, if (x, L) is an arbitrary point in $\tilde{X}$ then, again by minimality of (X, G), there is a sequence $g_n \in G$ with $\lim_n g_n x = x_0$. We can assume that the limit $\lim_n g_n(x, L) = (x_0, K)$ exists as well, and then the fact that $x_0 \in X_0$ implies that $K = G_{x_0}$. This shows that $(\tilde{X}, G)$ is minimal and then so is $Z = \alpha(\tilde{X})$.

(3) Given any $x \in X$ we argue, as in part (2), that (x_0, G_{x_0}) is in the orbit closure of (x, G_x). □

1.3. DEFINITION. Given a dynamical system (X, G) we call the system $Z \subset \mathcal{S}(G)$ the *stability system* of (X, G). We denote it by $Z = \mathcal{S}_X$. We say that (X, G) is *essentially free* if $Z = \{e\}$. Note that when $Z = \{N\}$ is a singleton then the subgroup $N < G$ is necessarily a normal subgroup of G and, by the upper-semi-continuity of the map $x \mapsto G_x$, it follows that $N < G_x$ for every $x \in X$, whence $N = \bigcap\{G_x : x \in X\}$. In this case then, the action reduces to an action of the group G/N and the latter is essentially free. In particular, if the action of G on X is *effective* (i.e. $\forall g \neq e, \exists x \in X, gx \neq x$) then $Z = \{N\}$ implies that $N = \{e\}$, so that the action is essentially free. Also note that part (3) of the proposition implies that if (X, G) is minimal and there is some point $x \in X$ with $G_x = \{e\}$ then (X, G) is necessarily essentially free.

1.4. PROPOSITION. *Let $X \subset \mathcal{S}(G)$ be a URS and let $X_0 \subset X$ denote the dense G_δ subset of X consisting of the continuity points of the map $x \mapsto G_x = N_G(x)$. Consider*
$$Z = \mathcal{S}_X = \mathrm{cls}\,\{G_x : x \in X_0\} = \mathrm{cls}\,\{N_G(x) : x \in X_0\},$$
the stability system of X.

(1) *Z is again a URS.*
(2) *If $N_G(x_0) = x_0$ for some $x_0 \in X_0$ then $Z = X$.*
(3) *Conversely, if $Z = X$ then $N_G(x) = x$ for every $x \in X_0$.*
(4) *If for some $x_0 \in X_0$ we have that $N = N_G(x_0)$ is a co-compact subgroup of G, then $Z = \mathcal{S}_X$ is a factor of the homogeneous G-space G/N.*
(5) *If, in addition, $N_G(N) = N$, then $Z \cong G/N$.*

PROOF. 1. The first part is a direct consequence of Proposition 1.2, and we also deduce that Z is a factor of an almost 1-1 extension of X, namely of $\tilde{X}$.

2. The second follows since by our assumption $x \in X \cap Z$.

3. Now suppose $Z = X$ and let x_0 be a point in X_0. We have $N = G_{x_0} = N_G(x_0) \in Z$ and thus $N \in X$. Let now $X_{max} = \{x \in X : y \supset x \,\&\, y \in X \Rightarrow y = x\}$. As in [18, Lemma 5.3] one shows that the set X_{max} is a dense G_δ subset of X. We now further assume that $x_0 \in X_0 \cap X_{max}$. But then the inclusion $x_0 \subset N$ implies $x_0 = N$ and we have
$$(x_0, N) = (x_0, G_{x_0}) = (x_0, x_0) \in \tilde{X}.$$
This implies $\tilde{X} = \{(x, x) : x \in X\}$ and, in particular
$$(x, G_x) = (x, N_G(x)) = (x, x),$$
for every $x \in X_0$.

4. For the fourth part note that, by assumption, $KN = G$ for some compact subset $K \subset G$. It follows that $G \cdot N = K \cdot N = \{kNk^{-1} : k \in K\}$, the G-orbit of N, is compact and therefore $Z = G \cdot N$. Moreover, the map $\psi : G/N \to Z$ defined by $\psi(gN) = gNg^{-1}$ is a homomorphism. Finally ψ is one-to-one when $N_G(N) = N$, whence follows part (5). □

The proof of the next proposition is straightforward.

1.5. PROPOSITION. (1) *A surjective group homomorphism $\eta : A \to B$ between two countable groups A and B induces an embedding $\eta_* : \mathcal{S}(B) \to \mathcal{S}(A)$ of the corresponding dynamical systems. Explicitly, for $H < B$ its image in $\mathcal{S}(A)$ is the subgroup $\tilde{H} = \eta^{-1}(H)$. Moreover we have*
$$\tilde{H}^a = \eta_*(H)^a = \eta_*(H^{\eta(a)}) = \widetilde{H^{\eta(a)}}$$
for every $a \in A$.

(2) *Let G be a group and $H < G$ a subgroup of finite index in G. If $Z_0 \subset \mathcal{S}(H)$ is a H-URS, then $Z = \cup\{gZ_0g^{-1} : g \in G\}$ is a G-URS in $\mathcal{S}(G)$.*

For further information on the space $\mathcal{S}(G)$ see [23].

Clearly for an abelian group G the conjugation action on $\mathcal{S}(G)$ is the identity action. If G is a finitely generated nilpotent group then every subgroup of G is finitely generated as well [**5**]. Thus G has only countably many subgroups and we conclude that every IRS (hence also every URS) of G is finite [1].

Let G be a connected semisimple Lie group with finite center and $\mathbb{R}$-rank ≥ 2, satisfying property (T) (e.g. $G = SL(3,\mathbb{R})$). It follows from the Stuck-Zimmer theorem [**25**] that any URS in $\mathcal{S}(G)$, which supports a G-invariant probability measure, is necessarily of the form $\{g\Gamma g^{-1} : g \in G\} \cong G/N_G(\Gamma) \cong \mathcal{S}_{G/\Gamma}$, where $\Gamma < G$ is a co-compact lattice and $N_G(\Gamma) = \{g \in G : g\Gamma g^{-1} = \Gamma\}$ its normalizer in G. Moreover, for each parabolic subgroup $Q < G$, the homogeneous space G/Q, with left multiplication, forms a minimal G action and clearly the corresponding stability system $\mathcal{S}_{G/Q} = \{gQg^{-1} : g \in G\} \cong G/Q$ is a URS. We don't know whether, up to isomorphism, these are the onlyURS's in $\mathcal{S}(G)$.

In this connection we have the following theorems of Stuck [**24**, Theorem 3.1 and Corollary 3.2]. (The action of a topological group G on a compact Hausdorff space X is *locally free* if for every $x \in X$ the stability subgroup G_x is discrete.)

1.6. THEOREM (Stuck). *Let G be a real algebraic group acting minimally on a compact Hausdorff space X. Fix $x \in X$ and let $N = N_G((G_x)^0)$. Then:*

(1) *N is co-compact in G.*
(2) *If H is any closed subgroup of G containing N then there is a compact minimal H-space Y such that X is G-equivariantly homeomorphic to $G \times_H Y$.*

1.7. THEOREM (Stuck). *Let G be a semisimple Lie group with finite center and without compact factors, acting minimally on a compact Hausdorff space X. Then either the action is locally free, or it is isomorphic to an induced action $G \times_Q Y$, where Q is a proper parabolic subgroup of G and Y is a compact Q-minimal space.*

1.8. PROBLEM. For a semisimple Lie group G as above, is it the case that every nontrivial URS of G is either of the form $\{g\Gamma g^{-1} : g \in G\}$, where $\Gamma < G$ is a co-compact lattice, or it admits G/Q as a factor with $Q < G$ a proper parabolic subgroup of G?[2]

We are currently working on that problem and have some indications that an affirmative answer is plausible. See [**24**] for information on minimal actions of semisimple Lie groups.

We next show that certain non-abelian infinite countable groups admit no nontrivial URS's. On the other hand, in Section 3 and 5 we will see many examples of nontrivial URS's.

1.9. EXAMPLE. A "Tarski monster" group G is a countable noncyclic group with the property that its only proper subgroups are cyclic (either all of a fixed prime order p, or all infinite cyclic). It is easy to see that such a group is necessarily simple. Since G is countable it follows that $\mathcal{S}(G)$ is a countable set. Moreover, the only URS's in $\mathcal{S}(G)$ are $\{e\}$ and $\{G\}$. In fact, if $X \subset \mathcal{S}(G)$ is an URS then, being a countable space, it must have an isolated point $H \in X$. Since X is minimal it must be finite (finitely many conjugates of the open set $\{H\}$ must cover X). If $H \in X$ and H is neither $\{e\}$ nor G, then H is a cyclic group and $X = \{gHg^{-1} : g \in G\}$ is a finite

---

[1]We thank Yair Glasner for this observation and for suggesting example 1.9 below.
[2]See also Problem 5.4 below.

set. Now G acts (by conjugation) on the finite set X as a group of permutations and the kernel of the homomorphism from G onto this group of permutations is a normal subgroup of G. As G is simple this kernel is either $\{e\}$ or G and both cases lead to contradictions.

1.10. EXAMPLE. It is not hard to see that $\{e\}$ and $\{G\}$ are the only URS's for $G = S_\infty(\mathbb{N})$, the countable group of finitely supported permutations on $\mathbb{N}$. This paucity of URS's is in sharp contrast to the abundance of IRS's of this group as described in [**26**].

2. Homogeneity properties of a URS

In this section we let G be a countable discrete group.

2.1. DEFINITION. We say that a property P of groups is *admissible* if

(1) P is preserved under isomorphisms.
(2) P is inherited by subgroups, i.e. if H has P and $K < H$ then K has P.
(3) P is preserved under increasing unions, i.e. if $H_n \nearrow H$, where $\{H_n\}_{n\in\mathbb{N}}$ is an increasing sequence of subgroups of H and each H_n has P, then so does H.

2.2. PROPOSITION. *Let P be an admissible property of groups. Then the subset*

$$\mathcal{S}(G)_P = \{H < G : H \text{ has } P\}$$

is a closed invariant subset of the dynamical system $\mathcal{S}(G)$.

PROOF. Suppose $H_n \to H$ in $\mathcal{S}(G)$ with $H_n \in \mathcal{S}(G)_P$ for every $n \in \mathbb{N}$. Let $H = \{h_0 = e, h_1, h_2, \dots\}$ be an enumeration of H. Given $k \geq 0$ there exits n_0 such that for $n \geq n_0$ we have $\{h_0 = e, h_1, h_2, \dots, h_k\} \subset H_n$. Let $H^{(k)}$ be the subgroup of H (and of H_n) generated by the set $\{h_0 = e, h_1, h_2, \dots, h_k\}$. It is now clear that each $H^{(k)}$ has P and that $H^{(k)} \nearrow H$. It thus follows that also H has P. $\square$

The assertions in the next proposition are well known and not hard to check.

2.3. PROPOSITION. *The following properties are admissible:*

(1) *Commutativity.*
(2) *Nilpotency of degree* $\leq d$.
(3) *Solvability of degree* $\leq d$
(4) *Having an exponent* $q \geq 2$ *(i.e. satisfying the identity* $x^q = e$*).*
(5) *Amenability.*

2.4. REMARK. The question whether, for a general locally compact topological group G, the collection of closed amenable subgroups of G forms a closed subset of $\mathcal{S}(G)$ is open. As is shown in [**10**] this is true for a very large class of groups (which includes the discrete groups). The discrete group case follows directly from Schochetman's work [**23**].

2.5. COROLLARY. *Let* $Z \subset \mathcal{S}(G)$ *be a URS and let P be an admissible property. Then either every element of Z has P, or none has P.*

2.6. REMARK. One can easily check that e.g. nilpotency and being perfect (i.e. the property: $[H, H] = H$) are not admissible properties.

We next consider topologically transitive subgroups (TTS) (i.e. closed invariant topologically transitive subsets of $\mathcal{S}(G)$). In the spirit of [**19**] let us say that a subset $\mathcal{L} \subset \mathcal{S}(G)$ is a *dynamical property* if it is Baire measurable, and G-invariant; i.e. invariant under conjugations. In view of Proposition 2.2 every admissible property of groups defines a corresponding dynamical property in $\mathcal{S}(G)$. The next proposition is just a special instance of the general "zero-one law" for topologically transitive dynamical systems, see e.g. [**19**].

2.7. PROPOSITION. *Let $Z \subset \mathcal{S}(G)$ be a TTS and $\mathcal{L} \subset \mathcal{S}(G)$ be a dynamical property. Then the set $Z \cap \mathcal{L}$ is either meager or comeager.*

2.8. COROLLARY. *Let $Z \subset \mathcal{S}(G)$ be a URS then the set of perfect elements in Z is either meager or comeager.*

For the corresponding measure theory zero-one law see Proposition 4.1 below.

3. URS's for Wreath products

Let Γ be an arbitrary countable infinite group and let $G = \mathbb{Z}_2 \wr \Gamma = \{1, -1\} \wr \Gamma$ be the corresponding (restricted) Wreath product. Recall that the group $\mathbb{Z}_2 \wr \Gamma$ is defined as the semidirect product $G = K \rtimes \Gamma$, where $K = \oplus_{\gamma \in \Gamma} \mathbb{Z}_2$ and Γ acts on K by permutations: $(\gamma \cdot k)(\gamma') = k(\gamma'\gamma)$, $\gamma, \gamma' \in \Gamma, k \in K$. Thus, for $(l, \gamma), (k, \gamma') \in K \wr \Gamma$,

$$(l, \gamma)(k, \gamma') = (l(\gamma \cdot k), \gamma\gamma').$$

Next let $\Omega = \{-1, 0, 1\}^\Gamma$ be the product space equipped with the compact product topology. We let G act on Ω as follows. The elements of K act by coordinatewise left multiplication (in the semigroup $\{-1, 0, 1\}$), while the action of $\gamma \in \Gamma$ is again via the corresponding permutation. Thus, for $\omega \in \Omega, k \in K, \gamma, \gamma' \in \Gamma$,

$$((k, \gamma)\omega)(\gamma') = k(\gamma)\omega(\gamma'\gamma).$$

Consider the dynamical system (Ω, G) and, as above, let $G_\omega = \{g \in G : g\omega = \omega\}$. As our goal is to realize our URS's as collections of subgroups of K, we would like to work with a G-action in which the Γ-subaction is free. Now by an old result of Ellis [**14**], the action of Γ on its universal minimal dynamical system $M(\Gamma)$ is free. From this it is easy to deduce that there exists a metrizable minimal free Γ system, say (W, Γ). We will consider W as a G-system (where K acts trivially) and we let (Ω', G) be the product G-system with $\Omega' = \Omega \times W$.

Now it is not hard to see that the set $\{G_{\omega'} : \omega' \in \Omega'\}$ is a closed invariant subset of $\mathcal{S}(G)$ which is isomorphic to the Γ 2-shift $(\Theta, G) := (\{0, 1\}^\Gamma, G)$ (where K acts trivially). In fact for a point $\omega' \in \Omega'$ the corresponding subgroup $G_{\omega'}$ is the subgroup $\oplus_A \{1, -1\}$ (considered as a subgroup of K), where $A = A(\omega') = \{\gamma \in \Gamma : \omega'(\gamma) = 0\}$. Thus the groups $\{G_{\omega'} : \omega' \in \Omega'\}$ are in one-to-one correspondence with the functions $\{\mathbf{1}_A : A \subset \Gamma\} = \{0, 1\}^\Gamma$. Moreover, we see that the map $\phi : \omega' \to G_{\omega'}$ can be viewed as the homomorphism $\omega' \to \mathbf{1}_{A(\omega')}$, $\Omega' \to \{0, 1\}^\Gamma$. This observation shows that the dynamical system $\mathcal{S}_{\Omega'} = \{G_{\omega'} : \omega' \in \Omega'\} \subset \mathcal{S}(G)$, the stability system of (Ω', G), is isomorphic to the dynamical system $(\Theta, G) = (\{0, 1\}^\Gamma, G) = (\{0, 1\}^\Gamma, \Gamma)$.

Next to every Γ-invariant closed subset $\Lambda \subset \Theta$ we associate a minimal G-system $\tilde{\Lambda} \subset \phi^{-1}(\Lambda) \subset \Omega$. Now $\tilde{\Lambda}$ is a G minimal system and its stability system $\mathcal{S}_{\tilde{\Lambda}} = \{G_\omega : \omega \in \tilde{\Lambda}\}$ is a G-URS which is isomorphic to Λ. Thus we have shown that *every Γ-minimal subset of the 2-shift $(\{0,1\}^{\Gamma}, \Gamma)$ appears as a stability system,* hence as a URS in $\mathcal{S}(G)$.

3.1. EXAMPLE. Let Γ be the free group on two generators, $\Gamma = \mathbb{F}_2 = \langle a, b \rangle$. Let Z be the space of right infinite reduced words on the letters $\{a, a^{-1}, b, b^{-1}\}$. Then Γ acts on Z by concatenation on the left and cancelation.

Let $C(a) = \{z \in Z : \text{ the first letter of } z \text{ is } a\}$ and let $\psi : Z \to \{0,1\}^{\Gamma}$ be defined by $\psi(z)(\gamma) = \mathbf{1}_{C(a)}(\gamma z)$, $z \in Z, \gamma \in \Gamma$. It is not hard to check that ψ is an isomorphism of the minimal system (Z, Γ) into the system $(\{0,1\}^{\Gamma}, \Gamma)$. Let $\Lambda = \psi(Z)$, then Λ is a minimal subsystem of $(\{0,1\}^{\Gamma}, \Gamma)$, and thus, via the construction described above, we obtain a realization of the system (Z, Γ) as a URS of the group $G = \mathbb{Z}_2 \wr \Gamma$, namely as the stability system of the dynamical system $(\tilde{\Lambda}, G)$.

Let m be the probability measure $m = \frac{1}{4}(\delta_a + \delta_b + \delta_{a^{-1}} + \delta_{b^{-1}})$ on Γ and let η be the probability measure on Z given by

$$\eta(C(\epsilon_1, \ldots, \epsilon_n)) = \frac{1}{4 \cdot 3^{n-1}},$$

where for $\epsilon_j \in \{a, a^{-1}, b, b^{-1}\}$, $C(\epsilon_1, \ldots, \epsilon_n) = \{z \in Z : z_j = \epsilon_j, \ j = 1, \ldots, n\}$. The measure η is m-stationary (i.e. $m * \eta = \eta$) and the m-system $\mathbf{Z} = (Z, \eta, \Gamma)$ is the Poisson boundary $\Pi(\mathbb{F}_2, m)$ (see e.g. [15]). The push forward $\psi_*(\eta)$ is an m-stationary measure on Λ. The fact that the system (Z, Γ) is strongly proximal (see [17]) shows that there is no Γ invariant measure on the URS (Λ, Γ). Thus we have the following:

3.2. PROPOSITION. *The URS $\mathcal{S}_{\tilde{\Lambda}} \cong (\Lambda, \Gamma)$ carries the probability measure η which is an ergodic NSRS, but it admits no IRS.*

4. Applications to ergodic systems

Our first application is a direct consequence of Proposition 2.7.

4.1. PROPOSITION. *Let $\mathbf{X} = (X, \mathcal{B}, \mu, G)$ be an ergodic nonsingular dynamical system. Let $\phi : X \to \mathcal{S}(\Gamma)$ be the map $x \mapsto G_x = \{g \in G : gx = x\}$. Let $\nu = \phi_*(\mu)$, the push forward probability measure on $\mathcal{S}(G)$. Finally, let $Z = \mathrm{supp}\,(\nu)$. Then Z is a TTS and for every dynamical property $\mathcal{L} \subset \mathcal{S}(\Gamma)$ the set $Z \cap \mathcal{L}$ is either meager or comeager. If moreover the dynamical property $\mathcal{L}$ is ν-measurable then $\nu(Z \cap \mathcal{L})$ is either 0 or 1.*

PROOF. The ergodicity of $\mathbf{X}$ implies the ergodicity of the factor nonsingular system (Z, ν, G). Now the latter is a topological system, where by construction $\mathrm{supp}\,(\nu) = Z$. In this situation ergodicity implies topological transitivity and the zero-one law applies. The last assertion follows directly from the ergodicity of $\mathbf{X}$. $\qquad\square$

Thus e.g. let us single out the following.

4.2. PROPOSITION. *For an ergodic nonsingular $\mathbf{X}$ either μ-a.e. G_x is amenable or μ-a.e. G_x is non amenable.*

For more results on the stability systems of ergodic actions of Lie groups we refer to [22].

For our next application consider a dynamical system (X, G) (either compact or Borel). For each $g \in G$ let $F_g = \{x \in X : gx = x\}$. We then have, for every $x \in X$,

$$G_x = \{g \in G : x \in F_g\}.$$

4.3. COROLLARY. *Let (Ω, μ, G) be a probability measure preserving system and suppose it admits a minimal model (X, ν, G). Thus, we assume the existence of a measure isomorphism $\rho : (\Omega, \mu, G) \to (X, \nu, G)$. Then for μ-a.e. $\omega \in \Omega$ the orbit closure $\text{cls}\{gG_\omega g^{-1} : g \in G\}$ must contain S_X as a unique minimal subset.*

PROOF. For each $g \in G$ let

$$A_g = \{\omega \in \Omega : g\omega = \omega\}.$$

Then for every $\omega \in \Omega$ we have

$$G_\omega = \{g \in G : \omega \in A_g\}.$$

Now for every $g \in G$ we have $\nu(\rho(A_g) \triangle F_g) = 0$ and it follows that for μ-a.e. ω, $G_\omega = G_{\rho(\omega)}$. Our claim now follows from Proposition 1.2.(3). □

4.4. EXAMPLE. As in Section 3, let $\Omega = \{-1, 0, 1\}^{\mathbb{Z}}$ and $\sigma : \Omega \to \Omega$ the shift map. Let μ denote the product measure $\{1/4, 1/2, 1/4\}^{\mathbb{Z}}$ on Ω. Set $\mathbf{L}_0 = \oplus_{\mathbb{Z}}\{-1, 1\}$ (the countable direct sum) and let $\mathbf{L}_0$ act on Ω by coordinate-wise multiplication. Let $\mathbf{L} = \{1, -1\} \wr \mathbb{Z}$, the *lamplighter group*. We view $\mathbf{L}$ as the group of homeomorphisms of Ω generated by $\mathbf{L}_0$ and σ. As in Section 3 we observe that the set $\{\mathbf{L}_\omega : \omega \in \Omega\}$ is a closed invariant subset of $S(\mathbf{L})$ which is isomorphic to the 2-shift $(\{0, 1\}^{\mathbb{Z}}, \sigma)$ (where $\mathbf{L}_0$ acts trivially). For a point $\omega \in \Omega$ the corresponding subgroup $\mathbf{L}_\omega$ is the subgroup $\oplus_A\{-1, 1\}$, where $A = A(\omega) = \{n \in \mathbb{Z} : \omega(n) = 0\}$. The groups $\{\mathbf{L}_\omega : \omega \in \Omega\}$ are in one-to-one correspondence with the functions $\{\mathbf{1}_A : A \subset \mathbb{Z}\} = \{0, 1\}^{\mathbb{Z}}$, and the map $\phi : \omega \to \mathbf{L}_\omega$ is a homomorphism $\omega \mapsto \mathbf{1}_{A_\omega}$, $\Omega \to \{0, 1\}^{\mathbb{Z}}$. Thus the dynamical system $S_\Omega = \{\mathbf{L}_\omega : \omega \in \Omega\} \subset S(\mathbf{L})$, the stability system of $(\Omega, \mathbf{L})$, is isomorphic to the full shift dynamical system $(\{0, 1\}^{\mathbb{Z}}, \sigma)$.

It was shown in [27] that for an arbitrary countable infinite group G, any **free** ergodic probability measure preserving G-system admits a minimal model. In contrast we show next that for the lamplighter group there is an ergodic measure preserving action which does not admit a minimal model.

4.5. THEOREM. *For the lamplighter group $\mathbf{L}$ there is an ergodic dynamical system for which no minimal model exits.*

PROOF. Consider the dynamical system $(\Omega, \mu, \mathbf{L})$ presented in Example 4.4. Suppose it admits a minimal model $(X, \nu, \mathbf{L})$. Thus, we assume the existence of a measure isomorphism $\rho : (\Omega, \mu, \mathbf{L}) \to (X, \nu, \mathbf{L})$. However, the topological system $(\{0, 1\}^{\mathbb{Z}}, \mathbf{L}) = (\{0, 1\}^{\mathbb{Z}}, \sigma)$ contains uncountably many distinct minimal sets and applying Corollary 4.3 we arrive at a contradiction. □

4.6. PROPOSITION. *Let (X, G) be a minimal system. Let μ be a G-invariant probability measure on X such that the action of G on the probability space (X, μ) is μ essentially free. Then the action (X, G) is essentially free.*

PROOF. For each $g \in G$ we set $F_g = \{x \in X : gx = x\}$, a closed subset of X. If for some $g \neq e$, int $F_g \neq \emptyset$ then, by minimality, $\mu(F_g) > 0$ which contradicts our assumption that the measure action is essentially free. Thus each set $F_g, g \neq e$ is nowhere dense, hence the set $F = \bigcup\{F_g : e \neq g \in G\}$ is meager. Since $G_x = \{e\}$ for every $x \in X \setminus F$, we conclude that the system (X, G) is indeed essentially free. $\square$

4.7. REMARK. The same proof works assuming only that μ is nonsingular.

5. Uncountably many URS's for $\mathbb{F}_2$

5.1. THEOREM. *For the free group* $G = \mathbb{F}_2$, *the space* $\mathcal{S}(G)$ *contains uncountably many non-isomorphic infinite URS's.*

PROOF. Our construction is based on Example 4.5 above and the following basic observation. The lamplighter group $\mathbf{L}$ is generated by two elements. Explicitly we can take these to be e_1 and σ, where $e_1 \in \mathbf{L}_0$ is defined by $e_1(n) = (-1)^{\delta_{1n}}$. Let $\eta : \mathbb{F}_2 \to \mathbf{L}$ be the surjective group homomorphism which is determined by $\eta(a) = e_1$ and $\eta(b) = \sigma$. Next define an action of $\mathbb{F}_2$ on Ω by letting $g\omega = \eta(g)\omega$, $g \in \mathbb{F}_2, \omega \in \Omega$. Clearly then $G_\omega = \eta^{-1}(\mathbf{L}_\omega)$, $(\omega \in \Omega)$ and again we see that the dynamical system $\mathcal{S}_\Omega = \{G_\omega : \omega \in \Omega\} \subset \mathcal{S}(G)$, the G-stability system of (Ω, G), is isomorphic to the dynamical system $(\{0, 1\}^{\mathbb{Z}}, \sigma)$ (see Proposition 1.5.(1)). As the latter system contains an uncountable family of pairwise non isomorphic minimal subsystems, our proof is complete. $\square$

5.2. REMARK. This is of course a much stronger assertion than the claim in [25, Lemma 3.9] that the action of $\mathbb{F}_2$ on $\mathcal{S}(\mathbb{F}_2)$ is not tame.

5.3. REMARK. By construction these URS's are noneffective. A more sophisticated construction due to Louis Bowen [7, Theorem 5.1] and the main result of [16] (actually [20, Theorem 3.1] will suffice here) yield an uncountable family of pairwise nonisomorphic effective URS's for $\mathbb{F}_2$.

Since $\mathbb{F}_2$ sits as a finite index subgroup in the group in $SL(2, \mathbb{Z})$ it follows that this latter group also admits an uncountable family of pairwise non isomorphic URS's. By the well known work of Stuck and Zimmer, for the group $G = SL(3, \mathbb{Z})$, any IRS on $\mathcal{S}(G)$ is finite (see [25, Corollaries 4.4 and 4.5]). We do not know whether the same holds for URS's on $\mathcal{S}(G)$.

5.4. PROBLEM. Are there infinite URS's in $\mathcal{S}(SL(3, \mathbb{Z}))$?

6. Realizations of URS's as stability systems

A basic result proved in [2] is that to every ergodic IRS ν (a G-invariant probability measure on $\mathcal{S}(G)$) there corresponds an ergodic probability measure preserving system $\mathbf{X} = (X, \mathcal{B}, \mu, G)$ whose stability system is $(\mathcal{S}(G), \nu)$. See the recent works of Creutz and Peterson [13, Theorem 3.3] and Creutz [12, Theorem 3.3] for continuous and NSRS versions of this theorem. We expected to be able to obtain an analogous statement for URS's. However, the question whether this desired analog holds remains open (see Problem 6.2 below) and we only have the following:

6.1. PROPOSITION. *Let G be a countable infinite group. For every URS $Z \subset \mathcal{S}(G)$ there is a topologically transitive system (X, G) with Z as its stability system.*

$105 2015 rectrine   R

1142190   rectrine   QA614   2014-021630   978-1-4704-0931-9
Recent Trends in Ergodic Theory and Dynamical Systems; proceedings
International Conference in Honor of S. G. Dani's 65th Birthday. Recent Trends in Ergodic
Theory and Dynamical System (2012: Vadodara, India) Edited by Siddhartha Bhattacharya,
Tarun Das, Anish Ghosh, and Riddhi Shah
(Contemporary Mathematics; Volume 631)
American Mathematical Society,   258 p. $105.00   (Paper)
c2014
ref
2/10/2015   MARC

PROOF. We begin with the case where some $H \in Z$ has finite index in G. It then follows that Z is finite and we let $Y = G/H = \{gH : g \in G\}$, the finite homogeneous G-space of right H-cosets. We now take X to be the G-orbit of the point $x_0 = (H, H)$ in the product system $(Z \times Y, G)$. Note that G acts on Z by conjugation, whereas it acts on Y by multiplication on the left. The stability subgroup G_{x_0} is H and it follows that indeed Z is the stability system of the (minimal) finite system (X, G).

We now assume that $[G : H] = \infty$ for every $H \in Z$.

Set $\Omega = \{0, 1\}^G$ and let G act on Ω by $g\omega(h) = \omega(g^{-1}h)$. For $H < G$ let

$$\Omega_H = \{\omega \in \Omega : h\omega = \omega, \ \forall h \in H\},$$

and

$$\Theta = \bigcup\{\{H\} \times \Omega_H : H \in Z\}.$$

It is easy to check that Θ is a closed G-invariant subset of $Z \times \Omega$. Also note that the map $\pi : (\Theta, G) \to (Z, G)$ is clearly a homomorphism of G-systems.

1. Our first observation is that for each $x = (H, \omega) \in \Theta$ we have $\pi(x) = H$ and

$$H \subset G_x = \{g \in G : gx = x\} \subset G_{\pi(x)} = G_H = \{g \in G : gHg^{-1} = H\} = N_G(H),$$

the normalizer of H in G. Also note that $H \lhd N_G(H)$, so that $\tilde{H} = N_G(H)/H$ is a group. Finally observe that for every $x \in \Theta$ the group $\tilde{H}$, with $H = \pi(x)$, acts on the fiber $\pi^{-1}(H) = \{H\} \times \Omega_H \cong \Omega_H$.

2. Given $H \in Z$ consider the map $\rho_H : \Omega_H \to \{0, 1\}^{\tilde{H}}$ which sends an element $\omega \in \Omega_H \subset \Omega = \{0, 1\}^G$ to its restriction to the subset $N_G(H) \subset G$. (We identify the image $\rho_H(\Omega_H)$ with $\{0, 1\}^{\tilde{H}}$.) Clearly ρ_H is a surjective homomorphism of $\tilde{H}$-systems.

3. We now fix some (arbitrary) point $H \in Z$. We claim that there exists a $\tilde{H}$-minimal subsystem of the symbolic system $(\{0, 1\}^{\tilde{H}}, \tilde{H})$, say $\tilde{Y}$, such that the action of $\tilde{H}$ on $\tilde{Y}$ is free. There are two cases we need to consider. The first case is when the group $\tilde{H}$ is finite. We then take $\tilde{Y}$ to be $\tilde{H}\chi_e$ where $\chi_e \in \{0, 1\}^{\tilde{H}}$ is the function $\chi_e(h) = \delta_{e,h}$. In the second case, where $\tilde{H}$ is infinite, we apply a recent result of Gao, Jackson and Seward [16], which ensures the existence of a free $\tilde{H}$-minimal subsystem $\tilde{Y}$ of $(\{0, 1\}^{\tilde{H}}, \tilde{H})$. In each of these cases choose Y to be any minimal subsystem of the system $(\Omega_H, \tilde{H})$ such that $\rho_H(Y) = \tilde{Y}$. From the fact that $\tilde{H}$ acts freely on Y it follows immediately that for every $x = (H, \omega) \in Y$ we have

$$N_G(H)_x = G_x = H.$$

4. Now pick any point $y_0 = (H, \omega_0) \in Y$ and let $X = \mathrm{cls}\, Gy_0 \subset \Theta$. If $x_1 = (H_1, \omega_1)$ is a continuity point of the map $x \mapsto G_x$ (from X to $\mathcal{S}(G)$), and $g_i y_0 \to x_1$ then $G_{g_i y_0} = g_i H g_i^{-1} \to G_{x_1} = H_1 = \pi(x_1) \in Z$. Thus on the dense G_δ set $X_c \subset X$ of continuity points of the map $x \mapsto G_x$, this map coincides with π. In other words Z is the stability system of X. $\square$

6.2. PROBLEM. Can Proposition 6.1 be strengthened to provide a minimal (X, G) with Z as its stability system ?

References

[1] Miklos Abert, Nicolas Bergeron, Ian Biringer, Tsachik Gelander, Nikolay Nikolov, Jean Raimbault, and Iddo Samet, *On the growth of Betti numbers of locally symmetric spaces* (English, with English and French summaries), C. R. Math. Acad. Sci. Paris **349** (2011), no. 15-16, 831–835, DOI 10.1016/j.crma.2011.07.013. MR2835886 (2012j:58033)

[2] Miklós Abért, Yair Glasner, and Bálint Virág, *Kesten's theorem for invariant random subgroups*, Duke Math. J. **163** (2014), no. 3, 465–488, DOI 10.1215/00127094-2410064. MR3165420

[3] M. Abert, Y. Glasner and B. Virag, *The measurable Kesten theorem*, a preprint; arXiv:1111.2080.

[4] Louis Auslander and Calvin C. Moore, *Unitary representations of solvable Lie groups*, Mem. Amer. Math. Soc. No. **62** (1966), 199. MR0207910 (34 #7723)

[5] Gilbert Baumslag, *Lecture notes on nilpotent groups*, Regional Conference Series in Mathematics, No. 2, American Mathematical Society, Providence, R.I., 1971. MR0283082 (44 #315)

[6] Gerald Beer, *Topologies on closed and closed convex sets*, Mathematics and its Applications, vol. 268, Kluwer Academic Publishers Group, Dordrecht, 1993. MR1269778 (95k:49001)

[7] Lewis Bowen, *Random walks on random coset spaces with applications to Furstenberg entropy*, Invent. Math. **196** (2014), no. 2, 485–510, DOI 10.1007/s00222-013-0473-0. MR3193754

[8] L. Bowen, *Invariant random subgroups of the free group*, a preprint, arXiv:1204.5939.

[9] L. Bowen, R. Grigorchuk and R. Kravchenko, *Random subgroups of the lamplighter group*, a preprint, arXiv:1206.6780.

[10] Pierre-Emmanuel Caprace and Nicolas Monod, *Relative amenability*, Groups Geom. Dyn. **8** (2014), no. 3, 747–774, DOI 10.4171/GGD/246. MR3267522

[11] G. Choquet, *Convergences*, Ann. Univ. Grenoble. Sect. Sci. Math. Phys. (N.S.) **23** (1948), 57–112. MR0025716 (10,53d)

[12] D. Creutz, *Stabilizers of Actions of Lattices in Products of Groups*, arXiv:1305.3648 [math.DS].

[13] D. Creutz and J. Peterson, *Stabilizers of Ergodic Actions of Lattices and Commensurators*, arXiv:1303.3949.

[14] Robert Ellis, *Universal minimal sets*, Proc. Amer. Math. Soc. **11** (1960), 540–543. MR0117716 (22 #8491)

[15] Hillel Furstenberg and Eli Glasner, *Stationary dynamical systems*, Dynamical numbers—interplay between dynamical systems and number theory, Contemp. Math., vol. 532, Amer. Math. Soc., Providence, RI, 2010, pp. 1–28, DOI 10.1090/conm/532/10481. MR2762131 (2012g:22007)

[16] Su Gao, Steve Jackson, and Brandon Seward, *A coloring property for countable groups*, Math. Proc. Cambridge Philos. Soc. **147** (2009), no. 3, 579–592, DOI 10.1017/S0305004109002655. MR2557144 (2010j:03056)

[17] Shmuel Glasner, *Proximal flows*, Lecture Notes in Mathematics, Vol. 517, Springer-Verlag, Berlin-New York, 1976. MR0474243 (57 #13890)

[18] Eli Glasner, *A topological version of a theorem of Veech and almost simple flows*, Ergodic Theory Dynam. Systems **10** (1990), no. 3, 463–482, DOI 10.1017/S0143385700005691. MR1074314 (92e:28012)

[19] Eli Glasner and Jonathan L. King, *A zero-one law for dynamical properties*, Topological dynamics and applications (Minneapolis, MN, 1995), Contemp. Math., vol. 215, Amer. Math. Soc., Providence, RI, 1998, pp. 231–242, DOI 10.1090/conm/215/02944. MR1603201 (99d:28039)

[20] Eli Glasner and Vladimir V. Uspenskij, *Effective minimal subflows of Bernoulli flows*, Proc. Amer. Math. Soc. **137** (2009), no. 9, 3147–3154, DOI 10.1090/S0002-9939-09-09905-5. MR2506474 (2010i:37020)

[21] E. Glasner and B. Weiss, *On the interplay between measurable and topological dynamics*, Handbook of dynamical systems. Vol. 1B, Elsevier B. V., Amsterdam, 2006, pp. 597–648, DOI 10.1016/S1874-575X(06)80035-4. MR2186250 (2006i:37005)

[22] Valentin Ya. Golodets and Sergey D. Sinel'shchikov, *On the conjugacy and isomorphism problems for stabilizers of Lie group actions*, Ergodic Theory Dynam. Systems **19** (1999), no. 2, 391–411, DOI 10.1017/S014338579913013X. MR1685400 (2001c:22006)

[23] I. Schochetman, *Nets of subgroups and amenability*, Proc. Amer. Math. Soc. **29** (1971), 397–403. MR0281837 (43 #7551)

[24] Garrett Stuck, *Minimal actions of semisimple groups*, Ergodic Theory Dynam. Systems **16** (1996), no. 4, 821–831, DOI 10.1017/S0143385700009123. MR1406436 (98a:57046)

[25] Garrett Stuck and Robert J. Zimmer, *Stabilizers for ergodic actions of higher rank semisimple groups*, Ann. of Math. (2) **139** (1994), no. 3, 723–747, DOI 10.2307/2118577. MR1283875 (95h:22007)

[26] A. M. Vershik, *Totally nonfree actions and the infinite symmetric group* (English, with English and Russian summaries), Mosc. Math. J. **12** (2012), no. 1, 193–212, 216. MR2952431

[27] Benjamin Weiss, *Minimal models for free actions*, Dynamical systems and group actions, Contemp. Math., vol. 567, Amer. Math. Soc., Providence, RI, 2012, pp. 249–264, DOI 10.1090/conm/567/11253. MR2931921

DEPARTMENT OF MATHEMATICS, TEL AVIV UNIVERSITY, TEL AVIV, ISRAEL
E-mail address: `glasner@math.tau.ac.il`

INSTITUTE OF MATHEMATICS, HEBREW UNIVERSITY OF JERUSALEM, JERUSALEM, ISRAEL
E-mail address: `weiss@math.huji.ac.il`

Contemporary Mathematics
Volume **631**, 2015
http://dx.doi.org/10.1090/conm/631/12597

Values of binary quadratic forms at integer points and Schmidt games

Dmitry Kleinbock and Barak Weiss

ABSTRACT. We prove that for any countable set A of real numbers, the set of binary indefinite quadratic forms Q such that the closure of $Q\left(\mathbb{Z}^2\right)$ is disjoint from A has full Hausdorff dimension.

Dedicated to S.G. Dani on the occasion of his 65th birthday

1. Introduction

We start with the following statement, conjectured by Oppenheim and Davenport in the 1930s and proved by Margulis in the 1980s:

THEOREM 1.1. Let Q be a real nondegenerate indefinite quadratic form of $n > 2$ variables which is not proportional to a rational form. Then

$$(1.1) \qquad 0 \text{ belongs to the closure of } Q(\mathbb{Z}^n \smallsetminus \{0\}).$$

Margulis used an approach which was suggested earlier by Raghunathan and implicitly used earlier by Cassels and Swinnerton-Dyer [**CaSD**]. Let Q_0 be a fixed quadratic form of the same signature as Q, then one can write $Q(\mathbf{v}) = aQ_0(g\mathbf{v})$ for some $g \in G = \mathrm{SL}_n(\mathbb{R})$ and $a \in \mathbb{R}$. Let F be the stabilizer of Q_0 and let $\Gamma = \mathrm{SL}_n(\mathbb{Z})$. Then (1.1) holds if and only if the orbit $Fg\mathbb{Z}^n$ in the space G/Γ of unimodular lattices in $\mathbb{R}^n$ is unbounded. The theorem proved by Margulis (in the case $n = 3$, to which the general case can be reduced) stated that any bounded orbit must be compact, from which it is not hard to derive Theorem 1.1.

Note that it was later proved by Dani and Margulis that for Q as above the set $Q(\mathbb{Z}^n)$, and even $Q\left(P(\mathbb{Z}^n)\right)$, is dense in $\mathbb{R}$ (here $P(\mathbb{Z}^n)$ stands for the set of primitive integer points in $\mathbb{Z}^n$).

When $n = 2$ it is well known the conclusion of Theorem 1.1 fails. Namely, take

$$(1.2) \qquad Q(p,q) = p^2 - \lambda^2 q^2 = q^2 \left(\frac{p}{q} - \lambda\right)\left(\frac{p}{q} + \lambda\right)$$

such that λ is *badly approximable*, that is,

$$\inf_{p \in \mathbb{Z},\, q \in \mathbb{N}} q^2 \left|\frac{p}{q} - \lambda\right| > 0\,;$$

2010 *Mathematics Subject Classification.* Primary: 37A17, 11J06, 11J83.

Key words and phrases. Oppenheim conjecture, the space of lattices, quadratic forms, nondense orbits, Schmidt games, hyperplane, absolute winning.

then the absolute value of $Q(p,q)$ is uniformly bounded away from 0 for any nonzero integer (p,q). It is known from the work of Jarník [**J**] that the set of such λ's, although null, is *thick*, that is, its intersection with any nonempty open set has full Hausdorff dimension. As in the reduction of the Oppenheim conjecture to a dynamical statement, choose $Q_0(x,y) \overset{\text{def}}{=} xy$, let

$$(1.3) \qquad\qquad G \overset{\text{def}}{=} \mathrm{SL}_2(\mathbb{R}),$$

and let F be the connected component of the identity in the stabilizer of Q_0, namely

$$(1.4) \qquad\qquad F = \{g_t : t \in \mathbb{R}\} \text{ where } g_t = \begin{pmatrix} e^t & 0 \\ 0 & e^{-t} \end{pmatrix}.$$

We also let

$$(1.5) \qquad\qquad X \overset{\text{def}}{=} G/\Gamma \text{ where } \Gamma \overset{\text{def}}{=} \mathrm{SL}_2(\mathbb{Z}), \text{ and } \mathcal{Q} \overset{\text{def}}{=} F \backslash G;$$

thus X can be identified with the space of unimodular lattices in $\mathbb{R}^2$, and $\mathcal{Q}$ can be identified with the space of binary indefinite quadratic forms, considered up to scaling. The set of integer values of a quadratic form is a (set-valued) function on $F\backslash G/\Gamma$ but since this double coset space has a complicated topological structure, it is more useful to consider it either as an F-invariant function on X, or dually, as a Γ-invariant function on $\mathcal{Q}$. This duality principle (already evident in the work of Cassels and Swinnerton-Dyer [**CaSD**] and developed explicitly by Dani [**D1**] in a related context) makes it possible to recast dynamical properties of the F-action on X as properties of quadratic forms. In particular (see e.g. [**Ma2**, Lemma 2.2.1]), it follows from Mahler's Compactness Criterion that for a quadratic form $Q(\mathbf{v}) = Q_0(g\mathbf{v}) \in \mathcal{Q}$, the set of values $Q(\mathbb{Z}^2)$ has a gap at zero if and only if the orbit $F(g\mathbb{Z}^2)$ is bounded in X. It is known [**KM**] that the set of points of X with bounded F-orbits is thick. Thus one has

THEOREM 1.2. *Let X and Q_0 be as above. Then the set*

$$\left\{ x \in X : 0 \notin \overline{Q_0(x \smallsetminus \{0\})} \right\}$$

is thick; dually, the set of binary indefinite quadratic forms whose set of values at integer points has a gap at 0 is thick in the space of all binary indefinite forms.

Note that the paper [**KM**] was motivated by the work of Dani [**D1, D2**], who used Schmidt's results and methods [**S1, S2**] to find thick sets of points in homogeneous spaces with bounded trajectories of one-parameter diagonal semigroups. See also [**Ar, AL, KW2**] for related results.

The goal of this paper is to strengthen Theorem 1.2 by considering even more complicated properties of the sets of values of quadratic forms at integer points, and, correspondingly, the behavior of F-orbits in X. Here is one of our main results:

THEOREM 1.3. *For any countable subset $A \subset \mathbb{R}$, the set*

$$\left\{ x \in X : \overline{Q_0(x \smallsetminus \{0\})} \cap A = \varnothing \right\}$$

is thick. Consequently, the set of binary indefinite quadratic forms whose set of values at integer points miss a given countable set is thick in the space of all binary indefinite forms.

Note that by a theorem of Lekkerkerker [**L**], for Q as in (1.2) the set of accumulation points of $Q(\mathbb{Z}^2)$ is discrete if and only if λ is quadratic irrational. See [**TV, DN**] for a precise description of the sets of accumulation points in these cases.

We will derive Theorem 1.3 from its more general dynamical counterpart. To state it, we need the following definition. Let H be a connected subgroup of G different from F, and let Z be a submanifold of X. Say that Z is (F, H)-*transversal* if the following two conditions hold:

(F) for any $x \in Z$, $T_x(Fx)$ is not contained in $T_x Z$;

(H, F) for any $x \in Z$, $T_x(Hx)$ is not contained in $T_x Z \oplus T_x(Fx)$.

For example, the above conditions are satisfied if Z consists of a single point. If $\dim(Z) = 1$, the (F, H)-transversality of Z is equivalent to saying that for each $x \in Z$ the lines $T_x(Hx)$, $T_x Z$ and $T_x(Fx)$ are in general position, i.e. they generate the space $T_x X$.

We permit Z to be a manifold with boundary; in such a case, smoothness of maps and definitions of tangent spaces at the boundary points are defined by positing the existence of smooth extensions (see e.g. [**GP**, Chap. 2]). We also note that in the application to quadratic forms, Z is an analytic manifold, but in all our arguments, it suffices to assume that Z is C^1-smooth.

Now let us denote by H^+ and H^- the *expanding and contracting horospherical subgroups* with respect to g_1:

$$(1.6) \qquad H^+ \stackrel{\text{def}}{=} \left\{ h_s \stackrel{\text{def}}{=} \begin{pmatrix} 1 & s \\ 0 & 1 \end{pmatrix} : s \in \mathbb{R} \right\}, \quad H^- \stackrel{\text{def}}{=} \left\{ \begin{pmatrix} 1 & 0 \\ s & 1 \end{pmatrix} : s \in \mathbb{R} \right\}.$$

A dynamical statement which will imply Theorem 1.3 is as follows:

THEOREM 1.4. *Let Z be a countable union of submanifolds of X which are both (F, H^+)- and (F, H^-)-transversal. Then the set*

$$(1.7) \qquad \left\{ x \in X : Fx \text{ is bounded and } \overline{Fx} \cap Z = \varnothing \right\}$$

is thick.

Note that for arbitrary homogeneous spaces $X = G/\Gamma$, non-quasiunipotent subgroups F and *finite* sets Z the thickness of the set (1.7) was conjectured by Margulis [**Ma1**]. Then in [**KM**] it was shown that, for mixing flows, the set of points with bounded orbits is thick, and in [**K**], for arbitrary homogeneous spaces and Z being a *finite* union of Z_i as in the above theorem[1]—that the set

$$\left\{ x \in X : \overline{Fx} \cap Z = \varnothing \right\}$$

is thick. Margulis' Conjecture was finally proved in [**KW2**] using the technique of Schmidt games, with the set Z upgraded from finite to countable. However the argument of [**KW2**] could not produce a result for Z being more than zero-dimensional, which, in particular, is needed for an application to quadratic forms.

Organization of the paper. In §2 we explain the reduction of Theorem 1.3 to Theorem 1.4. In §3 we describe a variant of Schmidt's (α, β)-game. This variant is very close to the *absolute game* and *hyperplane absolute game* introduced in [**Mc**] and [**BFKRW**] respectively, but adapted to a situation in which we want to play

---

[1]Technically the statement in [**K**, Corollary 4.4.2] is weaker than quoted here, namely Z_i are assumed to satisfy condition (F) and have dimension smaller than dimensions of H^+ and H^-—but the argument relies precisely on the combination of (F, H^+)- and (F, H^-)-transversality.

on a smooth manifold. In §4 we state our main technical result, Theorem 4.3 which implies Theorem 1.4, and is a result on the winning property of a certain set for the above game. We complete the proof of Theorem 4.3 in §5.

Acknowledgements. Thanks are due to Steffen Weil and the anonymous referee for useful suggestions. The authors gratefully acknowledge the support of BSF grant 2010428, ERC starter grant DLGAPS 279893, and NSF grant DMS-1101320.

2. Dynamics and gaps between values of quadratic forms

In this section we explain how to reduce Theorem 1.3 to Theorem 1.4. As was mentioned in the introduction, the equivalence between the set of values of the form at integer points being bounded away from 0 and the F-orbit of the corresponding lattice being bounded in X is well-known. We need to understand how to dynamically describe the set of binary quadratic forms whose values at $\mathbb{Z}^2$ miss a fixed $a \neq 0$. So fix a nonzero a and choose $\mathbf{v} \in \mathbb{R}^2$ such that $Q_0(\mathbf{v}) = a$. Not much will depend on this choice, yet the most obvious one seems to be choosing

$$(2.1) \qquad \mathbf{v} = \begin{cases} (\sqrt{a}, \sqrt{a}) & \text{if } a > 0 \\ (-\sqrt{|a|}, \sqrt{|a|}) & \text{otherwise} \end{cases}$$

which is what we will do. Now define

$$\tilde{Z}_{\mathbf{v}} \stackrel{\text{def}}{=} \{x \in X : \mathbf{v} \in x\},$$

that is, the set of unimodular lattices in $\mathbb{R}^2$ containing $\mathbf{v}$. The structure of this set is easy to describe: $\tilde{Z}_{\mathbf{v}} = \bigcup_{n \in \mathbb{N}} Z_{\frac{1}{n}\mathbf{v}}$ where

$$Z_{\mathbf{v}} \stackrel{\text{def}}{=} \{x \in X : \mathbf{v} \in P(x)\},$$

the set of unimodular lattices in $\mathbb{R}^2$ containing $\mathbf{v}$ as a *primitive* vector. The latter is simply a closed horocycle, namely a periodic orbit under the action of the subgroup V of G stabilizing $\mathbf{v}$, which is easily seen to be, for our choice of $\mathbf{v}$, equal to

$$(2.2) \qquad V \stackrel{\text{def}}{=} \begin{cases} \left\{ \begin{pmatrix} 1+s & -s \\ s & 1-s \end{pmatrix} : s \in \mathbb{R} \right\} & \text{if } a > 0 \\[2em] \left\{ \begin{pmatrix} -1-s & s \\ s & -1+s \end{pmatrix} : s \in \mathbb{R} \right\} & \text{otherwise} \end{cases}$$

Now let us state a proposition which relates escaping $\tilde{Z}_{\mathbf{v}}$ with a gap in values of quadratic forms at a:

PROPOSITION 2.1. *Suppose* $x \in X$ *is such that* Fx *is bounded, let* $a \neq 0$, *and define* $\mathbf{v}$ *by (2.1). Then* $\overline{Fx} \cap \tilde{Z}_{\mathbf{v}} = \varnothing$ *if and only if* $a \notin \overline{Q_0(x \smallsetminus \{0\})}$.

PROOF. For the 'if' direction, suppose that $x_0 \in \overline{Fx}$ contains $\mathbf{v}$. Then there are t_n such that $g_{t_n} x \to x_0$ and so there are $\mathbf{w}_n \in x$ with $g_{t_n} \mathbf{w}_n \to \mathbf{v}$. In particular $\mathbf{w}_n \neq 0$. Hence $Q_0(\mathbf{w}_n) = Q_0(g_{t_n} \mathbf{w}_n) \to Q_0(\mathbf{v}) = a$, so $a \in \overline{Q_0(x \smallsetminus \{0\})}$, a contradiction.

Suppose the converse does not hold, that is, there exist vectors $\mathbf{v}_k \in x$ such that $Q_0(\mathbf{v}_k) \to a$ as $k \to \infty$. For each k, choose $t_k \in \mathbb{R}$ such that $g_{t_k} \mathbf{v}_k$ belongs to the line passing through $\mathbf{v}$; then after passing to a subsequence, one can assume that

$$(2.3) \qquad\qquad g_{t_k}\mathbf{v}_k \to \mathbf{v} \text{ or } g_{t_k} \to -\mathbf{v} \text{ as } k \to \infty .$$

But Fx is relatively compact, hence there exists a limit point x_0 of the sequence $g_{t_k}x$ of lattices, and from (2.3) it follows that x_0 contains $\mathbf{v}$, or, equivalently, $x_0 \in \tilde{Z}_\mathbf{v}$, contrary to assumption. $\qquad\square$

REMARK 2.2. The boundedness of Fx was only used in proving the implication $\Longrightarrow$. If Fx is not bounded (or equivalently, $0 \in \overline{Q_0(x \smallsetminus \{0\})}$) then the situation is more interesting. A simple argument involving multiplication by integers, shows that when $Q_0(x \smallsetminus \{0\})$ contains sequences approaching 0 from both sides, then the set of values $Q_0(x \smallsetminus \{0\})$ is dense. However, as shown by Oppenheim [**O**], lattices x for which 0 is only a one-sided limit of $Q_0(x \smallsetminus \{0\})$ do exist.

We record the following:

LEMMA 2.3. For $\mathbf{v}$ as in (2.1) (in fact, the same is true for every $\mathbf{v}$ with $Q_0(\mathbf{v}) \neq 0$), the manifold $Z_\mathbf{v}$ is both (F, H^+)-transversal and (F, H^-)-transversal.

PROOF. It suffices to show that the Lie algebras of F, V and H^+, as well as those of F, V and H^-, span the Lie algebra of G. This is an easy computation using (1.4), (1.6) and (2.2). $\qquad\square$

We can now see that Theorem 1.3 follows from Theorem 1.4:

PROOF OF THEOREM 1.3 ASSUMING THEOREM 1.4. We have already mentioned that the boundedness of the F-orbit of x implies that 0 is not in the closure of the set $Q_0(x \smallsetminus \{0\})$. Now for each $a \in A \smallsetminus 0$ take $\mathbf{v} = \mathbf{v}(a)$ as in (2.1), then, in view of Lemma 2.3, the set

$$Z = \bigcup_{a \in A \smallsetminus \{0\}} \tilde{Z}_{\mathbf{v}(a)}$$

satisfies the assumption of Theorem 1.4, hence the set (1.7) is thick. On the other hand, Proposition 2.1 implies that for every x from the set, $\overline{Q_0(x \smallsetminus \{0\})} \cap A = \varnothing$. $\qquad\square$

3. Schmidt games

We first recall Schmidt's (α, β)-game, introduced in [**S1**]. The game is played by two players Alice and Bob on a complete metric space[2] X equipped with a target set S and two fixed parameters $\alpha, \beta \in (0, 1)$. Bob begins the (α, β)-game by choosing $x_1 \in X, r_1 > 0$, thus specifying the closed ball $B_1 \overset{\text{def}}{=} \bar{B}(x_1, r_1)$, where

$$\bar{B}(z, \rho) \overset{\text{def}}{=} \{x \in X : \text{dist}(x, z) \le \rho\}.$$

Then Alice and Bob continue by alternately choosing x'_i, x_{i+1} so that

$$\text{dist}(x_i, x'_i) \le (1 - \alpha)r_i, \quad \text{dist}(x'_{i+1}, x_i) \le (1 - \beta)r'_i, \quad \text{where } r'_i \overset{\text{def}}{=} \alpha r_i, \;\; r_{i+1} \overset{\text{def}}{=} \beta r'_i.$$

This implies that the closed balls

$$A_i \overset{\text{def}}{=} \bar{B}(x'_i, r'_i), \quad B_{i+1} \overset{\text{def}}{=} \bar{B}(x_{i+1}, r_{i+1})$$

---

[2]In this section, following tradition, we denote this metric space by X; elsewhere X continues to denote the space of two-dimensional unimodular lattices.

are nested, i.e.

$$B_1 \supset A_1 \supset B_2 \supset \cdots$$

The set S is said to be α-*winning* if for any $\beta > 0$ Alice has a strategy in the (α, β)-game guaranteeing that the unique point of intersection $\bigcap_{i=1}^{\infty} B_i = \bigcap_{i=1}^{\infty} A_i$ of all the balls belongs to S, regardless of the way Bob chooses to play. It is called *winning* if it is α-winning for some α.

In [**BFKRW**], inspired by ideas of McMullen [**Mc**], the *absolute hyperplane game* was introduced. This modification is played on $\mathbb{R}^d$. Let $S \subset \mathbb{R}^d$ be a target set and let $\beta \in \left(0, \frac{1}{3}\right)$. As before Bob begins by choosing a closed ball B_1 of radius r_1 and then Alice and Bob alternate moves. The sets B_i chosen by Bob are closed balls of radii r_i satisfying

$$r_{i+1} \geq \beta r_i .$$

The sets A_i chosen by Alice are r_i'-neighborhoods of affine hyperplanes, where the r_i' satisfy $r_i' \leq \beta r_i$. Additionally Bob's choices must satisfy

$$B_{i+1} \subset B_i \setminus A_i .$$

Then $S \subset \mathbb{R}^d$ is said to be β-*HAW* (where HAW is an acronym for *hyperplane absolute winning*) if Alice has a strategy which leads to

$$(3.1) \qquad \bigcap_{i=1}^{\infty} B_i \cap S \neq \varnothing$$

regardless of how Bob chooses to play; S is said to be *HAW* if it is β-HAW for all $0 < \beta < \frac{1}{3}$. It is easy to see that β-HAW implies β'-HAW whenever $\beta \leq \beta' < 1/3$; thus a set is HAW iff it is β-HAW for arbitrary small positive β. In the case $d = 1$ hyperplanes are points, and thus the HAW property coincides with the *absolute winning* property[3] introduced by McMullen in [**Mc**].

The following proposition summarizes properties of winning and HAW subsets of $\mathbb{R}^d$:

PROPOSITION 3.1. (a) Winning sets are thick.
(b) HAW implies winning.
(c) The countable intersection of α-winning (resp., HAW) sets is again α-winning (resp., HAW).
(d) The image of a HAW set under a C^1 diffeomorphism $\mathbb{R}^d \to \mathbb{R}^d$ is HAW.

For the proofs, see [**S1, Mc, BFKRW**].

REMARK 3.2. Note that in the hyperplane absolute version the radii of balls do not have to tend to zero, therefore $\cap_i B_i$ does not have to be a single point. However the outcome with radii not tending to 0 is clearly winning for Alice as long as S is dense. Thus in all the proofs of the HAW property it will be safe to assume that Bob plays so that the radii of his balls tend to zero as $n \to \infty$. Indeed, if Alice has a strategy which is guaranteed to win the game whenever $r_n \to 0$, then the target set must be dense, and hence she will automatically win if $r_n \not\to 0$.

---

[3]More generally, the paper [**BFKRW**] introduced k-dimensional absolute winning for every $0 \leq k < d$; the case $k = 0$ was considered earlier by McMullen.

We will be interested in playing variants of the two games described above on a differentiable manifold. Note that a manifold is not equipped with an intrinsic metric, nor with an intrinsic notion of affine submanifolds, and thus the definitions given above cannot be applied directly. Our approach will be to work in a given coordinate system and argue using Proposition 3.1 that the class of winning sets does not depend on the choice of a coordinate system. It will be technically simpler to work with the hyperplane absolute game and this is all that we require for the present paper. We proceed to the details.

We first define the absolute hyperplane game on an open subset $W \subset \mathbb{R}^d$. This is defined just as the absolute hyperplane game on $\mathbb{R}^d$, except for requiring that Bob's first move B_1 be contained in W. If Alice has a winning strategy, we say that S is HAW on W. Now suppose X is a C^1 d-dimensional manifold equipped with an atlas of charts $\mathcal{A} = (U_\alpha, \varphi_\alpha)$; that is, X is a separable topological space, the sets U_α are open subsets of X with $X = \bigcup_\alpha U_\alpha$, each $\varphi_\alpha : U_\alpha \to \mathbb{R}^d$ is a homeomorphism onto its image, and the transition functions $\varphi_\beta \circ \varphi_\alpha^{-1} : \varphi_\alpha(U_\alpha \cap U_\beta) \to \mathbb{R}^d$ are C^1. To define the absolute hyperplane game on $(X, \mathcal{A})$, we specify a target set $S \subset X$. Bob begins play by choosing one coordinate chart $(U, \varphi) = (U_\alpha, \varphi_\alpha)$ in $\mathcal{A}$ and a closed ball $B_1 \subset \mathbb{R}^d$ contained in $\varphi(U)$. From this point on the game continues as before, where Alice and Bob alternate moves in $\varphi(U) \subset \mathbb{R}^d$. To decide the winner they check whether the point of intersection $\bigcap_i B_i$ belongs to $\varphi(S)$. If Alice has a winning strategy, we say that S is HAW on $(X, \mathcal{A})$.

Note that when W is an open subset of $\mathbb{R}^d$, the definition of the game on W, given at the beginning of the previous paragraph, coincides with the definition of the game on $(X, \mathcal{A})$ if we take $X = W$ and take $\mathcal{A}$ to be the atlas consisting of one chart (W, Id). Also note that Bob has been given the additional prerogative of choosing a coordinate chart within which to work, at the start of play, and this appears to make the winning property very restrictive. However we have:

PROPOSITION 3.3. Suppose X is a C^1 manifold with an atlas $\mathcal{A}$ and (U_i, φ_i) is a system of coordinate charts in $\mathcal{A}$, such that $X = \bigcup U_i$, and $S \subset X$. Then S is HAW on $(X, \mathcal{A})$ if and only if for each i, $\varphi_i(S)$ is HAW on $\varphi_i(U_i)$.

PROOF. The direction $\Longrightarrow$ is immediate from the definitions, since Bob may select to work with each of the charts φ_i on his first move. For the direction $\Longleftarrow$, note first that by Proposition 3.1(a,b), $\varphi_i(S)$ is dense in each $\varphi_i(U_i)$ and hence S is dense in X. Now suppose Bob chose to work in some chart φ_α distinct from the φ_i. If the diameters of the balls B_n chosen by Bob do not decrease to zero, then $\bigcap B_n$ has interior. Since S is dense in X, each $\varphi_\alpha(S)$ is dense in $\varphi_\alpha(U_\alpha)$, so Alice wins. Otherwise, since B_1 is compact and is covered by the open sets $\varphi_\alpha(U_i)$, there is a Lebesgue number for this cover, that is $\delta > 0$ such that each subset of B_1 of diameter at most δ is contained in one of the $\varphi_\alpha(U_i)$. Thus there are n, i such that

$$(3.2) \qquad \varphi_\alpha^{-1}(B_n) \subset U_i.$$

In light of Proposition 3.1(d), $\varphi_\alpha(S) = \varphi_\alpha \circ \varphi_i^{-1}\big(\varphi_i(S)\big)$ is HAW on $\varphi_\alpha(U_\alpha \cap U_i)$. In light of (3.2), the latter set contains Bob's choice B_n and so (applying her strategy for the case that B_n is the first ball chosen by Bob), Alice wins in this case as well. $\square$

Note that our definition above depended on the choice of atlas $\mathcal{A}$. We now deduce from Proposition 3.3 that the HAW property in fact depends only on the manifold structure on X, and not on a specific atlas. Namely, recall that two atlases of charts $\mathcal{A}_1 \stackrel{\text{def}}{=} (U_\alpha, \varphi_\alpha)$, $\mathcal{A}_2 \stackrel{\text{def}}{=} (V_\beta, \psi_\beta)$ on the same manifold X are said to be C^1-*compatible* if $\mathcal{A}_1 \cup \mathcal{A}_2$ is also a C^1-atlas of in the above sense.

COROLLARY 3.4. Suppose $\mathcal{A}_1, \mathcal{A}_2$ are two compatible atlases of charts, and $S \subset X$. Then S is HAW on $(X, \mathcal{A}_1)$ if and only if it is HAW on $(X, \mathcal{A}_2)$.

PROOF. The atlases $\mathcal{A}_1, \mathcal{A}_2$ are compatible if and only if the maximal atlas (with respect to inclusion) $\mathcal{A}_{\text{max}}$ containing $(U_\alpha, \varphi_\alpha)$ coincides with the maximal atlas containing (V_β, ψ_β). So it is enough to show that S is HAW on $(X, \mathcal{A}_1)$ if and only if it is HAW on $(X, \mathcal{A}_{\text{max}})$. For this, apply Proposition 3.3 with $\mathcal{A} = \mathcal{A}_{\text{max}}$ and $\{(U_i, \varphi_i)\} = \mathcal{A}_1$. $\square$

In light of Corollary 3.4, we will be justified below in omitting the atlas from the terminology and saying that $S \subset X$ is *HAW* if it is HAW on $(X, \mathcal{A})$ for some atlas of charts $\mathcal{A}$ defining the manifold structure on X. It is clear that Proposition 3.1 immediately implies

PROPOSITION 3.5. Let X be a C^1 manifold. Then:
  (a) HAW subsets of X are thick.
  (b) The countable intersection of HAW subsets of X is again HAW.
  (c) The image of a HAW subset of X under a C^1 diffeomorphism $X \to X$ is HAW.

Now, and for the rest of the paper, let X be as in (1.5) and F as in (1.4). Also define $F^+ \stackrel{\text{def}}{=} \{g_t : t \geq 0\}$. It turns out that Theorem 1.4 can be reduced to the following statement about one-sided orbits:

THEOREM 3.6.   (a) The set

(3.3) $\{x \in X : F^+x \text{ is bounded}\}$

     is HAW.
     (b) Let Z be a compact (F, H^+)-transversal submanifold of X; then the set

(3.4) $\{x \in X : \overline{F^+x} \cap Z = \varnothing\}$

     is HAW.

PROOF OF THEOREM 1.4 ASSUMING THEOREM 3.6. It is clear that a statement analogous to Theorem 3.6 holds for the semigroup $F^- \stackrel{\text{def}}{=} \{g_t : t \leq 0\}$ in place of F^+ and with the roles of H^+ and H^- exchanged: namely, the sets

(3.5) $\{x \in X : F^-x \text{ is bounded}\}$

and

(3.6) $\{x \in X : \overline{F^-x} \cap Z = \varnothing\},$

where Z is a compact (F, H^-)-transversal submanifold of X, are HAW. The set

(3.7) $\{x \in X : Fx \text{ is bounded and } \overline{Fx} \cap Z = \varnothing\}$

is the intersection of sets (3.3)–(3.6); hence, in view of Proposition 3.5(b), it is HAW whenever $Z \subset X$ is compact and both (F, H^+)- and (F, H^-)-transversal. In Theorem 1.4 our set Z is a countable union of manifolds, and hence (replacing if necessary a manifold with a countable union of compact manifolds with boundary) a countable union of compact manifolds with boundary. Thus the set (1.7) is the countable intersection of sets of the form (3.7), and Theorem 1.4 follows by another application of Proposition 3.5(b). □

We will finish this section by proving part (a) of the above theorem, which is in fact a rather straightforward variation on some well-known results. In most of the earlier work concerning winning properties of sets of bounded trajectories, the games were actually played on expanding leaves for the F^+-action on X, which in our case can be parametrized as orbits of the expanding horospherical group H^+. An example is McMullen's strengthening [**Mc**, Theorem 1.3] of a theorem of Dani [**D2**] on the winning property of the set of directions in hyperbolic manifolds of finite volume with bounded geodesic rays, a special case of which can be restated as follows:

THEOREM 3.7. For any $y \in X$, the set

(3.8) $$\{s \in \mathbb{R} : F^+ h_s y \text{ is bounded}\}$$

is absolutely winning (which, for games played on $\mathbb{R}$, is the same as HAW).

To reduce Theorem 3.6(a) to Theorem 3.7, let us fix an atlas of coordinate charts for X as follows. Let $\mathfrak{g} \stackrel{\text{def}}{=} \text{Lie}(G)$, and for any $y \in X$ denote by $\exp_y$ the map

(3.9) $$\exp_y : \mathfrak{g} \to X, \quad \mathbf{x} \mapsto \exp(\mathbf{x})y.$$

For any $y \in X$ one can choose a neighborhood W_y of $0 \in \mathfrak{g}$ such that $\exp_y|_{W_y}$ is one-to-one. Denote

(3.10) $$U_y \stackrel{\text{def}}{=} \exp_y(W_y) \text{ and } \varphi_y \stackrel{\text{def}}{=} \exp_y^{-1}|_{U_y}.$$

We will identify $\mathfrak{g}$ with $\mathbb{R}^3$ and use the collection $\{(U_y, \varphi_y) : y \in X\}$ as an atlas of coordinate charts. Since for an arbitrary y we have $y \in U_y$, X is indeed covered by the union of the charts.

PROOF OF THEOREM 3.6(A). In view of Proposition 3.3 it suffices to show that for any $y \in X$, the set

(3.11) $$\varphi_y(\{x \in U_y : F^+x \text{ is bounded}\}) = \{\mathbf{x} \in W_y : F^+ \exp(\mathbf{x})y \text{ is bounded}\}$$

is HAW on W_y. We know from Theorem 3.7 that the set (3.8) is HAW. Note that conjugation by g_t, $t \geq 0$, does not expand elements of H^-F. Therefore, for any $x \in X$,

(3.12) $$F^+x \text{ is bounded} \iff F^+gx \text{ is bounded } \forall g \in H^-F.$$

The set U_y is foliated by connected components of orbits for the action of H^-F. By composing φ_y with a suitable diffeomorphism of W_y, which we are allowed to do by Proposition 3.1(d), we can assume that this foliation is mapped into the foliation of $\mathfrak{g}$ by translates of $\mathfrak{p} \stackrel{\text{def}}{=} \text{Lie}(H^-) \oplus \text{Lie}(F)$. Let us denote by $\mathfrak{h}$ the Lie algebra of H^+, by $\pi : \mathfrak{g} \to \mathfrak{h}$ the projection associated with the direct sum decomposition $\mathfrak{g} = \mathfrak{h} \oplus \mathfrak{p}$, and let $W_y^+ \stackrel{\text{def}}{=} \pi(W_y)$.

Fix $0 < \beta < 1/3$. Bob begins with a ball $B_1 \subset W_y$. Alice consults the strategy she, in view of Theorem 3.7, is assumed to have for playing on $\mathbb{R}$ for the chosen value of β with the target set (3.8) and Bob's first move being $B_1' \stackrel{\text{def}}{=} \pi(B_1)$ (here we identify $\mathfrak{h}$ with $\mathbb{R}$ via $s \leftrightarrow h_s$). The strategy specifies an interval (neighborhood of a point) $A_1' \subset \mathfrak{h}$, and in the game on W_y Alice chooses $A_1 \stackrel{\text{def}}{=} \pi^{-1}(A_1')$, which is a hyperplane neighborhood. Continuing iteratively, suppose Bob has chosen the ball $B_i \subset W_y$. The ball $B_i' \stackrel{\text{def}}{=} \pi(B_i)$ is a legal move for Bob playing on W_y^+, since the projection π does not affect the radii of balls and since

$$B_i \subset B_{i-1} \setminus A_{i-1} \implies B_i' \subset B_{i-1}' \setminus A_{i-1}'.$$

Thus Alice's strategy for playing on W_y^+ specifies a move $A_i' \subset \mathfrak{h}$, and in the new game Alice chooses $A_i \stackrel{\text{def}}{=} \pi^{-1}(A_i')$. This defines her strategy for playing on W_y and guarantees that $\bigcap B_i'$ belongs to the set (3.8). By (3.12), the point $\bigcap B_i$ belongs to the set (3.11).   □

4. Transversality and reduction to discrete time actions

It would seem natural to attempt to prove an analogue of Theorem 3.7 for orbits escaping Z, that is, show that for Z as in Theorem 3.6(b) and $y \in X$, the set $\{s \in \mathbb{R} : F^+ h_s y \cap Z = \varnothing\}$ is HAW. And indeed the above statement is true; however it would not be enough for proving Theorem 3.6(b). A reason for that is that one can state a version of the equivalence (3.12) for this situation, namely, that

$$\overline{F^+ x} \cap Z = \varnothing \iff \overline{F^+ \exp(\mathbf{p})x} \cap Z = \varnothing \quad \forall \mathbf{p} \in \mathfrak{p} \text{ with } \|\mathbf{p}\| \leq \varepsilon,$$

where $\varepsilon > 0$ depends on x. However to derive a winning property of the set (3.4) one would need a uniform lower bound on ε, which is not available here. Thus we will need to play on X itself.

Our first step will be a reduction to discrete time actions. The argument here loosely follows [**K**, §4]. For a parameter $\tau > 0$ to be defined later, denote by

$$F_\tau^+ \stackrel{\text{def}}{=} \{g_{n\tau} : n \in \mathbb{Z}_+\}$$

the cyclic subsemigroup of F generated by g_τ. We will make a reduction showing that we may replace the continuous semigroup F^+ with F_τ^+. For $Z \subset X$ we define

$$Z_{[t_1, t_2]} \stackrel{\text{def}}{=} \bigcup_{t_1 \leq t \leq t_2} g_t Z.$$

We have:

LEMMA 4.1. Suppose Z is a compact C^1 submanifold of X and H a connected subgroup of G. Then:
  (a) If condition (F) holds, then there exists $\sigma = \sigma(Z) > 0$ such that $Z_{[-\sigma, \sigma]}$ is a C^1 manifold.
  (b) If, in addition, condition (H, F) holds, then there exists positive $\tau = \tau(Z) \leq \sigma(Z)$ such that for any x in $Z_{[0, \tau]}$, $T_x(Hx)$ is not contained in $T_x(Z_{[0, \tau]})$.

Note that the conclusion of part (b) of the above lemma coincides with condition (F) with F replaced by H and Z replaced by $Z_{[0, \tau]}$. It will be convenient to introduce more notation and, for a subgroup H of G and a smooth submanifold Z of X, say that Z is H-transversal at $x \in Z$ if $T_x(Hx)$ is not contained in $T_x Z$, and that Z is H-transversal if it is H-transversal at every point of Z. In other words, if condition (F) holds with F replaced by H. This way, condition (F) says that Z is F-transversal, and the conclusion of Lemma 4.1(b) states that $Z_{[0,\tau]}$ is H-transversal.

PROOF. Let $\dim(Z) = k$. Using a finite covering of Z by appropriate coordinate charts of X, one can without loss of generality assume that Z is of the form $\psi\left(\overline{U}\right)$ for some bounded open $U \subset \mathbb{R}^k$, with ψ being a C^1, nonsingular embedding defined on an open $U' \subset \mathbb{R}^k$ strictly containing $\overline{U}$. Define $\tilde{\psi} : U' \times \mathbb{R} \to X$ by putting $\tilde{\psi}(u,t) = g_t\left(\psi(u)\right)$. From the F-transversality of Z it follows that $\tilde{\psi}$ is nonsingular at $t = 0$ and $u \in \overline{U}$. Hence $\tilde{\psi}$ is a nonsingular embedding of $U'' \times [-\sigma, \sigma]$ into X for some $\sigma > 0$ and an open set U'' strictly containing $\overline{U}$, and (a) is proved.

Clearly the tangent space to $Z_{[-\sigma,\sigma]}$ at $x \in Z$ is equal to $T_x Z \oplus T_x(Fx)$. Therefore condition (H, F) implies that $Z_{[-\sigma,\sigma]}$ is H-transversal at any point of Z. But the H-transversality is clearly an open condition, hence it holds at any point of $Z_{[-\sigma,\sigma]}$ which is close enough to Z. By compactness one can choose a positive $\tau \leq \sigma$ such that $Z_{[0,\tau]}$ is H-transversal. $\qquad \square$

Here is another way to express H-transversality: fix a Riemannian metric 'dist' on the tangent bundle of X, and for a C^1 submanifold Z of X consider the function $\theta_H : Z \to \mathbb{R}$,

$$\theta_H(x) \stackrel{\text{def}}{=} \sup_{v \in T_x(Hx), \|v\|=1} \text{dist}(v, T_x Z).$$

It is clear that $\theta_H(x) \neq 0$ iff Z is H-transversal at x, and that θ_H is continuous in $x \in Z$. Therefore the following holds:

LEMMA 4.2. A compact C^1 submanifold Z of X is H-transversal iff there exist $c = c(Z) > 0$ such that $\theta_H(x) \geq c$ for all $x \in Z$.

A right-invariant metric on G induces a well-defined Riemannian metric on X and we now change notation slightly, writing 'dist' for the resulting path metric on X. Now let Z be as in Theorem 3.6, that is, compact and (H^+, F)-transversal, and take positive $\tau \leq \tau(Z)$ as in Lemma 4.1 satisfying in addition:

$$(4.1) \qquad x, y \in X, 0 \leq t \leq \tau \implies \text{dist}(g_t x, g_t y) \leq 2 \text{dist}(x, y).$$

This can be done because $g_t, |t| \leq \tau$ is bounded and hence there is a uniform bound on the amount by which it distorts the Riemannian metric. Suppose that for some $x \in X$ and $\varepsilon > 0$ there exists $t \geq 0$ and $z \in Z$ such that the distance between $g_t x$ and z is less than ε. Choose $0 \leq t_1 < \tau$ such that $t + t_1 = n\tau$ for some $n \in \mathbb{N}$. It follows from (4.1) that $\text{dist}(g_{n\tau} x, g_{t_1} z) < 2\varepsilon$. This rather elementary argument proves that

$$\overline{F_\tau^+ x} \cap Z_{[0,\tau]} = \varnothing \implies \overline{F^+ x} \cap Z = \varnothing.$$

Thus to establish Theorem 3.6(b) it suffices to prove the following:

THEOREM 4.3. For any compact H^+-transversal submanifold Z of X and any $\tau > 0$ the set

$$(4.2) \qquad\qquad \left\{x \in X : F_\tau^+ x \cap Z = \varnothing\right\}$$

is HAW.

Theorem 3.6(b) can then be obtained from Theorem 4.3 with Z replaced by $Z_{[0,\tau]}$.

5. Completion of the proof: the percentage game

In this final section we prove Theorem 4.3. The argument below originates with an idea of Moshchevitin [**Mo**], which was developed further in the Ph.D. Thesis of Broderick [**B**] and in the papers [**BFK, BFS**]. We follow the streamlined presentation of [**BFS**], which consists of defining yet another game. In what follows, for a subset Y in a metric space and $\varepsilon > 0$ we will denote by $Y^{(\varepsilon)}$ the ε-neighborhood of Y.

Fix $\beta > 0$ and a target set $S \subset \mathbb{R}^d$. The *hyperplane percentage game* is defined as follows: Bob begins as usual by choosing a closed ball $B_1 \subset \mathbb{R}^d$. Then, for each $i \geq 1$, once B_i (of radius r_i) is chosen, Alice chooses finitely many affine hyperplanes $L_{i,j}$ and numbers $\varepsilon_{i,j}$, where $j = 1, \ldots, N_i$, satisfying $0 < \varepsilon_{i,j} \leq \beta r_i$. Here N_i can be any positive integer that Alice chooses. Bob then chooses a ball $B_{i+1} \subset B_i$ with radius $r_{i+1} \geq \beta r_i$ such that

$$B_{i+1} \cap L_{i,j}^{(\varepsilon_{i,j})} = \varnothing \text{ for at least } \frac{N_i}{2} \text{ values of } j\,.$$

Thus we obtain as before a nested sequence of closed balls $B_1 \supset B_2 \supset \cdots$ and declare Alice the winner if and only if (3.1) holds. If Alice has a strategy to win regardless of Bob's play, we say that S is *β-hyperplane percentage winning, or β-HPW*. Note that for large values of β it is possible for Alice to leave Bob with no available moves after finitely many turns. However, an elementary argument (see [**Mo**, Lemma 2] or [**BFK**, §2]) shows that Bob always has a legal move if β is smaller than some constant $\beta_0(d)$. In particular $\beta_0(1) = 1/5$. If S is β-HPW for each $0 < \beta < \beta_0(d)$, we say that S is *hyperplane percentage winning (HPW)*. It is clear that β-HPW implies β'-HPW if $\beta \leq \beta'$; thus HPW is equivalent to β-HPW for arbitrarily small values of β.

We remark that the game defined above is actually a special case of the (β, p)-game defined in [**BFS**], corresponding to the choice $p = 1/2$. Here p represents the percentage of hyperplanes that Bob is obliged to stay away from.

One sees that in the hyperplane percentage game the rules are more favorable to Alice than in the hyperplane absolute game, and so any HAW set is automatically HPW. Surprisingly, the converse is true, see [**BFS**, Lemma 2.1]:

LEMMA 5.1. *For any $\beta \in (0, 1/3)$ there exists $\beta' \in \bigl(0, \beta_0(d)\bigr)$ such that any set which is β'-HPW is β-HAW. In particular the HPW and HAW properties are equivalent.*

PROOF OF THEOREM 4.3. Recall that we are given $\tau, \beta > 0$ and a compact H^+-transversal submanifold Z of X. Our goal (after using Lemma 5.1 and rewriting β for β') is to show that the set (4.2) is β-HPW. We will assume, without loss of generality, that

$$(5.1) \qquad\qquad\qquad\qquad \beta < e^{-2\tau}\,.$$

Let us say that a map between two metric spaces is *C-bi-Lipschitz* if the ratio of dist$\bigl(f(x), f(y)\bigr)$ and dist(x, y) is uniformly bounded between $1/C$ and C.

The first step is to fix an atlas of coordinate charts for X. We will do it as in the proof of Theorem 3.6(a), that is, using charts (3.10) where $\exp_y$ is defined by (3.9) and $\exp_y |_{W_y}$ is one-to-one for all $y \in X$. As before, it suffices to show that for any $y \in X$, the set

$$\varphi_y \big(\{ x \in U_y : F_\tau^+ x \cap Z = \varnothing \} \big) = \{ \mathbf{x} \in W_y : F_\tau^+ \exp(\mathbf{x}) y \cap Z = \varnothing \}$$

is HAW on W_y.

The next step is to collect some information about Z. Since it is bounded, one can choose $\sigma_1 < 1$ such that the restriction of $\exp_y$ to $B_{\mathfrak{g}}(0, 4\sigma_1)$ is 2-bi-Lipschitz (in particular, injective) whenever $B_X(y, 2\sigma_1) \cap Z \neq \varnothing$ (here, as before, we work with a path-metric on X coming from a right-invariant Riemannian metric on G obtained from some inner product on $\mathfrak{g}$). In light of Corollary 3.4, we can (by replacing the U_y with smaller sets, depending on Z) also assume that $U_y \subset B_X(y, 2\sigma_1)$ for any $y \in Z^{(4\sigma_1)}$. Then, because of compactness and C^1-smoothness of Z, for every $b > 0$ there exists $0 < \sigma_2(b) \leq \sigma_1$ such that
(5.2)

$$\sigma \leq \sigma_2(b) \text{ and } B_X(y, \sigma) \cap Z \neq \varnothing \quad \Rightarrow \quad \text{there exists}$$

a $\dim(Z)$-dimensional subspace L of $\mathfrak{g}$ such that $\varphi_y \big(B_X(y, \sigma) \cap Z \big) \subset L^{(b\sigma)}$.

(More precisely, L is the tangent space to $\varphi_y(Z)$ at $\varphi_y(z)$, where z is some point in the intersection of $B_X(y, \sigma)$ and Z.) Also, recall that Z is H^+-transversal, and let $c = c(Z)$ be as in Lemma 4.2.

Now choose m large enough so that

(5.3) $$\beta^{-n} < e^{2m\tau}, \quad \text{where } n = \lfloor \log_2 m \rfloor + 1 .$$

This is possible since the left (resp., right) side of the inequality in (5.3) depends polynomially (resp., exponentially) on m.

Pick an arbitrary $y \in X$ and let us suppose that Bob chooses a ball $B_0 \subset W_y$ of radius r_0. Set

(5.4) $$b \overset{\text{def}}{=} \frac{c\beta^{n+2}e^{-2m\tau}}{16} ,$$

(5.5) $$\sigma \overset{\text{def}}{=} \min \left(\frac{1}{4}\sigma_2(b), e^{2m\tau} r_0 \right) ,$$

and let

(5.6) $$\delta \overset{\text{def}}{=} e^{-2m\tau} \sigma \quad \text{and} \quad \varepsilon \overset{\text{def}}{=} b\sigma = \frac{1}{16}c\beta^{n+2}\delta .$$

We will show that Alice can play the β-hyperplane percentage game in such a way that the intersection of all the balls belongs to the set

$$\{ \mathbf{x} \in W_y : F_\tau^+ \exp(\mathbf{x}) y \cap Z^{(\varepsilon)} = \varnothing \} .$$

Note that it follows from (5.5) and (5.6) that $\delta \leq r_0$. The game will start with Alice making moves in an arbitrary way until the first time Bob's ball has radius $r_1 \leq \delta$. (Recall that, in view of Remark 3.2, it suffices to assume that Alice has a winning strategy whenever the radii of Bob's balls tend to zero.) Re-indexing if necessary, let us call this ball B_1 and its radius r_1; note that we have $r_1 \geq \beta\delta$.

In order to specify Alice's strategy we will partition her moves into windows. For each $j, k \in \mathbb{N}$, we will say that k lies in the jth window if

(5.7) $$\beta^{-n(j-1)} \leq e^{2k\tau} < \beta^{-jn} .$$

By (5.3), for every $j \in \mathbb{N}$, there are at most m indices k lying in the jth window, and (5.1) guarantees that every $k \in \mathbb{N}$ lies in some window. We will call the indices i for which

$$(5.8) \qquad \beta^{nj} r_1 < r_i \leq \beta^{n(j-1)} r_1$$

the *jth stage of the game*. The first stage begins with Bob's initial ball B_1, and the rules of the game imply that each stage contains at least n indices. Loosely speaking, Alice will use her moves indexed by numbers in the jth stage, to ensure that the points $\mathbf{x}$ contained in the balls chosen by Bob satisfy $g_\tau^k \exp(\mathbf{x}) y \notin Z^{(\varepsilon)}$ for any k in the jth window. We now specify Alice's strategy in more detail.

Fix j and suppose that $i = i(j)$ is the first index of stage j. For any k belonging to the jth window, denote

$$A_{j,k} \overset{\text{def}}{=} \left\{ g_\tau^k \exp(\mathbf{x}) y : \mathbf{x} \in B_i \right\}.$$

In view of (5.8), the diameter of $A_{j,k}$ is at most

$$2 e^{2k\tau} \beta^{n(j-1)} r_1 \overset{(5.7)}{\leq} 2\beta^{-n} r_1 \overset{(5.3)}{\leq} 2 e^{2m\tau} r_1 \overset{(5.6)}{\leq} 2\sigma \overset{(5.5)}{\leq} \frac{1}{2} \sigma_2(b).$$

If $A_{j,k}$ does not intersect $Z^{(\varepsilon)}$ for any k in the jth window, Alice makes all her moves in the jth stage of the game in an arbitrary way (e.g. she could put $N_i = 0$, that is, decide not to specify any hyperplanes on her ith move). Suppose that one of them does; let $x \overset{\text{def}}{=} g_\tau^k y$. Then $A_{j,k} \cap B_X\big(x, \sigma_2(b)\big) \neq \varnothing$. Now let us use φ_x to map everything to $\mathfrak{g}$. In view of (5.2), there exists $z \in Z$ and a $\dim(Z)$-dimensional subspace $L = T_{\varphi_x(z)}\big(\varphi_x(Z)\big)$ of $\mathfrak{g}$ such that

$$\varphi_x\left(A_{j,k} \cap Z^{(\varepsilon)} \right) \subset \varphi_x\left(A_{j,k} \right) \cap \left(\varphi_x Z \right)^{(2\varepsilon)} \overset{(5.2)}{\subset} \varphi_x\left(A_{j,k} \right) \cap L^{(2\varepsilon + 2b\sigma)}$$

$$\overset{(5.4),\,(5.6)}{\subset} \varphi_x\left(A_{j,k} \right) \cap L^{\left(c\beta^{n+2} e^{-2m\tau} \sigma / 4 \right)}.$$

In view of Lemma 4.2 and the 2-bi-Lipschitz property of maps φ_x and φ_z, the intersection of $L^{\left(c\beta^{n+2} e^{-2m\tau} \sigma / 8 \right)}$ with any translate of $\mathfrak{h}$ has length at most $2\beta^{n+2} e^{-2m\tau} \sigma$. Consequently, the intersection of

$$(5.9) \qquad \left\{ \mathbf{x} \in B_i : g_\tau^k \exp(\mathbf{x}) y \in Z^{(\varepsilon)} \right\} = \varphi_y\left(g_\tau^{-k}(A_{j,k} \cap Z^{(\varepsilon)}) \right) = \mathrm{Ad}_{g_\tau^{-k}}\left(\varphi_x(A_{j,k} \cap Z^{(\varepsilon)}) \right)$$

with any translate of $\mathfrak{h}$ has length at most

$$2 e^{-2k\tau} \beta^{n+2} e^{-2m\tau} \sigma \overset{(5.7),\,(5.6)}{\leq} 2\beta^{n(j-1)} \beta^{n+2} \delta \leq 2\beta^{n(j-1)} \beta^{n+1} r_1 \leq 2\beta^n r_i.$$

This implies that $\left\{ \mathbf{x} \in B_i : g_\tau^k \exp(\mathbf{x}) y \in Z^{(\varepsilon)} \right\}$ is contained in a $\beta^n r_i$-neighborhood of some hyperplane, and hence the union of sets (5.9) over all indices k contained in the jth window is covered by at most m $\beta^n r_i$-neighborhoods of hyperplanes. Let us denote these hyperplanes by $L_{i,j}$ and let $\varepsilon_{i,j} = \beta^n r_i$. These will be Alice's choices in the ith move; the above discussion ensures that they constitute a valid move for Alice. In each of the remaining steps belonging to the jth stage, Alice will choose those neighborhoods which still intersect the ball chosen by Bob. That is, if we write the indices belonging to the jth stage as $i(j), i(j) + 1, \ldots, i(j+1) - 1$, Alice's choices in stage j will be those hyperplanes $L_{i,j}$ for which $L_{i,j}^{(\varepsilon_{i,j})} \cap B_i \neq \varnothing$, equipped with $\varepsilon_{i,j} = \beta^\ell r_{i(j)}$. This choice and (5.8) ensure that all of these moves are valid moves for Alice.

Since the jth stage contains at least n indices, and in every one of his moves Bob must choose a ball intersecting at most $1/2$ of the intervals chosen by Alice, we have that out of those neighborhoods, at most $2^{-n}m$ can intersect the first ball $B_{i(j+1)}$ in the first move of the $(j+1)$-th stage of the game. But $2^{-n}m < 1$ in view of (5.3). Consequently, for k in the jth window and $\mathbf{x} \in B_{i(j+1)}$, we have that $g_\tau^k \exp(\mathbf{x})y$ is not in $Z^{(\varepsilon)}$, and therefore, if $\mathbf{x}_\infty$ is the intersection point of all Bob's balls, it follow that $\exp(\mathbf{x}_\infty)y$ belongs to the set (4.2). The theorem is proved. $\square$

References

[Ar]   C. S. Aravinda, *Bounded geodesics and Hausdorff dimension*, Math. Proc. Cambridge Philos. Soc. **116** (1994), no. 3, 505–511, DOI 10.1017/S0305004100072777. MR1291756 (95f:53081)

[AL]   C. S. Aravinda and E. Leuzinger, *Bounded geodesics in rank-1 locally symmetric spaces*, Ergodic Theory Dynam. Systems **15** (1995), no. 5, 813–820, DOI 10.1017/S0143385700009640. MR1356615 (96h:53050)

[B]   R. Broderick, *Incompressibility and Fractals*, ProQuest LLC, Ann Arbor, MI, 2011. Thesis (Ph.D.)–Brandeis University. MR2942227

[BFK]   R. Broderick, L. Fishman, and D. Kleinbock, *Schmidt's game, fractals, and orbits of toral endomorphisms*, Ergodic Theory Dynam. Systems **31** (2011), no. 4, 1095–1107, DOI 10.1017/S0143385710000374. MR2818688 (2012e:11130)

[BFS]   R. Broderick, L. Fishman, and D. Simmons, *Badly approximable systems of affine forms and incompressibility on fractals*, Number Theory **133** (2013), no. 7, 2186–2205, DOI 10.1016/j.jnt.2012.12.004. MR3035957

[BFKRW]   R. Broderick, L. Fishman, D. Kleinbock, A. Reich, and B. Weiss, *The set of badly approximable vectors is strongly C^1 incompressible*, Math. Proc. Cambridge Philos. Soc. **153** (2012), no. 2, 319–339, DOI 10.1017/S0305004112000242. MR2981929

[CaSD]   J. W. S. Cassels and H. P. F. Swinnerton-Dyer, *On the product of three homogeneous linear forms and the indefinite ternary quadratic forms*, Philos. Trans. Roy. Soc. London. Ser. A. **248** (1955), 73–96. MR0070653 (17,14f)

[D1]   S. G. Dani, *Divergent trajectories of flows on homogeneous spaces and Diophantine approximation*, J. Reine Angew. Math. **359** (1985), 55–89, DOI 10.1515/crll.1985.359.55. MR794799 (87g:58110a)

[D2]   S. G. Dani, *Bounded orbits of flows on homogeneous spaces*, Comment. Math. Helv. **61** (1986), no. 4, 636–660, DOI 10.1007/BF02621936. MR870710 (88i:22011)

[DN]   S. G. Dani and Arnaldo Nogueira, *On orbits of SL(2,$\mathbb{Z}$)$_+$ and values of binary quadratic forms on positive integral pairs*, J. Number Theory **95** (2002), no. 2, 313–328. MR1924105 (2003j:11039)

[GP]   V. Guillemin and A. Pollack, *Differential topology*, Prentice-Hall, Inc., Englewood Cliffs, N.J., 1974. MR0348781 (50 #1276)

[J]   V. Jarník, *Zur metrischen Theorie der diophantischen Approximationen*, Prace mat. fiz. **36** (1928), 91–106.

[K]   D. Y. Kleinbock, *Nondense orbits of flows on homogeneous spaces*, Ergodic Theory Dynam. Systems **18** (1998), no. 2, 373–396, DOI 10.1017/S0143385798100408. MR1619563 (99e:58122)

[KM]   D. Y. Kleinbock and G. A. Margulis, *Bounded orbits of nonquasiunipotent flows on homogeneous spaces*, Sinaĭ's Moscow Seminar on Dynamical Systems, Amer. Math. Soc. Transl. Ser. 2, vol. 171, Amer. Math. Soc., Providence, RI, 1996, pp. 141–172. MR1359098 (96k:22022)

[KW1]   D. Kleinbock and B. Weiss, *Modified Schmidt games and Diophantine approximation with weights*, Adv. Math. **223** (2010), no. 4, 1276–1298, DOI 10.1016/j.aim.2009.09.018. MR2581371 (2011b:11103)

[KW2]   D. Kleinbock and B. Weiss, *Dirichlet's theorem on Diophantine approximation and homogeneous flows*, J. Mod. Dyn. **2** (2008), no. 1, 43–62. MR2366229 (2008k:11078)

[L]   C. G. Lekkerkerker, *Una questione di approssimazione diofantea e una proprietà caratteristica dei numeri quadratici. I* (Italian), Atti Accad. Naz. Lincei. Rend. Cl. Sci. Fis. Mat. Nat. (8) **21** (1956), 179–185. MR0086099 (19,124c)

[Ma1]   G. A. Margulis, *Dynamical and ergodic properties of subgroup actions on homogeneous spaces with applications to number theory*, Proceedings of the International Congress of Mathematicians, Vol. I, II (Kyoto, 1990), Math. Soc. Japan, Tokyo, 1991, pp. 193–215. MR1159213 (93g:22011)

[Ma2]   G. A. Margulis, *Oppenheim conjecture*, Fields Medallists' lectures, World Sci. Ser. 20th Century Math., vol. 5, World Sci. Publ., River Edge, NJ, 1997, pp. 272–327, DOI 10.1142/9789812385215_0035. MR1622909 (99e:11046)

[Mc]   C. T. McMullen, *Winning sets, quasiconformal maps and Diophantine approximation*, Geom. Funct. Anal. **20** (2010), no. 3, 726–740, DOI 10.1007/s00039-010-0078-3. MR2720230 (2012a:30061)

[Mo]   N. G. Moshchevitin, *A note on badly approximable affine forms and winning sets* (English, with English and Russian summaries), Mosc. Math. J. **11** (2011), no. 1, 129–137, 182. MR2808214 (2012f:11127)

[O]   A. Oppenheim, *Values of quadratic forms. I*, Quart. J. Math., Oxford Ser. (2) **4** (1953), 54–59. MR0054650 (14,955a)

[S1]   W. M. Schmidt, *On badly approximable numbers and certain games*, Trans. Amer. Math. Soc. **123** (1966), 178–199. MR0195595 (33 #3793)

[S2]   W. M. Schmidt, *Badly approximable systems of linear forms*, J. Number Theory **1** (1969), 139–154. MR0248090 (40 #1344)

[TV]   G. Troessaert and A. Valette, *On values at integer points of some irrational, binary quadratic forms*, Essays on geometry and related topics, Vol. 1, 2, Monogr. Enseign. Math., vol. 38, Enseignement Math., Geneva, 2001, pp. 597–610. MR1929341 (2004e:11033)

DEPARTMENT OF MATHEMATICS, BRANDEIS UNIVERSITY, WALTHAM MASSACHUSETTS 02454-9110
E-mail address: kleinboc@brandeis.edu

DEPARTMENT OF MATHEMATICS, TEL AVIV UNIVERSITY, TEL AVIV, ISRAEL
E-mail address: barakw@post.tau.ac.il

Contemporary Mathematics
Volume **631**, 2015
http://dx.doi.org/10.1090/conm/631/12598

Conformal families of measures for general iterated function systems

Manfred Denker and Michiko Yuri

ABSTRACT. We introduce some thermodynamic methods for the dynamics of
families of partially defined maps on metric spaces, a general form of iterated
function systems. We show the existence of conformal families of probability
measures for this type of iterated functions systems. It extends the well known
theory of conformal measures.

1. Introduction

The study of non-invertible transformations is often reduced to the investigation of its inverse branches, the main feature being a type of open set condition for
the overlap of images of the inverse branches. The picture changes fundamentally
when considering iteration of some family of maps, for example in the case of random dynamics, semigroups of transformations or, more recently, place dependent
dynamics where maps are only defined locally. In this case the overlap of 'inverse
branches' is no longer a matter of extension of well known methods. The purpose
of this note is to develop ideas for how these systems can be treated in terms of a
thermodynamic formalism.

We will do this for a special class of such systems, called iterated function
systems (see Definition 2.6). They consist of a (mostly compact) metric space X
and a family $\mathfrak{V}$ of homeomorphisms $v : D(v) \to v(D(v)) \subset X$, where $D(v) \subset X$
is closed, and where infinite orbits exist. This class contains the case of maps
$T : X \to X$ which have continuous inverse branches, and stretches over a variety of
other examples (see Section 2 below) to topological versions of position dependent
random dynamics ([**1**]) where at each point $x \in X$ there are several choices of
locally defined maps which can be applied to x (of course, each of these maps has
to have continuous inverse branches). One particular example we have in mind is
given by sofic systems on shift spaces over countable alphabets (this will be treated
in a subsequent paper).

2010 *Mathematics Subject Classification.* Primary 28D99, 28D20, 37A40, 37A30, 37C30,
37D35, 37F10, 37A45.
Key words and phrases. Iterated function system, algebraic pressure, conformal measures,
weak bounded variation.
The research was supported by JSPS Grant-in-Aid for Scientific Research (B) 21340018 and
by the National Science Foundation under Grant Number DMS-1008538.
The research was supported by JSPS Grant-in-Aid for Scientific Research (B) 21340018.

In general, the iterated function systems considered here are not Markovian, as is the case in other work (e.g. [8], [11], [12], [15], [2]). Accordingly, some of our results have similar proofs, some extend earlier research. However, there is a significant difference, since our model covers the Markov graph structure introduced in [8] but the dynamics is not a factor (even in some loose sense) of the corresponding symbolic dynamics. Moreover, we do not rely on the Markov property.

In Section 2.2 we introduce (locally) self-homeomorphic sets which extend the notion of self-similar sets, and show in Theorem 2.5 that locally self-homeomorphic sets exist if an infinite iteration procedure is possible on minimal invariant and nonempty sets. These sets are unique if the maps are uniformly contractive (this may be considered as an extension of Hutchinson's theorem [7]). It permits to define a general notion of an iterated function system (Definition 2.6), and we obtain as a consequence that every iterated function system contains at least one locally self-homeomorphic set which itself forms an iterated function system (which is self-homeomorphic in some important cases, see Theorem 2.7).

In Section 2.3 we study limit sets in analogy of those for Fuchsian groups (see [9]). This has some similarities with the work in [12], and - as in there - we also give conditions when a point is directly accessible through an iterated application of maps from the dynamics. In Theorem 2.10 we state conditions that a self-homeomorphic iterated function system is equal to its limit set.

Section 3.2 on conformal measures has some new aspects for the general theory (see [4] for a survey of results on conformal measures). The problem of the existence of conformal measures can be turned into a deficiency statement. We show that an iterated function system always carries conformal families (see Definition 3.5) for any potential with eigenvalues whose logarithm equals the transition parameter (Theorem 3.6). If the potential is of weak bounded variation and the iterated function system is aperiodic, then it is shown in Theorem 3.19 that the logarithm of the eigenvalue is the algebraic pressure (see the definition in Lemma 3.1). The tool here is the Patterson construction of obtaining limit measures ([9], [4], see also [10] for a recent work). As said before, a conformal family has a deficiency which originates from the behavior of the iterated function systems on boundaries of the domains of definition. This deficiency vanishes if the potential vanishes on boundaries (including the case when the boundaries are empty), or the measures in the conformal family vanish on boundaries. This then leads to the usual existence statements of conformal measures. We also show (under some mild conditions) that there are conformal measures supported by the radial limit set (Theorem 3.11).

This paper is a first part on our investigation of general iterated function systems. The second part contains the notions of and results for factors, powers and products, as well as for systems with finite domain structure. Moreover it contains the analog of Bowen's theorem on the existence of conformal measures as a fixed point and the representation of the algebraic pressure as Gurevic pressure. Moreover, sofic systems over countable alphabets will be considered.

2. Iterated Function Systems

2.1. Preliminaries. Let $X \neq \emptyset$ be a complete metric space with metric d and $\mathfrak{V} = \mathfrak{V}_1$ be a family of homeomorphisms $v : D(v) \to v(D(v))$ defined on some closed nonempty subset $D(v) \subset X$. The pair $(X, \mathfrak{V})$ will be called a *function system*. If $\mathfrak{V}$ is a finite family and $D(v) = X$ for all $v \in \mathfrak{V}$, the system $(X, \mathfrak{V})$ is called an iterated function system (IFS) in the literature ([8]), and, moreover, if $X = \mathbb{R}^d$ and each v is affine it is called an IFS of similarities. More generally the graph directed Markov systems in [8] and the pseudo Markov systems in [12] are special cases of such a function system, where the domains have a Markov structure and satisfy the open set condition, and where the maps are contractions. Below we will consider systems without the Markov structure.

Let us mention a few examples. The Gauss map $x \mapsto \{\frac{1}{x}\}$ ($x \in X = (0, 1]$) defines a pseudo Markov system since the map restricted to $(\frac{1}{n}, \frac{1}{n-1}]$ is a homeomorphism, extends continuously to the closure and the inverse a contraction. For the β-transformation the maps are defined by $v_l : [0, 1] \rightarrow [\frac{l-1}{\beta}, \frac{l}{\beta}]$, $v_l(x) = \beta^{-1}(l - 1 + x)$ for $l = 1, ..., k$ where $k \leq \beta < k + 1$, and $v_{k+1} : [0, \beta - k] \rightarrow [\frac{k}{\beta}, 1]$, $v_{k+1}(x) = \beta^{-1}(k + x)$. If β is not an integer, then this function system is not Markov. Other motivations originate from the action of a Fuchsian or Kleinian group on its limit set (see [12]) and from random dynamics: Let $\mathfrak{V} \subset \{f : D(f) \rightarrow f(D(f)) \subset X : f \text{ homeomorphism on some closed domain } D(f)\}$ be a countable family of partially defined homeomorphisms. Let Σ be the space of all infinite sequences $(f_1, f_2, ...)$ of maps from $\mathfrak{V}$ such that for each n the map $f_n \circ f_{n-1} \circ ... \circ f_1$ is well defined on some domain. If m is a shift invariant probability on Σ, the random maps $\Pi \circ \sigma^n$, $n \geq 0$, define a stationary process of random iterated maps on (random) domains, where σ denotes the shift and $\Pi((f_1, f_2, ...)) = f_1$ is the projection onto the first coordinate. The topological version of this defines a system which we are interested in. Such examples have some meaning in network dynamics if the state of the network is position dependent, for example in neuronal dynamics when the network connections depend on the actual state of the system (so far only fixed networks were considered). As one of the basic investigations on position dependent dynamics we refer to [1]. The connection to dynamics describing the cortex in neural science is described in [13], and it is immediate to extend this to position dependent random dynamics and the iterated function systems considered here.

We need to impose some general conditions on the function system $(X, \mathfrak{V})$ in order to make an iteration process meaningful. We let $\mathfrak{V}_n$ be the family of n-fold concatenations of maps from $\mathfrak{V}$, that is for $u \in \mathfrak{V}_n$ there are $v \in \mathfrak{V}$ and $w \in \mathfrak{V}_{n-1}$ with $v \circ w = u$, and we denote by $D(u) = D(w) \cap w^{-1}(D(v))$ the (closed) domain of u (which always is assumed to be nonempty). We denote by $int A$ the interior and by $\overline{A}$ the closure of a set $A \subset X$. Moreover, for a subset $A \subset X$ we set

$$A^\circ_{[n]} = \bigcup_{v \in \mathfrak{V}_n} v(D(v) \cap A) \quad \text{and} \quad A_{[n]} = \overline{\bigcup_{v \in \mathfrak{V}_n} v(D(v) \cap A)} \quad (n \geq 1),$$

$$A^\circ_{[\infty]} = \bigcap_{n \geq 1} A^\circ_{[n]} \quad \text{and} \quad A_{[\infty]} = \bigcap_{n \geq 1} A_{[n]}.$$

By definition we obtain

LEMMA 2.1. *Let A be a subset of X satisfying $A^\circ_{[1]} \subseteq A$. Then $\{A^\circ_{[n]} : n \geq 1\}$ and $\{A_{[n]} : n \geq 1\}$ are nested sequences of subsets of X.*

2.2. Self-Homeomorphisms. Recall that a self-similar set for an IFS of similarities is a nonempty compact set K satisfying $K = K^\circ_{[1]}$. This generalizes to

DEFINITION 2.2. A non-empty closed set K is called *locally self-homeomorphic* with respect to $(X, \mathfrak{V})$ if

$$K = K_{[1]}.$$

It is called *self-homeomorphic* if in addition $K \subseteq \bigcup_{v \in \mathfrak{V}} D(v)$.

The existence of locally self-homeomorphic sets extends Hutchinson's result in [7] and is easily proved by Zorn's lemma.

PROPOSITION 2.3. *Let* $(X, \mathfrak{V})$ *be a function system. If there exists a compact and non-empty set* $C \subset X$ *such that* $C^{\circ}_{[1]} \subset C$, *then there are minimal (by set inclusion) compact sets* K *satisfying* $K \neq \emptyset$ *and* $K_{[1]} \subset K$. *Each of these minimal sets is either locally self-homeomorphic or* $K \subset X \setminus \bigcup_{v \in \mathfrak{V}} D(v)$ *with* $K_{[1]} = \emptyset$.

The existence of locally self-homeomorphic sets does not guarantee that some point can be iterated infinitely often. This follows from each of the next two results. The first lemma is straight forward.

LEMMA 2.4. *Let* K *be a locally self-homeomorphic set. Define for all* $n \geq 1$,

$$K^{\circ}_{(n)} = \bigcup_{v \in \mathfrak{V}_n} v(int(D(v)) \cap K) \quad K_{(n)} = \overline{K^{\circ}_{(n)}}.$$

If $K = K_{(1)}$, *then*

$$K_{(n)} = K_{(1)} = K$$

for all $n \in \mathbb{N}$.

In case that for all domains $D(v) = X$, no $K_{[1]}$ is empty as long as $K \neq \emptyset$, hence self-homeomorphic sets always exist. The set K as in Proposition 2.3 may fail to be locally self-homeomorphic; and if it is so, it even may fail to allow infinite iteration of any point. We therefore refine Proposition 2.3 and Lemma 2.4 in the next theorem. We shall use the following terminology: Let $A \subset X$ and $v \in \mathfrak{V}$. A point $x \in D(v) \setminus \overline{A \cap D(v)}$ with a finite orbit is called an (A, v)-*isolated finite orbit*, and x is called an A-isolated finite orbit if it is an (A, v)-isolated finite orbit for some $v \in \mathfrak{V}$.

THEOREM 2.5. *Let* $C \subset X$ *be compact, non-empty and satisfying* $C^{\circ}_{[1]} \subset C$. *Then there are pairs* $(K, \widetilde{K}) \subset C \times C$ *such that*

(1) $\emptyset \neq K \subset C$ *is compact and* $\widetilde{K} \subset K$.

(2) $(K, \widetilde{K})$ *is minimal with respect to the partial ordering*

$$(L, \widetilde{L}) < (J, \widetilde{J}) \text{ if } L \subset J \text{ and } L \cap \widetilde{J} \subset \widetilde{L}.$$

(3) $K_{[1]} \subset K$.

(4) $\widetilde{K} \subset \bigcup_{v \in \mathfrak{V}} D(v)$ *and* $\widetilde{K}^{\circ}_{[1]} \subset \widetilde{K}$.

(5) *If* $\widetilde{K} \neq \emptyset$ *and if there is no* $\widetilde{K}^{\circ}_{[1]}$-*isolated finite orbit, then*

   a): $\widetilde{K} \subset K$ *is dense.*

   b): K *is locally self-homeomorphic.*

PROOF. Let

$$\mathfrak{K} := \left\{ (L, \widetilde{L}) : \emptyset \neq L \subset C; L \text{ compact}; \widetilde{L} \subset \bigcup_{v \in \mathfrak{V}} D(v) \cap L; \widetilde{L}^{\circ}_{[1]} \subset \widetilde{L}; L^{\circ}_{[1]} \subset L \right\}.$$

$\mathfrak{K}$ is partially ordered by the partial ordering defined in 2. Since $\mathfrak{K}$ contains $(C, \emptyset)$, by Zorn's lemma there are minimal pairs $(K, \widetilde{K}) \in \mathfrak{K}$. By definition of $\mathfrak{K}$, for such a minimal pair $(K, \widetilde{K})$, K is compact, non-empty, $\subset C$, $K^{\circ}_{[1]} \subset K$, $\widetilde{K}^{\circ}_{[1]} \subset \widetilde{K}$ and $\widetilde{K} \subset \bigcup_{v \in \mathfrak{V}} D(v) \cap K$. This proves statements 1.–4.

Assume now that $\widetilde{K} \neq \emptyset$ and that there is no $\widetilde{K}^{\circ}_{[1]}$-isolated finite orbit. Let $\widetilde{L} = \widetilde{K}$ and $L = \overline{\widetilde{K}}$. We show that $(L, \widetilde{L})$ belongs to $\mathfrak{K}$.

Clearly, $\emptyset \neq L \subset K \subset C$ is compact. Since $\widetilde{L} = \widetilde{K} \subset L$, we have $\widetilde{L} \subset \bigcup_{v \in \mathfrak{V}} D(v) \cap L$ and $\widetilde{L}^{\circ}_{[1]} \subset \widetilde{L}$. Let $x \in D(v) \cap L$ for some $v \in \mathfrak{V}$. If $x \in \widetilde{L}$, then $v(x) \in \widetilde{L} \subset L$, so also if $x \in \overline{D(v) \cap \widetilde{L}}$, $v(x) \in L$. The remaining case is when $x \notin \overline{D(v) \cap \widetilde{L}}$. Since there is no $\widetilde{K}$-isolated finite orbit (which follows from the assumption and since $\widetilde{K}^{\circ}_{[1]} \subset \widetilde{K}$), the orbit

$$O(x) = \{w(x) : w \in \bigcup_{n \in \mathbb{N}} \mathfrak{V}_n\}$$

has the following properties: $O(x) \subset \bigcup_{u \in \mathfrak{V}} D(u)$, $\quad O(x) \subset K$ (use induction), and $O(x)^{\circ}_{[1]} \subset O(x)$.

Therefore $M = \widetilde{K} \cup O(x)$ satisfies $M \subset \bigcup_{v \in \mathfrak{V}} D(v) \cap K$ and $M^{\circ}_{[1]} \subset M$. Therefore, since (K, M) is smaller than $(K, \widetilde{K})$, $\widetilde{K} = M$ which implies $O(x) = \emptyset$, and there is no point $x \in D(v) \cap L$ which is not in the closure of $D(v) \cap \widetilde{L}$. Thus we have

$$L^{\circ}_{[1]} \subset L.$$

It follows now that $(L, \widetilde{L})$ belongs to $\mathfrak{K}$ and is smaller than $(K, \widetilde{K})$, therefore $K = L = \overline{\widetilde{K}}$. This proves a).

In order to show b), consider the pair $(L, \widetilde{L})$ with

$$L = \widetilde{K}_{[1]} \quad \text{and} \quad \widetilde{L} = \widetilde{K} \cap L.$$

Note that L is compact and non-empty, and by definition, $\widetilde{L} \subset \bigcup_{v \in \mathfrak{V}} D(v) \cap L$. Moreover, for all $v, w \in \mathfrak{V}$ we have

$$v(D(v) \cap w(D(w) \cap \widetilde{K})) \subset v(D(v) \cap \widetilde{K}) \subset \widetilde{K} \cap \widetilde{K}^{\circ}_{[1]}.$$

Therefore

$$v(D(v) \cap \widetilde{K}^{\circ}_{[1]}) \subset \widetilde{K}^{\circ}_{[1]} \subset L,$$

$$\overline{v(D(v) \cap \widetilde{K}^{\circ}_{[1]})} \subset \widetilde{K}_{[1]} = L,$$

and by the same reasoning as in b) there is no point in $L \cap D(v) \setminus \overline{D(v) \cap \widetilde{K}^{\circ}_{[1]}}$. It follows that

$$v(D(v) \cap L) \subset L.$$

Finally observe that for any $v \in \mathfrak{V}$

$$v(D(v) \cap \widetilde{L}) \subset \widetilde{K} \cap v(D(v) \cap L) \subset \widetilde{K} \cap L = \widetilde{L}.$$

It follows that the pair $(L, \widetilde{L})$ belongs to $\mathfrak{K}$ and is dominated by $(K, \widetilde{K})$. Hence by minimality

$$K = L = \widetilde{K}_{[1]} \subset K_{[1]} \subset K,$$

and K is locally self-homeomorphic. $\qquad \square$

DEFINITION 2.6. A function system $(X, \mathfrak{V})$ is called an iterated function system (IFS) if there is a subset $O \subset X$ such that

$$\emptyset \neq \bigcup_{v \in \mathfrak{V}} v(D(v) \cap O) \subset \bigcup_{v \in \mathfrak{V}} D(v) \cap O.$$

By definition, an IFS always has an infinite orbit. Any point in $\bigcup_{v \in \mathfrak{V}} D(v) \cap O$ has no finite orbit. Removing all isolated finite orbits from the domain of definition, allows to apply Theorem 2.5 in its full strength.

Note that for a locally self-homeomorphic set K we can define the action of $\mathfrak{V}$ by restriction: $v \in \mathfrak{V}$ defines a map $v : D(v) \cap K \to X$ by restriction. If $D(v) \cap K = \emptyset$ we delete this map from the family $\mathfrak{V}$, hence $(K, \mathfrak{V})$ also defines a systems considered in this section. We specialize the foregoing results in

THEOREM 2.7. *Let $(X, \mathfrak{V})$ be a functional system with compact set $\bigcup_{v \in \mathfrak{V}} D(v)$ and*

$$\bigcup_{v \in \mathfrak{V}} v(D(v)) \subset \bigcup_{v \in \mathfrak{V}} D(v).$$

Then there is a self-homeomorphic set K such that the subsystem $(K, \mathfrak{V})$ is an IFS.

PROOF. Apply Theorem 2.5 with $C = \bigcup_{v \in \mathfrak{V}} D(v)$. Then the set $\mathfrak{K}$ in the proof of that theorem contains the pair $(C, \widetilde{L})$ with $\widetilde{L} = \bigcup_{v \in \mathfrak{V}} v(D(v))$. Since $\bigcup_{v \in \mathfrak{V}} v(D(v))$ is nonempty and there are no finite orbits, there is a minimal pair $(K, \widetilde{K})$ with $\widetilde{K} \neq \emptyset$ and K is a locally self-homeomorphic set satisfying $\widetilde{\widetilde{K}} = K \subset \bigcup_{v \in \mathfrak{V}} D(v)$, hence K is even self-homeomorphic. $\qquad \square$

2.3. Limit Sets. Following [12] limit points of an iterated function system can be distinguished in the following way.

DEFINITION 2.8. Let $A \subset X$. A point $x \in X$ is said to belong to the extended limit set $\Lambda_e(A)$ of A if $\forall n \in \mathbb{N} \; \exists v_n \in \mathfrak{V} \; \exists z_n \in A$ with $x = \lim_{n \to \infty} v_1 \circ \ldots \circ v_n(z_n)\}$. If $z_n = z$ for all n, then x is said to belong to the limit set $\Lambda(A)$.

An IFS $(X, \mathfrak{V})$ is said to be of *finite multiplicity for $A \subset X$* if for each $x \in A$ the number of maps $v \in \mathfrak{V}$ with $x \in v(D(v))$ is bounded (independently of $x \in A$). It has the *strong generator property for A* if $\lim_{n \to \infty} \sup_{v \in \mathfrak{V}_n} \operatorname{diam} v(D(v) \cap A) = 0$. It is said to have the *finite intersection property for A* if for all sequences $\{v_j \in \mathfrak{V} : j \geq 1\}$ with $D(v_1 \circ \ldots \circ v_n) \cap A \neq \emptyset$ for all $n \geq 1$ also $\bigcap_{k=1}^n D(v_1 \circ v_2 \circ \ldots \circ v_k) \cap A \neq \emptyset$.

PROPOSITION 2.9. (1) *For any set $A \subset X$*

$$\Lambda(A) \subset \Lambda_e(A).$$

(2) *Let $K \subset X$ be compact and $(X, \mathfrak{V})$ have the strong generator and the finite intersection properties for K. Then*

$$\Lambda_e(K) = \Lambda(K).$$

(3) *If $(X, \mathfrak{V})$ is of finite multiplicity for the $\mathfrak{V}$-invariant set $K \subset X$, then*

$$K^{\circ}_{[\infty]} \subset \Lambda_e(K) \subset K_{[\infty]}.$$

PROOF. (1) and (2) have standard proofs (see [12]). So we only prove (3).

Let $x \in K^{\circ}_{[\infty]}$. Then for every $n \in \mathbb{N}$ there are $v_n = v_n^1 \circ v_n^2 \circ \ldots \circ v_n^n \in \mathfrak{V}_n$ and $z_n \in D(v_n) \cap K$ such that $x = v_n(z_n)$.

By assumption, the set $E(z)$ of pairs (w, y) with $w \in \mathfrak{V}$, $y \in D(w) \cap K$ and $z = w(y)$ is finite and its cardinality is uniformly in n bounded by M for some $M \in \mathbb{N}$. Define

$$E_n = \{(u_1, u_2, \ldots) : u_j \in \mathfrak{V}; \exists z \in D(u_1 \circ \ldots \circ u_n) \cap K \ni x = u_1 \circ \ldots \circ u_n(z)\}.$$

Then $E_n \neq \emptyset$ for every $n \in \mathbb{N}$, and by invariance for $(u_1, ..., u_n, ...) \in E_{n+1}$ and some $z \in D(u_1 \circ ... \circ u_{n+1}) \cap K$ one has $x = u_1 \circ ... \circ u_n(u_{n+1}(z))$ with $u_{n+1}(z) \in D(u_1 \circ ... \circ u_n) \cap K$. It follows that $E_{n+1} \subset E_n$. Now $E_1 \subset \bigcup_{u \in \mathfrak{V}:u^{-1}(x) \in D(u)}[u]$, where $[u] = \{(w_1, w_2, ...) : w_1 = u\}$. The number of such u is bounded by M, hence there exists u_1 such that for all $n \geq 1$ $E_n \cap [u_1] \neq \emptyset$. By induction one defines a sequence u_n $(n \geq 1)$ such that $E_m \cap [u_1, u_2, ..., u_n] \neq \emptyset$ for all $n \geq 1$ and $m \geq n$. Hence for all $n \geq 1$ there exists $z_n \in D(u_1 \circ ... \circ u_n) \cap K$ such that $x = u_1 \circ ... \circ u_n(z_n)$.

Finally, let $x \in \Lambda_e(K)$ and $v_1, v_2, ... \in \mathfrak{V}$, $z_n \in D(v_1 \circ ... \circ v_n) \cap K$ with $x = \lim v_1 \circ ... \circ v_n(z_n)$. Then $x_n = v_1 \circ ... \circ v_n(z_n)$ belongs to $K_{[n]}^\circ \subset K_{[m]}$ for all $m \leq n$ by Lemma 2.1, hence $x_n \in K_{[m]}$ for all $n \geq m$, and since $x = \lim x_n$, also $x \in K_{[m]}$. $\qquad\square$

THEOREM 2.10. *Let $(K, \mathfrak{V})$ be a self-homeomorphic IFS having the strong generator and the finite intersection properties, and being of finite multiplicity. If K is compact and minimal in the sense of Proposition 2.3, if $K_{[\infty]}^\circ \neq \emptyset$ and if $(K, K_{[\infty]}^\circ)$ is minimal in the sense of Theorem 2.5, then the limit set $\Lambda(K)$ is dense in K.*

PROOF. Since K is a self-homeomorphic set and $(K, \mathfrak{V})$ an IFS we have that $K = \bigcup_{v \in \mathfrak{V}} D(v)$. It follows that $K_{[n]}^\circ = \bigcup_{v \in \mathfrak{V}_n} v(D(v))$ and therefore for $v \in \mathfrak{V}$

$$v(D(v) \cap K_{[n]}^\circ) \subset K_{[n+1]}^\circ.$$

Taking intersections over n yields

(2.1) $$v(D(v) \cap K_{[\infty]}^\circ) \subset K_{[\infty]}^\circ.$$

Since there are no finite isolated orbits, as in the proof of Theorem 2.5, parts a) or b), one can show that for $L = \overline{K_{[\infty]}^\circ}$, (2.1) implies that $L_{[1]} \subset L$. Since L is nonempty and contained in K it must be equal to K by minimality in the sense of Proposition 2.3. Therefore

$$K = K_{[\infty]} = \overline{K_{[\infty]}^\circ}.$$

By the property of finite multiplicity, Proposition 2.9 (3) shows that

$$K_{[\infty]}^\circ \subset \Lambda_e(K) \subset K.$$

The theorem follows since the strong generator and finite intersection properties hold as well. $\qquad\square$

3. Conformal Measures

3.1. Transfer Operator And Algebraic Pressure. In this subsection let $(X, \mathfrak{V})$ be an iterated function system.

A *potential* for the system $(X, \mathfrak{V})$ is a family $\phi = \{\phi_v : v \in \mathfrak{V}\}$ of equi-uniformly continuous functions $\phi_v : D(v) \to \mathbb{R}_+$. Let $\Phi = \Phi(X, \mathfrak{V})$ denote the set of all potentials. A potential $\phi = \{\phi_v : v \in \mathfrak{V}\}$ is called *proper* if each function ϕ_v vanishes on the boundary $\partial D(v)$ of $D(v)$. A special case of a potential is obtained setting $\phi_v = \psi \circ v$ where $\psi \in C(X)$ is non-negative and $v \in \mathfrak{V}$. In case each set $D(v)$ $(v \in \mathfrak{V})$ is closed and open, every non-negative function $\psi \in C(X)$ defines a proper potential. We define for $w = v_n \circ v_{n-1} \circ ... \circ v_1 \in \mathfrak{V}_n$ and $x \in D(w)$

$$\phi_w(x) = \prod_{i=1}^n \phi_{v_i}(v_{i-1} \circ ... \circ v_0(x)) \qquad v_0 \text{ identity.}$$

The *Perron Frobenius operator* for $\mathfrak{V}$ and a potential ϕ is defined by

$$\mathfrak{L}_\phi f(x) = \sum_{v \in \mathfrak{V}; x \in D(v)} f(v(x))\phi_v(x),$$

where the operator acts on functions f, whenever the sum is well defined. Throughout the paper, measurability is understood by Borel-measurability and we denote by $\mathcal{F}$ the Borel σ-field of X. Note that the operator is always well defined on bounded measurable functions if ϕ satisfies the condition $\|\mathfrak{L}_\phi 1\|_\infty < \infty$. We let $\Phi^{uc} = \Phi^{uc}(X, \mathfrak{V})$ denote the set of all potentials ϕ such that for each $x \in X$ and $n \in \mathbb{N}$, $\mathfrak{L}_\phi^n 1$ converges uniformly in some neighborhood of x, and let Φ^p denote the subset of all proper potentials.

Let $\phi \in \Phi$. Define for each $n \geq 1$ the n-th partition function by

$$Z_n(\phi) := \sum_{v \in \mathfrak{V}_n} \sup_{x \in D(v)} \phi_v(x),$$

including the case when $Z_n(\phi) = \infty$. Likewise, for a potential $\phi \in \Phi^{uc}$ and for $x \in X$ we define the n-th partition function at x by

$$Z_n(\phi, x) := \sum_{v \in \mathfrak{V}_n; x \in D(v)} \phi_v(x) = \mathfrak{L}_\phi^n 1(x).$$

Here, due to the assumption of the potential, $Z_n(\phi, x)$ is always finite.

LEMMA 3.1. *Let $(X, \mathfrak{V})$ be an iterated function system and $\phi \in \Phi$. Then the limit $P_{alg}(\phi) := \lim_{n \to \infty} \frac{1}{n} \log Z_n(\phi)$ exists, or $P_{alg}(\phi)$ is infinite. If $Z_n(\phi)$ is finite for some $n \in \mathbb{N}$, then $P_{alg}(\phi) < \infty$. This limit will be called the algebraic pressure of ϕ.*

PROOF. $\log Z_n(\phi)$ is subadditive. $\qquad\qquad\qquad\qquad\qquad\qquad\square$

The definition is similar to the notion of pressure for shift spaces with finite or countable alphabets. Also note, the corresponding statement for $Z_n(\phi, x)$ does not hold without additional assumptions. We therefore set

$$P(\phi, x) = \limsup_{n \to \infty} \frac{1}{n} \log Z_n(\phi, x),$$

and call it the *transition parameter* of ϕ at x.

As an immediate consequence of the definition and Hölder's inequality we obtain the following lemma where we use the notation $\phi^p = \{\phi_v^p : v \in \mathfrak{V}\}$ and $\phi^\alpha \psi^\beta = \{\phi_v^\alpha \psi_v^\beta : v \in \mathfrak{V}\}$. The proof is standard and omitted.

LEMMA 3.2.   (1) *If $1 < p, q < \infty$ form a dual pair (i.e. $\frac{1}{p} + \frac{1}{q} = 1$), then for any $\phi, \psi \in \Phi$*

$$P_{alg}(\phi) \leq \frac{1}{p} P_{alg}(\psi^p) + \frac{1}{q} P_{alg}\left(\left(\frac{\phi}{\psi}\right)^q\right).$$

(2) *If $\psi, \phi \in \Phi$, and if $\log \phi := \{\log \phi_v : v \in \mathfrak{V}\}$ and $\log \psi := \{\log \psi_v : v \in \mathfrak{V}\}$ are well defined families of uniformly continuous functions, then*

$$|P_{alg}(\phi) - P_{alg}(\psi)| \leq \sup_{v \in \mathfrak{V}} \sup_{x \in D(v)} |\log \phi_v(x) - \log \psi_v(x)|.$$

(3) *If $\psi, \phi \in \Phi$ and $\alpha + \beta = 1$, $\alpha, \beta \geq 0$, then*

$$P_{alg}(\phi^\alpha \psi^\beta) \leq \alpha P_{alg}(\phi) + \beta P_{alg}(\psi).$$

Let $C(\Phi^+)$ denote the set of all potentials $\phi \in \Phi$ satisfying

$$\sup_{v \in \mathfrak{V}} \| \log \phi_v \|_\infty < \infty.$$

The next theorem follows immediately from the foregoing lemma.

THEOREM 3.3. *The algebraic pressure is finite for all $\phi \in C(\Phi^+)$ if and only if it is finite for the unit potential $\{\phi_v \equiv 1 : v \in \mathfrak{V}\}$.*

3.2. Conformal Families. We assume here that X is compact and $\mathfrak{V}$ a finite or countable family of maps, unless stated otherwise. We let $\phi = (\phi_v)_{v \in \mathfrak{V}} \in \Phi^{uc}$ be a potential with $\phi_v \neq 0$ for all $v \in \mathfrak{V}$. If follows that the transition parameter of the potential $\phi \in \Phi^{uc}$ at $x \in X$ is well defined and finite:

$$P(\phi, x) = \limsup_{n \to \infty} \frac{1}{n} \log Z_n(\phi, x) = \limsup_{n \to \infty} \frac{1}{n} \log \mathcal{L}_\phi^n 1(x).$$

For $p > P(\phi, x)$ we shall be using the Patterson construction defining $M(p, x) = \sum_{n=1}^{\infty} b(n) \exp(-np) Z_n(\phi, x)$ and

$$m(p, x) = M(p, x)^{-1} \sum_{n=1}^{\infty} b(n) \exp(-np) \sum_{v \in \mathfrak{V}_n; x \in D(v)} \phi_v(x) \epsilon_{v(x)},$$

where ϵ_z denotes the point mass in $z \in X$ and $b(n)$ is a slowly varying sequence independent of p, for which the sequence $M(p, x)$ diverges for $(p \leq P(\phi, x))$, converges for $p > P(\phi, x)$ and $\lim_{p \downarrow P(\phi, x)} M(p, x) = \infty$ (see [**9**] or [**3**]).

A probability measure m on X is called a *limit measure* for the potential ϕ if it is a weak accumulation point of $\{m(c, x) : c > P(\phi, x)\}$ for some $x \in X$ as $c \downarrow P(\phi, x)$. We say that a sequence of measures $\{m_n : n \in \mathbb{N}\}$ converges $\mathfrak{V}$-*weakly* if there exists a measure m on X such that m_n converges weakly to m and for any $v \in \mathfrak{V}$ the measures $\{m_n^{(v)} : n \in \mathbb{N}\}$, defined by $m_n^{(v)}(A) = m_n(A \cap D(v))$, converge weakly to some measure m^v, where $D(v)$ is equipped with the induced topology (note that $m^{(v)}$ and m^v are in general different measures).

Obviously the following extension of weak compactness of measures holds:

LEMMA 3.4. *The space of probability measures on X is $\mathfrak{V}$-weakly sequentially compact. Moreover, for the limit measures we have the relation*

$$m^v_{|intD(v)} = m_{|intD(v)} \qquad \forall v \in \mathfrak{V}.$$

For a general result on existence of conformal measures we need the notion of a conformal family.

DEFINITION 3.5. Let $(X, \mathfrak{V})$ be an iterated function system and $\phi \in \Phi$ be a potential. A family of measures $\mathcal{M} = \{m\} \cup \{m^v : v \in \mathfrak{V}\}$ is called conformal for ϕ if there exists $\lambda_\phi \geq 0$ such that
1. For every continuous function $f \in C(X)$

$$\lambda_\phi \int f dm = \int \mathcal{L}_\phi f dm + \sum_{v \in \mathfrak{V}} \int_{\partial D(v)} f(v(x)) \phi_v(x) (m^v - m)(dx),$$

where $\partial D(v) = D(v) \setminus intD(v)$.

2. $m^v = m$ on $\operatorname{int} D(v)$ for all $v \in \mathfrak{V}$.

The function $\Delta(\phi, \mathcal{M}, \cdot) : C(X) \to \mathbb{R} \cup \{\pm\infty\}$,

$$\Delta(\phi, \mathcal{M}, f) = \sum_{v \in \mathfrak{V}} \int_{\partial D(v)} f(v(x)) \phi_v(x)(m^v - m)(dx)$$

will be called the deficiency of the conformal family. If the deficiency of a conformal family vanishes, the measure m is called conformal.

The existence of conformal families is the next result.

THEOREM 3.6. *Let $(X, \mathfrak{V})$ be an iterated function system, and let $\phi \in \Phi^{uc}$ be a potential with $Z_1(\phi) < \infty$. Then there exists a conformal family $\mathcal{M} = \{m\} \cup \{m^v : v \in \mathfrak{V}\}$ for ϕ. Its deficiency satisfies*

$$(3.1) \qquad\qquad \Delta(\phi, \mathcal{M}, f) \leq 0$$

for every positive $f \in C(X)$.

PROOF. Apply Lemma 3.4 to the measures in the Patterson construction and a subsequences $p_k \downarrow P(\phi, x)$ to obtain a family $\mathcal{M} = \{m, m^v : v \in \mathfrak{V}\}$ such that m is a limit measure, $m^v(D(v)) = m^v(X)$ and $m^v = m$ on $\operatorname{int} D(v)$.

Let $g \in C(X)$. By definition of $\mathfrak{L}_\phi$ one obtains

$$\int \mathfrak{L}_\phi g(y) m(p_k, x)(dy)$$

$$= \frac{1}{M(p_k, x)} \sum_{n=1}^{\infty} b(n) e^{-np_k} \sum_{w \in \mathfrak{V}_{n+1}; x \in D(w)} g(w(x)) \phi_w(x)$$

$$= e^{p_k} \int g(y) m(p_k, x)(dy)$$

$$+ \frac{1}{M(p_k, x)} \sum_{n=2}^{\infty} e^{p_k} (b(n-1) - b(n)) e^{-np_k} \sum_{v \in \mathfrak{V}_n} g(v(x)) \phi_v(x)$$

$$- M(p_k, x)^{-1} b(1) e^{-p_k} \sum_{v \in \mathfrak{V}, x \in D(v)} g(v(x)) \phi_v(x).$$

As $k \to \infty$ the right hand side tends to $e^{P(\phi, x)} \int g(y) m(dy)$.

It is left to show that the left hand side tends to

$$\int \mathfrak{L}_\phi g(y) m(dy) + \sum_{v \in \mathfrak{V}} \int_{\partial D(v)} g \circ v \phi_v d(m^v - m).$$

First note that for a finite subfamily $\mathfrak{V}_0 \subset \mathfrak{V}$ we have

$$\left| \int \mathfrak{L}_\phi g(y) m(p_k, x)(dy) - \sum_{v \in \mathfrak{V}_0} \int g(v(y)) \phi_v(y) m(p_k, x)^{(v)}(dy) \right|$$

$$\leq \left| \int \sum_{v \notin \mathfrak{V}_0} g(v(y)) \phi_v(y) m(p_k, x)(dy) \right|$$

$$\leq \|g\|_\infty \sum_{v \notin \mathfrak{V}_0} \sup_{z \in D(v)} \phi_v(z) m(p_k, x)(X).$$

Given $\epsilon > 0$, take $\mathfrak{V}_0$ so large that for any $k \geq 1$

$$\left| \int \mathcal{L}_\phi g(y) m(p_k, x)(dy) - \sum_{v \in \mathfrak{V}_0} \int g(v(y)) \phi_v(y) m(p_k, x)^{(v)}(dy) \right| < \epsilon,$$

then k_0 so large that for $k \geq k_0$

$$\sum_{v \in \mathfrak{V}_0} \left| \int g(v(y)) \phi_v(y) m(p_k, x)^{(v)}(dy) - \int g(v(y)) \phi_v(y) m^v(dy) \right| < \epsilon.$$

We find that

$$\lim_{k \to \infty} \int \mathcal{L}_\phi g(y) m(p_k, x)(dy) = \sum_{v \in \mathfrak{V}} \int g(v(y)) \phi_v(y) m^v(dy)$$

$$= \int \mathcal{L}_\phi g(y) m(dy) + \sum_{v \in \mathfrak{V}} \int_{\partial D(v)} g \circ v \phi_v d(m^v - m).$$

In order to show (3.1) let $A \subset D(v)$ be a closed set which is a m^v continuity set (boundary in the induced topology has measure zero). Then

$$m^v(A) = \lim_{k \to \infty} m(p_k, x)^{(v)}(A) = \limsup_{k \to \infty} m(p_k, x)(A) \leq m(A).$$

It follows that $m^v(C) \leq m(C)$ for all Borel sets, and this proves (3.1). $\qquad\square$

COROLLARY 3.7. *If Φ is a proper potential or if the measure m in Theorem 3.6 vanishes on each boundary $\partial D(v)$ for $v \in \mathfrak{V}$ then m is a conformal measure.*

EXAMPLE 3.8. The following simple example illustrates how to use the previous result for dynamical systems. Let $T_k : S^2 \to S^2$, $T_k(z) = z^k$ be the quadratic, resp. cubic map on the Riemann sphere. Define $T(z) = z^2$ for $|z| > 1$ and $T(z) = z^3$ for $|z| \leq 1$. Now consider the iterated function system $\mathfrak{V}$ defined by 5 maps: v_l for $l = 1, 2, 3$ denote inverse branches of T_3 on the unit disk $\{|z| \leq 1\}$, while v_4 and v_5 are those for T_2 on $D(v_4) = \{z : |z| \geq 1, Im(z) \geq 0\}$ and $D(v_5) = \{z : |z| \geq 1, Im(z) \leq 0\}$. Taking an initial point in the unit disk and the potential family given by the derivatives of the v_i's we obtain a conformal family. Since the disk and its complement stay invariant under T and S^1 is attracting for the inverse branches, we obtain a conformal family with deficiency

$$\Delta(\phi, \mathcal{M}, f) = -\sum_{k=4,5} \int_{\partial D(v_k)} f(v_k(x)) \phi_{v_k}(x) m(dx).$$

Note that the operator $\mathcal{L}$ also sums over the branches v_4 and v_5, so that the measure m will be conformal for the transformation $z \mapsto z^3$ on S^1, but it is not conformal for the operator $\mathcal{L}$. If the initial point is chosen in the complement of the disk the resulting conformal family contains the measure m as a conformal measure for the quadratic map on S^1. Of course, it is Lebesgue measure in each case. To see this one applies Theorem 3.6 and the known fact that the limiting procedure cannot produce point masses.

EXAMPLE 3.9. A more complicated example is given by the following piecewise monotonic transformation on $[0, 1]$.

$$T(x) = \begin{cases} \beta x & \text{if } 0 \leq x \leq \frac{1}{\beta} \\ \frac{\beta}{(\beta-1)^2}(x - \frac{1}{\beta})^2 & \text{if } \frac{1}{\beta} < x \leq 1, \end{cases}$$

where $\beta > 1$. There are two inverse branches. The iterated functions system becomes $v_1(x) = \frac{x}{\beta}$ with domain $D(v_1) = [0,1]$, and $v_2(x) = (\beta-1)\sqrt{\frac{x}{\beta}} + \frac{1}{\beta}$ with domain $D(v_2) = [0, \frac{1}{\beta}]$. It is easy to see that for all small $\eta > 0$

$$v_2(v_1([1-\eta, 1])) = [1 - \eta', 1]$$

for some $\eta' < \lambda\eta$ and $\lambda < 1$, independent of η. If E_n denotes the set of images of a point x_0 under all maps in $\mathfrak{V}_n$ then only a fraction $p_n(\eta)$ of these points will be mapped into a neighborhood of the form $[\frac{1}{\beta} - \eta, \frac{1}{\beta}]$ and the fraction tends uniformly in n to zero as $\eta \to 0$. This means that the measure m^{v_2} which is a measure on $D(v_2)$ has no mass on its boundary $\{\frac{1}{\beta}\}$. Since $D(v_1)$ has no boundary (in $[0,1]$) the deficiency is

$$\Delta(\phi, \mathcal{M}, f) = -f(1)\phi_{v_2}(1)m(\{\frac{1}{\beta}\}).$$

This vanishes since $m(\{\frac{1}{\beta}\}) = 0$, and the measure m is conformal.

COROLLARY 3.10. For any potential with $Z_1(\phi) < \infty$ there exists a weakly conformal measure m, that is: For any positive continuous function $f \in C(X)$

$$\lambda_\phi \int f dm \leq \int \mathcal{L}_\phi f dm.$$

Another variant of the proof of Theorem 3.6 is this result where we do not assume that X is compact.

THEOREM 3.11. Let $(X, \mathfrak{V})$ be an iterated function system. If ϕ is a proper potential with $P_{alg}(\phi) < \infty$ such that each $D(v)$ is compact, and if there is $x \in X$ with $P(\phi, x) > 0$, then there exists a conformal measure with support

$$X^\circ_{[\infty]} = \bigcap_{n=1}^{\infty} \bigcup_{v \in \mathfrak{V}_n} v(D(v)).$$

PROOF. Pick $x \in X$ with $P(\phi, x) > 0$. Let $p_\ell > P(\phi, x)$ $(\ell \in \mathbb{N})$ be a sequence converging to $P(\phi, x)$. We shall show that the family $\{m(p_\ell, x) : \ell \in \mathbb{N}\}$ is tight on $X^\circ_{[\infty]}$ by constructing compact sets K with $m(p_\ell, x)(K^c) < \epsilon$ for any given $\epsilon > 0$. Then the argument of the previous theorem remains valid replacing the compactness of X by tightness of $\{m(p_\ell, x) : \ell \in \mathbb{N}\}$.

Let $\epsilon > 0$. We first construct compact sets of the form $K_\ell = \bigcup_{v \in \mathfrak{V}^{(\ell)}} v(D(v))$ with a finite set $\mathfrak{V}^{(\ell)} \subset \mathfrak{V}_\ell$, $\ell \geq 1$, and then set $K = \bigcap_{\ell=1}^{\infty} K_\ell$.

Since $P_{alg}(\phi) < \infty$, there exist $\ell_0 \in \mathbb{N}$ with $Z_\ell(\phi) < \infty$ for $\ell \geq \ell_0$. Let $\ell \geq \ell_0$, and take $\epsilon_\ell > 0$ such that $\sum_{\ell \geq \ell_0} \epsilon_\ell < \frac{\epsilon}{C+1}$ where C is a constant determined below. Then, for each $\ell \geq \ell_0$, choose a finite set $\mathfrak{V}^{(\ell)} \subset \mathfrak{V}_\ell$ such that

$$\sum_{v \notin \mathfrak{V}^{(\ell)}} \sup_{y \in D(v)} \phi_v(x) \leq \epsilon_\ell.$$

Then for $n \geq \ell$ define $\mathfrak{V}_n^{(\ell)} = \{wu \in \mathfrak{V}_n, w \in \mathfrak{V}^{(\ell)}; u \in \mathfrak{V}_{n-\ell}\} \subset \mathfrak{V}_n$. It follows that

$$\sum_{v = wu \notin \mathfrak{V}_n^{(\ell)}} \phi_w(u(x))\phi_u(x) \leq \epsilon_\ell Z_{n-\ell}(\phi, x).$$

If $p > P(\phi, x)$ we conclude that, writing $M = M(p_k, x)$ for short,

$$m(p, x)(K_l^c) = \frac{1}{M} \sum_{n=1}^{\infty} b(n) e^{-np} \sum_{v \in \mathfrak{V}_n} \phi_v(x) 1_{K_\ell^c}(v(x))$$

$$\leq \frac{1}{M} \sum_{n=\ell+1}^{\infty} b(n) e^{-np} \sum_{v=wu \notin \mathfrak{V}_n^{(\ell)}} \phi_w(u(x)) \phi_u(x) + \frac{1}{M} \sum_{n=1}^{\ell} b(n) e^{-np} Z_n(\phi, x)$$

$$\leq \frac{\epsilon_\ell e^{-\ell p}}{M} \sup_{n \geq 1} \frac{b(n+\ell)}{b(n)} \sum_{n=1}^{\infty} b(n) e^{-np} Z_n(\phi, x) + \frac{1}{M} \sum_{n=1}^{\ell} b(n) e^{-np} Z_n(\phi, x).$$

Next observe that the sequence $b(n)$ can be constructed in such a way that for some sequence $\eta_k \to 0$ and a subsequence n_k ($k \in \mathbb{N}$),

$$\eta_k \leq \frac{1}{n} \log b(n) \leq \eta_{k-1} \qquad (n_{k-1} \leq n < n_k),$$

hence there is a constant C such that

$$\sup_{p > P(\phi, x)} e^{-\ell p} \sup_{n \geq 1} \frac{b(n+\ell)}{b(n)} = C < \infty$$

since $P(\phi, x) > 0$.

There exists κ_0 such that for $\kappa \geq \kappa_0$

$$\frac{1}{M(p_\kappa, x)} \sum_{n=1}^{\ell} e^{-np_\kappa} Z_n(\phi, x) \leq \epsilon_\ell.$$

It follows that for $\kappa \geq \kappa_0$

$$m(p_\kappa, x)(K_\ell^c) \leq \epsilon_\ell(C+1),$$

and enlarging K_ℓ by including more (but finitely many) sets into $\mathfrak{V}^{(\ell)}$ we may assume that this estimate holds for each p_κ. Setting $K = \bigcap_{\ell=\ell_0}^{\infty} K_\ell$ we find that

$$m(p_\kappa, x)(K) = 1 - m(p_\kappa, x)(K^c) \geq 1 - \sum_{\ell=\ell_0}^{\infty} m(p_\kappa, x)(K_\ell^c) \geq 1 - \epsilon.$$

$\square$

COROLLARY 3.12. Let

$$C_\partial(X) := \{f \in C(X) : \forall v \in \mathfrak{V} \Rightarrow f(v(x)) = 0 \text{ for } x \in \partial D(v)\}.$$

Then

$$\int \mathfrak{L}_\phi f dm = \lambda_\phi \int f dm \qquad \forall f \in C_\partial(X).$$

Let $(X, \mathfrak{V})$ be an iterated function system. A measure m is called *non-singular* if $m(A) = 0$ implies $m(v^{-1}(A)) = 0$ for all $v \in \mathfrak{V}$ and conversely.

LEMMA 3.13. *Every conformal measure is non-singular provided that* $m(\phi_v = 0) = 0$ *for all* $v \in \mathfrak{V}$.

PROOF. If the eigenvalue $\lambda = 0$, nothing has to be shown. Therefore let $\lambda > 0$. First note that the conformality equation

$$\int \mathfrak{L}_\phi f dm = \lambda_\phi \int f dm \qquad f \in C(X)$$

extends to indicator functions $f = 1_A$ of closed sets $A \subset X$ by dominated convergence, so that

$$\lambda_\phi m(A) = \int \mathfrak{L}_\phi 1_A(x) m(dx) = \int \sum_{v \in \mathfrak{V}; x \in D(v)} 1_{v^{-1}(A)}(x) \phi_v(x) m(dx).$$

If $m(A) = 0$ but $m(v^{-1}(A)) > 0$, the integrand on the right hand side is positive on a set of positive measure. Conversely, the right hand side vanishes if every $v^{-1}(A)$ is a null set. General sets are approximated from inside by closed sets using the regularity of Borel measures on X. □

3.3. Weak Bounded Variation. A potential ϕ is called *bounded* if

$$\sup_{v \in \mathfrak{V}} \|\phi\|_\infty = \sup_{v \in \mathfrak{V}} \max_{x \in D(v)} |\phi_v(x)| < \infty.$$

We denote by Φ^+ the subset of all potentials $\phi = \{\phi_v : \inf_{x \in D(v)} \phi_v(x) > 0 \ (v \in \mathfrak{V})\}$ and recall that

$$C(\Phi^+) = \{\phi \in \Phi^+ : \|\phi\| := \sup_{v \in \mathfrak{V}} \|\log \phi_v\|_\infty < \infty\}.$$

Φ^+ is a vector space and $C(\Phi^+)$ is a Banach space with norm

$$\|\phi\| = \sup_{v \in \mathfrak{V}} \|\log \phi_v\|_\infty.$$

LEMMA 3.14. *The map $P_{alg} : C(\Phi^+) \to \mathbb{R} \cup \{\pm\infty\}$ has the following properties:*
1. If $P_{alg}(\phi) = \infty$ (or $-\infty$) for some potential $\phi \in C(\Phi^+)$ then $P_{alg}(\psi) = \infty$ (or $-\infty$) for all potentials $\psi \in C(\Phi^+)$.
2. P_{alg} is Lipschitz-continuous if $|P_{alg}(\phi)| < \infty$ for some $\phi \in C(\Phi^+)$:

$$|P_{alg}(\phi) - P_{alg}(\psi)| \leq \|\phi/\psi\|$$

where $\phi/\psi = \{\phi_v/\psi_v : v \in \mathfrak{V}\}$.

PROOF. This follows from Theorem 3.3 and Lemma 3.2 2. □

DEFINITION 3.15. A potential $\phi \in \Phi$ is said to be of weak bounded variation (WBV) if there exists a sequence of positive numbers $\{C_n\}$ satisfying $\lim_{n \to \infty}(1/n) \log C_n = 0$ and $\forall n \geq 1, \forall v \in \mathfrak{V}_n$

$$\frac{\sup_{x \in D(v)} \phi_v(x)}{\inf_{y \in D(v)} \phi_v(y)} \leq C_n.$$

The constants C_n are called the WBV-constants.

REMARK 3.16. A potential ϕ is called a Bowen potential (of bounded variation) if $Var_n(\phi) = \sup_{v \in \mathfrak{V}_n} \sup_{x,y \in D(v)} \frac{\phi_v(x)}{\phi_v(y)}$ satisfies $\sum_{n=1}^\infty \log Var_n(\phi) < \infty$. Clearly, a Bowen potential is of weak bounded variation.

Let Φ^b denote the set of all potentials in Φ of weak bounded variation which do not contain functions identically zero. Note that each WBV potential is strictly positive (i.e. $\phi_v(x) > 0$ for all $v \in \mathfrak{V}$ and $x \in D(v)$) unless some ϕ_v vanishes, hence $\Phi^b \subset \Phi^+$.

DEFINITION 3.17. An iterated function system $(X, \mathfrak{V})$ is called aperiodic for $x \in X$ if for any $m \in \mathbb{N}$ there is a natural number $n(x, m) = o(m)$ such that $\forall v \in \mathfrak{V}_m, \exists k \leq n(x, m)$ and $w \in \mathfrak{V}_k$ with $x \in D(w) \cap w^{-1}(D(v))$.

PROPOSITION 3.18. 1. Φ^b is a linear subspace of Φ^+.

2. The map $P_{alg} : \Phi^b \cap C(\Phi^+) \to \mathbb{R}$ is Lipschitz continuous.

3. Let $(X, \mathfrak{V})$ be aperiodic for $x \in X$. If $\phi \in \Phi^b$ and $\inf_{v \in \mathfrak{V}; y \in D(v)} \phi_v(y) > 0$, then

$$P_{alg}(\phi) = P(\phi, x).$$

In particular, if $(X, \mathfrak{V})$ is an iterated function system with $\mathfrak{V}$ finite and $D(v) = X$ compact for all $v \in \mathfrak{V}$ then for every $\phi \in \Phi^b$, $P_{alg}(\phi) = P(\phi, x)$ ($\forall x \in X$).

4. Let m be a finite conformal measure for the potential $\phi \in \Phi^b$. Then

$$P_{alg}(\phi) = 0 \quad \text{if and only if} \quad \lambda_\phi = 1.$$

PROOF. 1. Let $\phi, \psi \in \Phi^b$ and $\alpha, \beta \in \mathbb{R}$ with WBV-constants $C_n(\phi)$ and $C_n(\psi)$. Then for $v \in \mathfrak{V}_n$ and $x, y \in D(v)$, $\phi_v^\alpha(x)\psi_v^\beta(x) \le C_n(\phi)^\alpha C_n(\psi)^\beta \phi_v^\alpha(y)\psi_v^\beta(y)$ and Φ^b is a linear subspace of Φ^+.

2. This is Lemma 3.2 2.

3. Clearly $P(\phi, x) \le P_{alg}(\phi)$, hence we need to show the converse inequality.

Let $v \in \mathfrak{V}_m$. Choose $w_v \in \mathfrak{V}_k$ and $k \le n := n(x, m)$ such that $w_v(x) \in D(v)$. Then by weak bounded variation

$$Z_m(\phi) \le C_m \sum_{v \in \mathfrak{V}_m} \phi_v(w_v(x)) \le C_m \sum_{v \in \mathfrak{V}_m} \frac{\phi_{vw_v}(x)}{\phi_{w_v}(x)}$$
$$\le C_m Z_{m+n}(\phi, x) \sup_{u \in \mathfrak{V}} \|1/\phi_u\|_\infty^n.$$

The claim follows from $n = o(m)$.

4. Let $n \ge 1$ and let $x_v \in D(v)$ for $v \in \mathfrak{V}_n$ such that $\phi_v(x_v) = \sup_{y \in D(v)} \phi_v(y)$. Then by weak bounded variation and conformality

$$C_n^{-1} m(X) \sum_{v \in \mathfrak{V}_n} \phi_v(x_v) = C_n^{-1} m(X) Z_n(\phi) \le \int \mathfrak{L}_\phi^n 1 dm$$
$$= \lambda_\phi^n m(X) \le m(X) \sum_{v \in \mathfrak{V}_n} \sup_{x \in D(v)} \phi_v(x) = m(X) Z_n(\phi)$$

The claim follows using Lemma 3.1. $\square$

A final result of our discussion is this:

THEOREM 3.19. *Let $(X, \mathfrak{V})$ be an iterated function system with an aperiodic point. Then for every non-vanishing proper potential $\phi \in \Phi^b \cap \Phi^p$ of bounded variation there exists a conformal measure with eigenvalue*

$$\lambda_\phi = e^{P_{alg}(\phi)}.$$

REMARK 3.20. A similar statement can be made for conformal families and general potentials in Φ^{uc}.

PROOF. Let x be an aperiodic point. By 3. in Proposition 3.18 we have $P_{alg}(\phi) = P(\phi, x)$ and by Corollary 3.7 there exists a conformal measure for ϕ with $\lambda_\phi = e^{P(\phi, x)} = e^{P_{alg}(\phi)}$. $\square$

References

[1] Wael Bahsoun, Christopher Bose, and Anthony Quas, *Deterministic representation for position dependent random maps*, Discrete Contin. Dyn. Syst. **22** (2008), no. 3, 529–540, DOI 10.3934/dcds.2008.22.529. MR2429852 (2009m:37149)

[2] Vaughn Climenhaga and Daniel J. Thompson, *Equilibrium states beyond specification and the Bowen property*, J. Lond. Math. Soc. (2) **87** (2013), no. 2, 401–427, DOI 10.1112/jlms/jds054. MR3046278

[3] Manfred Denker and Mariusz Urbański, *On the existence of conformal measures*, Trans. Amer. Math. Soc. **328** (1991), no. 2, 563–587, DOI 10.2307/2001795. MR1014246 (92k:58155)

[4] Manfred Denker and Bernd O. Stratmann, *The Patterson measure: classics, variations and applications*, Contributions in analytic and algebraic number theory, Springer Proc. Math., vol. 9, Springer, New York, 2012, pp. 171–195, DOI 10.1007/978-1-4614-1219-9_7. MR3060460

[5] Manfred Denker and Michiko Yuri, *A note on the construction of nonsingular Gibbs measures*. part 2, Colloq. Math. **84/85** (2000), no. part 2, 377–383. Dedicated to the memory of Anzelm Iwanik. MR1784215 (2001k:37013)

[6] Doris Fiebig, Ulf-Rainer Fiebig, and Michiko Yuri, *Pressure and equilibrium states for countable state Markov shifts*, Israel J. Math. **131** (2002), 221–257, DOI 10.1007/BF02785859. MR1942310 (2004b:37055)

[7] John E. Hutchinson, *Fractals and self-similarity*, Indiana Univ. Math. J. **30** (1981), no. 5, 713–747, DOI 10.1512/iumj.1981.30.30055. MR625600 (82h:49026)

[8] R. Daniel Mauldin and Mariusz Urbański, *Graph directed Markov systems*, Cambridge Tracts in Mathematics, vol. 148, Cambridge University Press, Cambridge, 2003. Geometry and dynamics of limit sets. MR2003772 (2006e:37036)

[9] S. J. Patterson, *The limit set of a Fuchsian group*, Acta Math. **136** (1976), no. 3-4, 241–273. MR0450547 (56 #8841)

[10] F. Paulik, M. Pollicott, B. Shapira: Equilibrium states in negative curvature. Nov. 2013. ArXiv: 1211.6242

[11] Omri M. Sarig, *Thermodynamic formalism for countable Markov shifts*, Ergodic Theory Dynam. Systems **19** (1999), no. 6, 1565–1593, DOI 10.1017/S0143385799146820. MR1738951 (2000m:37009)

[12] Bernd O. Stratmann and Mariusz Urbański, *Pseudo-Markov systems and infinitely generated Schottky groups*, Amer. J. Math. **129** (2007), no. 4, 1019–1062, DOI 10.1353/ajm.2007.0028. MR2343382 (2008j:37050)

[13] Y. Yamaguti, S. Kuroda, Y. Fukushima, M. Tsukada, I. Tsuda: A mathematical model for Cantor coding in the hippocampus. Neural Network, **24**(1) (2011), 43–53.

[14] Michiko Yuri, *Zeta functions for certain non-hyperbolic systems and topological Markov approximations*, Ergodic Theory Dynam. Systems **18** (1998), no. 6, 1589–1612, DOI 10.1017/S0143385798117972. MR1658607 (2000j:37024)

[15] Michiko Yuri, *Thermodynamic formalism for countable to one Markov systems*, Trans. Amer. Math. Soc. **355** (2003), no. 7, 2949–2971 (electronic), DOI 10.1090/S0002-9947-03-03269-0. MR1975407 (2004g:37012)

MATHEMATICS DEPARTMENT, THE PENNSYLVANIA STATE UNIVERSITY, STATE COLLEGE, PENNSYLVANIA 16802

E-mail address: denker@math.psu.edu

DEPARTMENT OF MATHEMATICS, HOKKAIDO UNIVERSITY, KITA 10, NISHI 8, KITA-KU, SAPPORO, HOKKAIDO 060-0810, JAPAN

E-mail address: yuri@math.sci.hokudai.ac.jp

Contemporary Mathematics
Volume **631**, 2015
http://dx.doi.org/10.1090/conm/631/12599

Dani's work on probability measures on groups

François Ledrappier and Riddhi Shah

ABSTRACT. We describe S. G. Dani's contributions in the area of probability measures on groups, especially on the embedding problem; other topics discussed include convergence-of-types theorems, concentration functions, Levy's measures.

1. Introduction to the embedding problem

Let G be a locally compact (Hausdorff) group and let $P(G)$ denote the convolution semigroup of probability measures endowed with the weak* topology. Let $\mu \in \mathbb{P}(G)$. For $\lambda \in P(G)$, let $\mu\lambda$ denote the convolution of μ and λ in $P(G)$. For $x \in G$, let δ_x denote the Dirac measure at x, and write $x\mu = \delta_x\mu$ and $\mu x = \mu\delta_x$ for the left and right shifts (or translates) of μ in $P(G)$. Let μ^n denote the n-fold convolution of μ with itself in $P(G)$. Let $G(\mu)$ denote the smallest closed subgroup containing $\operatorname{supp}\mu$, the support of μ in G and $N(\mu)$ (resp. $Z(\mu)$) be the normaliser (resp. the centraliser) of $G(\mu)$ in G. Let $F(\mu)$ denote the set of two sided convolution factors of μ, i.e. $F(\mu) = \{\lambda \in P(G) \mid \lambda\nu = \nu\lambda = \mu$ for some $\nu \in P(G)\}$. Then $F(\mu)$ is closed and $F(\mu) \subset P(N(\mu))$.

A probability measure μ on G is said to be *infinitely divisible* (i.d.) if for all $n \geq 1$, one can solve

$$(1) \qquad\qquad \nu^n = \mu,$$

for $\nu \in P(G)$; i.e. μ has convolution roots of all orders in $P(G)$. The solution of (1) is called an nth (convolution) root of μ.

For probabilists, this means that for every $n \in \mathbb{N}$ the random variable with law μ can be written as a product of n independent random variables with the same law.

Infinitely divisible probabilities on $\mathbb{R}$ were classified in late 1930s by Lévy and Khintchine in terms of their Fourier transforms. It turns out that any i.d. probability μ on $\mathbb{R}$ is embeddable in a continuous one-parameter convolution semigroup, i.e. there exist $\{\nu_t\}_{t\geq 0} \subset P(G)$ such that $t \mapsto \nu_t$ is continuous,

$$\nu_{t+s} = \nu_t\nu_s \quad \text{and} \quad \nu_1 = \mu.$$

There has been a long-standing conjecture that every infinitely divisible probability measure on a connected Lie group is embeddable in the sense as above.

2010 *Mathematics Subject Classification.* Primary 60B15; Secondary 60B10, 22E15, 22D45.
Key words and phrases. Infinitely divisible measures, embeddable measures, convergence-of-types theorems, concentration functions.

This is known as the embedding problem. It was raised as a question by K.R. Parthasarathy in [**P67**] and gradually came to be viewed as a conjecture, when some evidence was gathered. S. G. Dani has made major contributions towards resolution of the embedding problem.

We say that the group G has the *embedding property* if every infinitely divisible probability measure on G is embeddable.

A simple reason for which this may not hold for a group G is that there could be elements $g \in G$ admitting roots (in the group) of all orders, but which are not contained in one-parameter subgroups (e.g. 1 on $\mathbb{Q}$, the latter being viewed as a discrete group) and then δ_g is i.d. but not embeddable. This does not happen in a connected Lie group (not obvious, but true - a proof of this was written down by McCrudden in an unpublished preprint), which makes a case for focussing on connected Lie groups. (Dani comments in one of his papers that in some sense, the above phenomenon accounts for failure of the embedding property in non-connected groups; in particular one expects a group to have the embedding property if all point masses are embeddable.)

It was shown by K. R. Parthasarathy (cf. [**P67**]) that all (connected) compact Lie groups have the embedding property. By the 1970's all connected nilpotent Lie groups, and more generally all connected solvable Lie groups whose Lie algebra admits only real roots, were proved to have the embedding property (see [**H77**] for details). McCrudden also showed that any i.d. probability measure on a connected Lie group G is embeddable if supp μ is large, namely, if $G(\mu) = G$ (see [**Mc81**]).

2. The embedding problem on connected Lie groups

Here we discuss the techniques used in solving the embedding problem and Dani's results with McCrudden. The results mentioned above are based on considering primarily measures $\mu \in P(G)$ for which the set of all roots of μ is contained in a compact subset of $P(G)$. The thrust of the work of Dani and McCrudden is on groups in which this would not hold for all measures, for structural reasons. For instance if α is an automorphism that preserves μ, then, for any $n \in \mathbb{N}$, if ν is an nth root of μ, $\alpha(\nu)$ is also an nth root, and the set of such roots may not be contained in a compact set of measures on G.

The approach of Dani and McCrudden involves finding a set of nth roots, supported on a certain subgroup, which are contained in a compact set. As a first step, given a probability measure μ on a certain class of groups, they show that it is *almost factor compact*, i.e. the factor set $F(\mu)$ is relatively compact modulo $Z(\mu)$, the centraliser of the support of μ. More precisely,

THEOREM 2.1 ([**DMc88a**]). *Let G be an almost algebraic group and let $\mu \in P(G)$. Given a sequence $\{\lambda_n\}$ in $F(\mu)$, there exists a sequence $\{x_n\}$ in $Z(\mu)$, such that $\{\lambda_n x_n\}$ and $\{x_n \lambda_n\}$ are relatively compact. i.e. μ is almost factor compact.*

They also show that to embed a probability measure μ on a connected Lie group, it is enough to find a rational embedding of the measure i.e. a semigroup homomorphism $f : \mathbb{Q}_+^* \to P(G)$ such that $f(1) = \mu$ (cf. [**DMc88b**]). In particular, they show that such an embedding is necessarily locally tight, i.e. $f(]0,1[\cap\mathbb{Q})$ is relatively compact. As is well-known, the existence of such an f leads to a real embedding of μ on a (connected) Lie group (see [**H77**]).

Their major result on the embedding problem is for a connected Lie group G which admits a finite dimensional representation with discrete kernel (see [**DMc92**]).

They later refer to the class of such groups as class $\mathcal{C}$ in [**DMc07**]. For a probability measure μ on a group G in class $\mathcal{C}$, they show that the set of nth roots (solutions of (1)) is relatively compact up to equivalence by an inner automorphism $g \mapsto zgz^{-1}$, for $z \in Z^0(\mu, [G, G])$, where $Z(\mu, [G, G])$ is the centralizer of the support of μ within $[G, G]$ and $Z^0(\mu, [G, G])$ is its connected component of the identity. They also characterise class $\mathcal{C}$ as follows: Let G be a connected Lie group, let R be the solvable radical and let K be the maximal compact (connected) central subgroup contained in the closure of $[R, R]$. Then G is in class $\mathcal{C}$ if and only if K is trivial. In particular, a connected Lie group G belongs to class $\mathcal{C}$ if it does not contain a compact central subgroup of positive dimension; we will see that Dani has proven many other results for this latter class of groups.

THEOREM 2.2 ([**DMc92**], [**DMc07**]). *Let G be a connected Lie group in class $\mathcal{C}$, i.e. G admits a finite dimensional representation with discrete kernel. Then G has the embedding property.*

The theorem in particular implies the embedding property for all connected linear Lie groups, connected semisimple groups and also for simply connected Lie groups. For a general connected Lie group the following technical strengthening of the theorem is also noted.

THEOREM 2.3. *Let G be a connected Lie group and μ an i.d. probability measure on G. Let K be the maximal compact connected central subgroup contained in $\overline{[R, R]}$, where R is the solvable radical of G. Then there exists a sequence $\{x_n\}$ in G such that $x_n g K x_n^{-1} = gK$ in G/K, $n \in \mathbb{N}$, for all $g \in G(\mu)$, and $x_n \mu x_n^{-1} \to \lambda$ such that λ is embeddable. In particular, $\mu \omega_K = \omega_K \mu$ is embeddable, where ω_K is the normalised Haar measure supported on K.*

We refer the reader to two surveys by McCrudden [**Mc98**] and [**Mc06**] for further details on the work of Dani and McCrudden on the embedding problem, until about 2005, described above. Some more recent work of Dani on the problem, involving other coauthors, is discussed in § 4.

3. A basic technique and its applications

A general feature involved in the discussion on the embedding problem, is to understand the behavior of probability measures on a locally compact group under the application of a sequence of group automorphisms. Firstly, we want to understand how automorphisms can go to infinity.

There is an introduction by Dani in [**D06**] where he discusses this point and connects it with other problems of probabilities on groups.

Consider two real finite dimensional vector spaces V_1 and V_2. How does a sequence $\{\alpha_k\}$ of linear maps from V_1 to V_2 go to infinity? Possibly after passing to a subsequence, the answer is given by two subspaces $U \subset W \subset V_1$:

- $v \in U \iff \{\alpha_k(v)\}$ converges, and $\alpha_k(v) \to \infty$ when $v \notin U$,
- $w \in W \iff \exists \{v_k\}$ such that $v_k \to w$ and $\{\alpha_k(v_k)\}$ converges.

This translates for probabilities on V_1 as follows:

PROPOSITION 3.1 ([**D92b**]). *Assume there are $\{\lambda_k\}_{k \in \mathbb{N}}$ and λ in $P(V_1)$ such that $\lambda_k \to \lambda$ and $\{\alpha_k(\lambda_k)\}$ converges. Then the support of λ is contained in W. If the support of λ spans V_1, then $\{\alpha_k\}$ is bounded in $\mathrm{Hom}(V_1, V_2)$.*

There are several consequences of this basic result of Dani to probabilities on Lie groups as noted below:

A. Convergence-of-types theorems

Applying Proposition 3.1 to $\{\alpha_k\}$ and $\{\alpha_k^{-1}\}$, Dani gets:

THEOREM 3.2 ([**D92b**]). *Let V be a real finite dimensional vector space and $\{\alpha_k\} \subset GL(V)$. Assume there are $\{\lambda_k\}_{k \in \mathbb{N}}$, λ and μ in $P(V)$ such that $\lambda_k \to \lambda$ and $\alpha_k(\lambda_k) \to \mu$. Assume moreover that the support of λ, as well as the support of μ, spans V. Then $\{\alpha_k\}$ is bounded in $GL(V)$.*

Dani also proves a similar convergence-of-types theorem for connected Lie groups without any compact central subgroup of positive dimension (cf. [**D92b**]). As an application Dani obtains the following result which in particular plays a useful role in studying the embedding problem.

PROPOSITION 3.3 ([**D92b**]). *Let G be a connected Lie group such that the center of G has no compact central subgroup of positive dimension. Let $\lambda \in P(G)$ and set*

$$
\begin{aligned}
I(\lambda) &:= \{\alpha \in \mathrm{Aut}(G) \mid \alpha(\lambda) = \lambda\} \\
J(\lambda) &:= \{\alpha \in \mathrm{Aut}(G) \mid \alpha(g) = g, \forall g \in \mathrm{supp}\,\lambda\}.
\end{aligned}
$$

Then, $I(\lambda)/J(\lambda)$ is compact.

B. Concentration functions

Let μ be a probability measure on a group G and let $K \subset G$ be compact. The *concentration functions* of μ are defined as:

$$
c_K(\mu^n) := \sup_{x \in G} \mu^n(Kx), \ n \in \mathbb{N}.
$$

A probability measure μ on G is said to be *scattering* if $c_K(\mu^n) \to 0$ as $n \to \infty$, for all compact subsets K of G. Clearly, if $G(\mu)$ admits a compact normal subgroup H and μ is supported on a coset of H, then μ is not scattering. More generally, with R. Shah, Dani gets the following:

THEOREM 3.4 ([**DSh97**]). *Let G be a real algebraic group and $\mu \in P(G)$. Suppose that μ is not scattering. Then there exist a closed subgroup C containing $\mathrm{supp}\,\mu$, and a simply connected nilpotent normal subgroup N of C such that C/N is a direct product of a compact subgroup K and a (possibly trivial) one parameter subgroup ϕ and for $x \in \mathrm{supp}\,\mu$, the conjugation action of x on N is contracting.*

Using the above, R. Shah proved the following result which is useful in finding an embedding for certain i.d. probability measures on Lie groups which are not in class $\mathcal{C}$ (see § 4). For a measure μ on a group G, set $K(\mu) = \{g \in G \mid g\mu = \mu = \mu g\}$. It is a compact group.

THEOREM 3.5 ([**Sh99**]). *Let G be a connected Lie group and let $\mu \in P(G)$. Suppose $\lambda \in P(G)$ is such that $\lambda^n \in F(\mu)$ for all $n \in \mathbb{N}$. Then $\mathrm{supp}\,\nu \subset xK(\mu) = K(\mu)x$.*

C. Levy's measures

A probability measure μ on G is called a *Levy's measure* if it is a weak* limit of a sequence $\{\mu_i\}$ of probability measures, of the form

$$
\mu_i = g_i \alpha_i (\lambda_1 \cdots \lambda_{n_i}),
$$

where $g_i \in G$, $\alpha_i \in \mathrm{Aut}(G)$, $i \in \mathbb{N}$, and for all $j \in \mathbb{N}$, $\lambda_j \in P(G)$ are such that $\alpha_i(\lambda_j) \to \delta_e$ and $n_i \to \infty$, as $i \to \infty$.

It is well-known that on abelian divisible groups, Levy's measures are in fact i.d., and hence embeddable if any connected compact subgroup of G is arcwise connected; this holds in particular if G^0 is a Lie group.

THEOREM 3.6 ([**DSh00**]). *Let G be a connected Lie group and let C be the maximal compact connected central subgroup. Let $\{\alpha_i\}$ be a sequence in $\mathrm{Aut}(G)$. Then there exists a maximal connected Ad-unipotent subgroup N of G such that for all $x \in N$, $\alpha_i(xC) \to C$ in G/C. Moreover, for any $\mu \in P(G)$ if $\alpha_i(\mu) \to \delta_e$, then $\mathrm{supp}\,\mu \subset N$. In particular, any Levy's measure on G is supported on a coset of a closed connected nilpotent subgroup.*

D. Orbits of unipotent subgroups and almost factor compactness

For any measure λ on G and any subgroup T of $\mathrm{Aut}(G)$, the T-orbit of λ is defined as $T\lambda = \{\alpha(\lambda) \mid \alpha \in T\}$. With C. R. E. Raja, Dani proves the following result about certain orbits of a measure.

THEOREM 3.7 ([**DR98a**]). *Let G be a connected Lie group and T be an almost algebraic subgroup of $\mathrm{Aut}(G)$. Then for any measure λ in $P(G)$, $T\lambda$ is open in its closure. Moreover, $T\lambda$ is closed if T is unipotent.*

A connected Lie group is *W-algebraic* if $\mathrm{Ad}(G)$ is almost algebraic in $GL(\mathcal{G})$, where $\mathcal{G}$ is the Lie algebra of G, and for any $x \in G$, the intersection $\{xyx^{-1}y^{-1} \mid y \in G\} \cap C$ is finite where C is the maximal compact central subgroup of G. Note that any almost algebraic subgroup is W-algebraic. Dani and Raja generalise Theorem 2.1 to W-algebraic groups.

THEOREM 3.8 ([**DR98a**]). *Any probability measure λ on a W-algebraic group is almost factor compact.*

E. Tortrat groups

A locally compact group G is said to be a *Tortrat group* if for any probability measure λ on G which is not an idempotent, the closure of $\{g\lambda g^{-1} \mid g \in G\}$ does not contain any idempotent. Note that all idempotent probability measures are Haar measures of compact subgroups. A locally compact group is said to be *distal* if the identity e is not contained in the closure of $\{xgx^{-1} \mid x \in G\}$ for any nontrivial g in G. A result of Rosenblatt (in [**Ro79**]) asserts that a connected Lie group is distal if and only if it has polynomial growth, if and only if it is of type $\mathcal{R}$ (i.e. all the eigenvalues of $\mathrm{Ad}\,g$ are of absolute value 1 for all $g \in G$). Using similar techniques as above, Dani, with C. R. E. Raja, proves that for connected Lie groups the above class is same as that of Tortrat groups, namely:

THEOREM 3.9 ([**DR98b**]). *A connected Lie group is a Tortrat group if and only it is of type $\mathcal{R}$.*

From Rosenblatt's result it follows that, for connected Lie groups, the classes of Tortrat groups, distal groups and groups of polynomial growth are same.

Let us note here a result of Dani about the structure of the automorphism group of a Lie group which plays a crucial role in proving many of the above results.

THEOREM 3.10 ([**D92a**]). *Let G be a connected Lie group which has no compact central subgroup of positive dimension. Then $\mathrm{Aut}(G)$ is almost algebraic as a* *subgroup of $GL(\mathcal{G})$ where $\mathcal{G}$ is the Lie algebra of G. In particular, it has finitely many connected components.*

4. Recent results on the embedding problem on Lie groups

From all the previous results, we know that the most troublesome objects are those groups for which the center contains a torus of positive dimension. The first example is the *Walnut group*. Let $\mathbb{H} = \{(x, y, z) \mid x, y, z \in \mathbb{R}\}$ be the 3-dimensional Heisenberg group, whose center is $Z = \{(0, 0, z) \mid z \in \mathbb{R}\}$ and $D = \{(0, 0, z) \mid z \in \mathbb{Z}\}$ is a discrete co-compact subgroup in Z. Set $N = \mathbb{H}/D$. The Walnut group is the semidirect product of $SO(2)$ and N, where $SO(2)$ acts on N by rotations in the (x, y) plane. It is a 4-dimensional solvable group.

Dani, with Y. Guivarc'h and R. Shah, uses many of the techniques developed by him jointly with McCrudden, along with results on concentration functions and Fourier analysis to get the following results for embedding i.d. probability measures on Lie groups which are not in class $\mathcal{C}$.

THEOREM 4.1 ([**DGSh12**]). *Let G be a Lie group admitting a surjective continuous homomorphism $p : G \to \widetilde{G}$ onto an almost algebraic group $\widetilde{G}$, such that $\ker p$ is a compactly generated subgroup contained in the center of G and $(\ker p)^0$ is compact. Let $T = (\ker p)^0$ and $q : G \to G/T$ be the quotient homomorphism. Let $\mu \in P(G)$ be such that $q(\mu)$ has no nontrivial idempotent factor in $P(G/T)$. If μ is i.d., then it is embeddable.*

Using the theorem and previously known results, they also deduce following two corollaries.

COROLLARY 4.2 ([**DGSh12**]). *Let G be a connected Lie group which is a semidirect product of $SL(2, \mathbb{R})$ and $N = \mathbb{H}/D$ described above. Let H be any closed subgroup of G. Then H has the embedding property. In particular, the Walnut group has the embedding property.*

COROLLARY 4.3 ([**DGSh12**]). *Let G be a connected Lie group and let $\mu \in P(G)$ be i.d.. Then μ is embeddable if μ is supported on the nilradical of G.*

Moreover, they also prove the following result.

THEOREM 4.4 ([**DGSh12**]). *Let G be a connected Lie group. Then any i.d. $\mu \in P(G)$ is embeddable if $G(\mu)/\overline{[G(\mu), G(\mu)]}$ is compact.*

Note that the conditions in Corollary 4.3 and Theorem 4.4 are complementary to each other unless $G(\mu)$ is compact. The class of groups H such that $H/\overline{[H, H]}$ is compact, is quite large, as it is closed under taking finite products, compact extensions and taking quotients. It also contains all connected reductive Lie groups and also groups with the Kazhdan property T. Note that lattices in the group with the Kazhdan property T also have the Kazhdan property T. There are also amenable groups for e.g. $H = SO(2) \ltimes \mathbb{R}^2$ or the Walnut group such that $H/\overline{[H, H]}$ is compact. In particular, any i.d. probability measure μ on a connected locally compact group G is embeddable if $G(\mu)$ has the Kazhdan property T.

5. The embedding problem on general locally compact groups

Dani has also worked on the embedding problem on general locally compact groups. It was proven in [**Sh91**], that on a linear p-adic Lie group, any infinitely divisible probability measure μ has a shift $x\mu$, for some $x \in Z(\mu)$, which is embeddable (see also [**McW99**]). Using some results on real and p-adic algebraic groups, Dani, with R. Shah, shows the following:

THEOREM 5.1 ([**DSh93**]). *Any infinitely divisible probability measure on a finitely generated subgroup of $GL(n, A)$ is embeddable, where A is the field of algebraic numbers.*

Dani also considered the embedding problem on a subgroup H of a connected Lie group. For example, with McCrudden, he proved the following:

THEOREM 5.2 ([**DMc96**]). *Let G be a discrete linear group and let μ in $P(G)$ be i.d. Then μ has a shift $x\mu$, with $x \in Z(\mu)$, which is embeddable.*

They also construct an example of a measure on such a G which is i.d. but not embeddable in $P(G)$. Note that any i.d. probability measure μ on $G \subset GL(V)$ is embeddable on $GL(V)$.

For a general locally compact group G, with K. Schmidt, Dani considered measures μ supported on a connected abelian subgroup and they prove the following:

THEOREM 5.3 ([**DSc02**]). *Let G be a locally compact group and let $\mu \in P(G)$ be i.d.. Suppose μ is supported on a connected abelian subgroup A. Then μ is i.d. in $P(A)$, and hence embeddable, if one of the following holds:*
  (i) *A is isomorphic to $\mathbb{R}^n$ for some n.*
  (ii) *G is a connected Lie group and $G(\mu) = A$.*

COROLLARY 5.4. *If G is a connected locally compact group and a $\mu \in P(G)$ is such that μ is i.d. and $G(\mu)$ is a connected abelian subgroup, then μ is embeddable.*

The approach in solving the embedding problem for a given measure μ on a Lie group in class $\mathcal{C}$ is to show that there exists a certain nilpotent group N in $Z(\mu)$ such that a particular set of nth roots are relatively compact modulo N and hence one can find nth roots which are in a compact set. In [**D10**], Dani has generalised this approach to locally compact groups and described conditions suitable to find a rational embedding.

The embedding problem, and the related probabilistic problems in general, turned out to raise deep questions about general structure of groups and about convolutions of probability measures. S. G. Dani led the group of scholars who are exploring this domain and turning it into one that is rich and engrossing.

The second named author would like to take this opportunity to express her deep gratitude to Professor S. G. Dani, from whom she learned a lot of mathematics, first as a student and later as a colleague. She had the privilege to collaborate with him on many occasions in the area of Probabilities on Groups. Apart from his deep knowledge and phenomenal mathematical achievements, she admires his humility, patience, compassion and many other qualities which make him a fine human being.

References

[D92a] S. G. Dani, *On automorphism groups of connected Lie groups*, Manuscripta Math. **74** (1992), no. 4, 445–452, DOI 10.1007/BF02567680. MR1152505 (92k:22012)

[D92b] S. G. Dani, *Invariance groups and convergence of types of measures on Lie groups*, Math. Proc. Cambridge Philos. Soc. **112** (1992), no. 1, 91–108, DOI 10.1017/S030500410007078X. MR1162935 (93f:60008)

[D06]   S. G. Dani, *Asymptotic behaviour of measures under automorphisms*, Probability mea-
          sures on groups: recent directions and trends, Tata Inst. Fund. Res., Mumbai, 2006,
          pp. 149–178. MR2213478 (2007j:60007)

[D10]   S. G. Dani, *Convolution roots and embeddings of probability measures on locally com-
          pact groups*, Indian J. Pure Appl. Math. **41** (2010), no. 1, 241–250, DOI 10.1007/s13226-
          010-0016-y. MR2650110 (2011d:60013)

[DGSh12] S. G. Dani, Yves Guivarc'h, and Riddhi Shah, *On the embeddability of certain infinitely
          divisible probability measures on Lie groups*, Math. Z. **272** (2012), no. 1-2, 361–379,
          DOI 10.1007/s00209-011-0937-0. MR2968229

[DMc88a] S. G. Dani and M. McCrudden, *Factors, roots and embeddability of measures on Lie
          groups*, Math. Z. **199** (1988), no. 3, 369–385, DOI 10.1007/BF01159785. MR961817
          (89j:43001)

[DMc88b] S. G. Dani and M. McCrudden, *On the factor sets of measures and local tightness of
          convolution semigroups over Lie groups*, J. Theoret. Probab. **1** (1988), no. 4, 357–370,
          DOI 10.1007/BF01048725. MR958243 (90d:43005)

[DMc92]   S. G. Dani and M. McCrudden, *Embeddability of infinitely divisible distributions on
          linear Lie groups*, Invent. Math. **110** (1992), no. 2, 237–261, DOI 10.1007/BF01231332.
          MR1185583 (94d:60010)

[DMc96]   S. G. Dani and M. McCrudden, *Infinitely divisible probabilities on discrete linear
          groups*, J. Theoret. Probab. **9** (1996), no. 1, 215–229, DOI 10.1007/BF02213741.
          MR1371077 (97e:60010)

[DMc07]   S. G. Dani and M. McCrudden, *Convolution roots and embeddings of proba-
          bility measures on Lie groups*, Adv. Math. **209** (2007), no. 1, 198–211, DOI
          10.1016/j.aim.2006.05.002. MR2294221 (2008g:60016)

[DR98a]   S. G. Dani and C. R. E. Raja, *Asymptotics of measures under group automorphisms
          and an application to factor sets*, Lie groups and ergodic theory (Mumbai, 1996), Tata
          Inst. Fund. Res. Stud. Math., vol. 14, Tata Inst. Fund. Res., Bombay, 1998, pp. 59–73.
          MR1699358 (2000i:22026)

[DR98b]   S. G. Dani and C. R. E. Raja, *A note on Tortrat groups*, J. Theoret. Probab. **11** (1998),
          no. 2, 571–576, DOI 10.1023/A:1022600326181. MR1622587 (99g:60012)

[DSc02]   S. G. Dani and Klaus Schmidt, *Affinely infinitely divisible distributions and
          the embedding problem*, Math. Res. Lett. **9** (2002), no. 5-6, 607–620, DOI
          10.4310/MRL.2002.v9.n5.a4. MR1906064 (2004c:60017)

[DSh93]   S. G. Dani and Riddhi Shah, *On infinitely divisible measures on certain finitely
          generated groups*, Math. Z. **212** (1993), no. 4, 631–636, DOI 10.1007/BF02571681.
          MR1214051 (94m:43001)

[DSh97]   S. G. Dani and Riddhi Shah, *Collapsible probability measures and concentration func-
          tions on Lie groups*, Math. Proc. Cambridge Philos. Soc. **122** (1997), no. 1, 105–113,
          DOI 10.1017/S0305004196001223. MR1443590 (98e:60015)

[DSh00]   S. G. Dani and Riddhi Shah, *Contractible measures and Levy's measures on Lie groups*,
          Probability on algebraic structures (Gainesville, FL, 1999), Contemp. Math., vol. 261,
          Amer. Math. Soc., Providence, RI, 2000, pp. 3–13, DOI 10.1090/conm/261/04129.
          MR1787869 (2001j:60008)

[H77]   Herbert Heyer, *Probability measures on locally compact groups*, Springer-Verlag, Berlin-
          New York, 1977. Ergebnisse der Mathematik und ihrer Grenzgebiete, Band 94.
          MR0501241 (58 #18648)

[Mc81]   Michael McCrudden, *Factors and roots of large measures on connected Lie groups*,
          Math. Z. **177** (1981), no. 3, 315–322, DOI 10.1007/BF01162065.   MR618198
          (82m:60014)

[Mc98]   Mick McCrudden, *An introduction to the embedding problem for probabilities on locally
          compact groups*, Positivity in Lie theory: open problems, de Gruyter Exp. Math., vol. 26,
          de Gruyter, Berlin, 1998, pp. 147–164. MR1648701 (99i:60015)

[Mc06]   Mick McCrudden, *The embedding problem for probabilities on locally compact groups*,
          Probability measures on groups: recent directions and trends, Tata Inst. Fund. Res.,
          Mumbai, 2006, pp. 331–363. MR2213481 (2007c:60011)

[McW99]   M. McCrudden and S. Walker, *Infinitely divisible probabilities on linear p-adic
          groups*, Proc. Indian Acad. Sci. Math. Sci. **109** (1999), no. 3, 299–302, DOI
          10.1007/BF02843532. MR1709335 (2000i:22009)

[P67]   K. R. Parthasarathy, *On the imbedding of an infinitely divisible distribution in a one-parameter convolution semigroup* (English, with Russian summary), Teor. Verojatnost. i Primenen **12** (1967), 426–432. MR0216540 (35 #7371)

[Ro79]   Joseph Rosenblatt, *A distal property of groups and the growth of connected locally compact groups*, Mathematika **26** (1979), no. 1, 94–98, DOI 10.1112/S0025579300009669. MR557132 (81c:22014)

[Sh91]   Riddhi Shah, *Infinitely divisible measures on p-adic groups*, J. Theoret. Probab. **4** (1991), no. 2, 391–405, DOI 10.1007/BF01258744. MR1100241 (92k:60016)

[Sh99]   Riddhi Shah, *The central limit problem on locally compact groups*, Israel J. Math. **110** (1999), 189–218, DOI 10.1007/BF02808181. MR1750437 (2000m:60007)

DEPARTMENT OF MATHEMATICS, UNIVERSITY OF NOTRE DAME, INDIANA 46556-4618, USA & LPMA, BOÎTE COURRIER 188, 4, PLACE JUSSIEU, 75252 PARIS CEDEX 05, FRANCE
E-mail address: fledrapp@nd.edu

SCHOOL OF PHYSICAL SCIENCES, JAWAHARLAL NEHRU UNIVERSITY, NEW DELHI 110067, INDIA
E-mail address: rshah@jnu.ac.in

Contemporary Mathematics
Volume **631**, 2015
http://dx.doi.org/10.1090/conm/631/12600

On the homogeneity at infinity of the stationary probability for an affine random walk

Y. Guivarc'h and E. Le Page

ABSTRACT. We consider an affine random walk on $\mathbb{R}$. We assume the existence of a stationary probability ν on $\mathbb{R}$ and we describe the shape at infinity of ν, if ν has unbounded support. We discuss the connections of the result with geometrical or probabilistic problems.

1. Introduction

Let G be the affine group of the line. For $g \in G$, $x \in \mathbb{R}$, we write $gx = a(g)x + b(g)$ with $a(g) \in \mathbb{R}^*, b(g) \in \mathbb{R}$. Let μ be a probability on G. We denote by P the Markov operator on $\mathbb{R}$ defined by $P\varphi(x) = \int \varphi(gx)\mu(dg)$ where φ is a bounded Borel function. Our hypothesis H_μ is stated below and we observe that $H_\mu(1)$ and $H_\mu(2)$ imply that P has a unique stationary probability ν (see [**16**]) ; if $H_\mu(4)$ is also valid, then *suppν* is unbounded. Here we are interested in the "shape at infinity" of ν ; we show that for some $\alpha > 0$, the quantities $|t|^\alpha \nu(t, \infty)$ and $|t|^\alpha \nu(-\infty, t]$ have limits at infinity, we discuss their positivity and we illustrate the possible uses of this result by two corollaries in two different contexts. This "homogeneity at infinity" of ν plays an essential role in extreme value theory (see [**19**]), for random variables associated with the Markov chain X_n^x with kernel P on $\mathbb{R}$. On the other hand, in the geometrical context of excursions of geodesic flows on manifolds of negative curvature the "logarithm law" is well known (see [**22**], [**18**]) and we discuss analogous properties for the Markov chain X_n^x. Also, for random walk in a random medium on $\mathbb{Z}$ (see [**21**]), the slow diffusion property is closely related to this homogeneity (see [**6**], [**17**]). Furthermore the construction of ν given here provides a natural construction of a large class of heavy tailed measures which generates "anomalous" random walks on the additive group $\mathbb{R}$. This class of measures appears now to be of great interest from the physical point of view (see [**2**]).

We assume that μ satisfies the following set of conditions H_μ.

$H_\mu(1) : \int (|\ell n|a(g)|| + |(\ell n|b(g)||))\mu(dg) < \infty$.

$H_\mu(2) :$ For some $\alpha > 0$ $\int |a(g)|^\alpha \mu(dg) = 1$.

$H_\mu(3) : \int |a(g)|^\alpha \ell n^+|a(g)|\mu(dg) < \infty$, $\int |b(g)|^\alpha \mu(dg) < \infty$.

$H_\mu(4) :$ The elements of *suppμ* have no common fixed point in $\mathbb{R}$.

$H_\mu(5) :$ The set $\{\ell n|a(g)| \; ; \; g \in supp\mu\}$ generates a dense subgroup of $\mathbb{R}$.

2010 *Mathematics Subject Classification*. Primary: 37Hxx, 60Bxx.
Key words and phrases. Random walk, stationary measure, affine group, extreme value theory, heavy tail.

Then we have the

THEOREM 1.1. *Assume that μ satisfies H_μ. Then*
1) There exists $c_+ \geq 0$, $c_- \geq 0$ such that $\lim\limits_{t\to\infty} |t|^\alpha \nu(t,\infty) = c_+$, $\lim\limits_{t\to-\infty} |t|^\alpha \nu(-\infty,t) =$
c_-. Moreover $c = c_+ + c_- > 0$.
2) If $\mu\{g \in G \; ; \; a(g) < 0\} > 0$, then $c_+ = c_- > 0$.
3) If $\mu\{g \in G \; ; \; a(g) > 0\} = 1$, then $c_+ > 0$ (resp $c_- > 0$) if and only if the action
of $supp\mu$ on $\mathbb{R}$ has no invariant half-line of the form $]-\infty, k]$, (resp $[k, \infty[$) .
4) If $\mu\{g \in G \; ; \; a(g) < 0\} > 0$, then $supp\nu = \mathbb{R}$. Otherwise the set $supp\nu$ is a
half-line or the whole line if $supp\mu$ do not preserve a half-line.

We denote by $\mathbb{P}$ the product probability $\mu^{\otimes\mathbb{N}}$ on $\Omega = G^\mathbb{N}$, where $\mathbb{N}$ is the set
of positive integers. For $\omega \in \Omega$, we write $g_k = (a(g_k), b(g_k)) = (a_k, b_k)$. Then X_n^x
satisfies the stochastic recursion :
$$X_n^x = a_n X_{n-1}^x + b_n, \; X_0^x = x.$$
The Markov chain X_n^x on $\mathbb{R}$ will be called "affine random walk". It is well known
that, for the existence and uniqueness of the stationary measure ν, it is sufficient
to assume $H_\mu(1)$ and $\int \ln|a(g)|\mu(dg) < 0$; then X_n^x converges in law to $R =$
$$\sum_1^\infty a_1 \ldots a_{k-1} b_k$$ and the law of R is ν. Also if $\int |b(g)|^\alpha \mu(dg) < \infty$ and $0 < \beta < \alpha$,
then $\int |x|^\beta d\nu(x) < \infty$. We observe that R can be interpreted as the sum of a
"random geometric series", hence its interest for collective risk theory ([19]).

The validity of 1) and 2) was proved in [10], [16] using the renewal theorem
(see [7]) ; in particular implicit expression for c_+, c_- were given in [10] and the
relation $c_+ + c_- > 0$ was obtained. Here we restrict our study to 3) and 4), a
result which is new under hypothesis H_μ. A different proof was sketched in [12],
where a survey of the multidimensional situation was also given. We observe that
the main difficulty of the proof occurs when $supp\mu$ do not preserve a half line and
$a(g) > 0$ $\mu - a.e$; in this case we have $c_+ > 0, c_- > 0$. In this situation an essential
role is played by an ergodic method (Lemma 3) applied to an associated fibered
Markov chain. If $supp\mu$ has compact support, a short complex analytic proof of
the positivy of c_+, c_- following [11] and depending of a Lemma of E. Landau,
well known in analytic number theory, is given in Appendix 2. Furthermore, if
$\int(|a(g)| + |b(g)|)^{\alpha+\delta}\mu(dg) < \infty$, we give also in Appendix 2, a new proof of the
asymptotic formula $\lim\limits_{t\to\infty} t^\alpha \nu(t,\infty) = c_+$, using Wiener-Ikehara theorem instead of
the renewal theorem.

We recall that Fréchet's law with parameter γ is the probability Φ_α^γ on $\mathbb{R}_+$
given by $\Phi_\alpha^\gamma(0,t) = e^{-\gamma t^{-\alpha}}$ where $\gamma > 0$, $\alpha > 0$. This family of laws is one of the
three families of max-infinitely divisible laws of extreme value theory ([9], [19]).
The following is shown in [14].

COROLLARY 1.2. *For $x \in \mathbb{R}$ we denote*
$M_n^x = sup\{|X_k^x| \; ; \; 1 \leq k \leq n\}$, $^+M_n^x = sup\{X_k^x \; ; \; 1 \leq k \leq n\}$.
Then the sequence $n^{-1/\alpha}M_n^x$ (resp $n^{-1/\alpha}$ $^+M_n^x$) converges in $\mathbb{P}$-law to $\Phi_\alpha^{c\theta}$ with
$0 < \theta < 1$ (resp $\Phi_\alpha^{c_+\theta_+}$ with $0 < \theta_+ < 1$, if $c_+ > 0$).

Closely related properties have been intensively studied in the context of extreme value theory (see [19]). The positive number θ is the so-called extremal index

of the stochastic process X_n^x ; its inverse θ^{-1} gives a measure of the clustering of the exceptionally large values of the process. If the random variables X_n^x were i.i.d. with law ν, one would have $\theta = 1$ (see [9]). If $a(g) > 0$, $b(g) > 0$ for $g \in supp\mu$, the above corollary is proved in [15]. It is also known (see [13]) that, under hypothesis H_μ, the normalized Birkhoff sum of X_n^x converges in law to a stable law of index α if $\alpha < 2$. As mentioned in ([3], remark 4.8), this convergence is a consequence of extreme value properties of X_n^x, at least for $\alpha < 1$. The analysis of random walk in a random medium on $\mathbb{Z}$ developed in [6] is closely related to such properties for the sojourn time of the particle at a site in $\mathbb{Z}$, instead of its hitting time as in [17], where Birkhoff sums as above played a dominant role.

The following logarithm law is an easy consequence of Corollary 2.

COROLLARY 1.3. *For any $x \in \mathbb{R}$, we have the following $\mathbb{P} - a.e$ convergences :*
$$\limsup_{n \to \infty} \frac{\ell n|X_n^x|}{\ell n(n)} = \frac{1}{\alpha}, \quad \limsup_{n \to \infty} \frac{\ell n(X_n^x)^+}{\ell n(n)} = \frac{1}{\alpha} \text{ if } c_+ > 0.$$

The so-called "logarithm law" for excursions of geodesic flow around the cusps on hyperbolic manifolds was proved in [22] and extended to locally symmetric spaces, of finite volume in [18]. It was observed in [20] that in case of the modular surface, it is a simple consequence of Fréchet's law for geodesic flow which follows from already known extreme value properties of the continuous fraction expansion of a number x uniformly distributed on $[0, 1]$ (see [8]).

2. Calculation of invariant measures on $\mathbb{R}$ in a special case

In this section we consider the situation of stationary measures, finite or not, in a special case which allows explicit calculations.

The Lie algebra of G is generated by the vector fields $X = a\frac{\partial}{\partial a}$, $Y = \frac{\partial}{\partial b}$. We consider the left convolution semi-group of probability measures on G with infinitesimal generator $D = X^2 + Y^2 - (\beta + 1)X$. This operator is elliptic and we denote by $p^t (t \geq 0)$ the associated semi-group of probability measures.

We have $\int \ln a(g)p^t(dg) = -t(\beta+1)$ in particular $\int \ln a(g)p(dg)$ is negative if $\beta > -1$, hence p^t has a stationary probability ν on $\mathbb{R}$ in this case. We consider more generally, for any β, the action of p^t and X, Y, D on positive measures of the form $\nu = f(x)dx$ on the line. We denote by X^*, Y^*, D^* the operators adjoint to X, Y, D. Then the extremal solutions of the equation $D^* f = 0$ ($f \geq 0$) are described by the

PROPOSITION 2.1. *With the above notations, the equation $D^* f = 0$ has the following normalized positive extremal solutions :*
$\beta \geq -1 : f(x) = (1 + x^2)^{-(1+\beta/2)}$,
$\beta < -1 : f_+(x) = (1 + x^2)^{-(1+\beta/2)} \int_{-\infty}^{x} (1 + t^2)^{\beta/2} dt$,
and $f_-(x) = (1 + x^2)^{-(1+\beta/2)} \int_{x}^{\infty} (1 + t^2)^{\beta/2} dt$.
If $\beta > -1$, then $\int f(x)dx < \infty$. If $\beta \leq -1$ then $\int f(x)dx = \int f_+(x)dx = \int_- f(x)dx = \infty$.

PROOF. We calculate the action of X, Y on the measure $\nu = fdx$ as follows. Since dx is translation-invariant and the action of the one parameter group $x \to x+b$ is by translation we get $Y^* f = -f'$.

Since $X\varphi(x) = x\varphi'(x)$, we get also $X^*f(x) = -(xf(x))'$. It follows $D^*f(x) = (x(xf)')' + f'' + (\beta+1)(xf)'$, so that the equation $D^*f = 0$ implies :
$$x(xf)' + f'(x) + (\beta+1)(xf) = k,$$
for a certain constant k, i.e :
$$(1+x^2)f' + (\beta+2)(xf) = k.$$
With $u(x) = (1+x^2)^{-(1+\beta/2)}$ we have $(1+x^2)u'(x) + (\beta+2)xu(x) = 0$, hence the above differential equation has the solutions : $f = u(d+kv)$ with $v(x) = \int_0^x (1+t^2)^{\beta/2}dt$ and d is a constant.

For $\beta \geq -1$, we have $\lim_{x\to-\infty} v(x) = \infty$, hence the condition $f \geq 0$ implies $k = 0$. In this case the equation $D^*f = 0$ has only positive extremal solutions of the form $f(x) = d(1+x^2)^{-(1+\beta/2)}$. For $\beta = 0$, D is the hyperbolic Laplacian and we recover the Cauchy law on $\mathbb{R}$ with density $\frac{1}{\pi}\frac{1}{1+x^2}$. For $\beta > -1$, we get a probability law with density proportional to $(1+x^2)^{-(1+\beta/2)}$.

We verify that for $\beta < -1$, the equation $D^*f = 0$ has two basic extremal solutions :
$$f_+(x) = (1+x^2)^{-(1+\beta/2)} \int_{-\infty}^x (1+t^2)^{\beta/2}dt,$$
$$f_-(x) = (1+x^2)^{-(1+\beta/2)} \int_x^\infty (1+t^2)^{\beta/2}dt.$$
The measure ν corresponding to f_+ has infinite mass and satisfies :
$$\lim_{t\to-\infty} |t|^{2+\beta}\nu(-\infty, t) = c- > 0$$
At $+\infty$ $f_+(x)$ is asymptotic to c_+x^{-1} with $c_+ > 0$. Analogous properties are valid fo f_-.

Also, if $\beta = -1$ at $\pm\infty$, $f(x)$ is asymptotic to $|x|^{-1}$ $\square$

REMARK 2.2. The case $\beta > -1$ corresponds to the situation of the theorem with $\alpha = \beta + 1$.

The case $\beta = -1$ corresponds to the (critical) situation of [1], [4]. Then the unique basic extremal solution behaves at infinity like multiplicative Lebesgue measure on $\mathbb{R}^*$.

The situation $\beta < -1$, with two extremal solutions, corresponds to a so-called phase transition in $P.D.E$ theory, for example in the context of non linear Schroëdinger equations.

One can ask for a general corresponding picture for a Lie group acting on a non compact boundary.

3. Proof of Theorem 1.1

The proofs of 1) and 2) in [10] are based on the first renewal equation in Lemma 1 below. A delicate point in [10] for the use of the renewal theorem (see [7]) is solved by replacing $^\alpha f(t) = e^{\alpha t}f(t)$ by a related directly Riemann-integrable function. Here we give only the proofs of 3) and 4). We will now assume $\mu\{g ; a(g) > 0\} = 1$ and we will only study the non vanishing of c_+. To do that we need some preliminary notations and results.

Let T be the stopping time on Ω defined by :
$$T = Inf\{n \geq 1 ; g_1g_2 \cdots g_n \in G_+\}, \quad T = \infty \text{ if } \{n \geq 1 ; g_1g_2 \cdots g_n \in G_+\} = \phi,$$
where $G_+ = \{b(g) > 0\}$.

We denote by $\bar\mu$ the probability on the additive group $\mathbb{R}$ given by $\bar\mu(A) = \mu\{\ell n(a(g)) \in A\}$

Moreover we denote by μ_T the positive measure on $\mathbb{R}$ defined by :
$$\mu_T(A) = \mathbb{P}\{T < +\infty \; ; \; ln(a_1 a_2 \cdots a_T) \in A\},$$
where A is a Borel subset of $\mathbb{R}$. We have $\mu_T(\mathbb{R}) = \mathbb{P}(T < +\infty) \leq 1$, and we denote by μ_T^n the nth convolution power of μ_T on the additive group $\mathbb{R}$. Define f by
$$f(t) = \mathbb{P}\{R > e^t\} = \nu(]e^t, +\infty[) \quad t \in \mathbb{R},$$
and write $R_n = \overset{n}{\underset{k=1}{\Sigma}} a_1 a_2 \cdots a_{k-1} b_k$, $S_n = \overset{n}{\underset{k=1}{\Sigma}} ln(a_k)$.
Then we have the :

LEMMA 3.1. *1) For every real t, we have $f(t) = \bar{\mu} * f(t) + f_1(t) = \mu_T * f(t) + h_1(t)$ where : $f_1(t) = \mathbb{P}\{R > e^t\} - \mathbb{P}\{R - b_1 > e^t\}$, $h_1(t) = \mathbb{E}\{1_{[T<+\infty]}\nu(]e^{-S_T}(e^t - R_T), \; e^{t-S_T}])\}$*

2) For every real t, we have $f(t) = \overset{+\infty}{\underset{n=0}{\Sigma}} \mu_T^n * h_1(t) = \overset{\infty}{\underset{n=0}{\Sigma}} \mu^n * f_1(t)$.

If p is a bounded measure on $\mathbb{R}$ and φ is a positive Borel function, we write
$$p * \varphi(t) = \int \varphi(t - x) p(dx), \quad t \in \mathbb{R}.$$
We denote by $^\alpha\mu$, the probability measure on G defined by : $^\alpha\mu(dg) = a^\alpha(g)\mu(dg)$.

We define the probability $^\alpha\mathbb{P}$ on $G^{\mathbb{N}}$ by $^\alpha\mathbb{P} = {}^\alpha\mu^{\otimes\mathbb{N}}$ and we write $^\alpha\mathbb{E}$ for the corresponding expectation.

The measure $^\alpha\mu_T$ on $\mathbb{R}$ is defined by $^\alpha\mu_T(A) = {}^\alpha\mathbb{E}(1_A(ln(a_1 \cdots a_T)))$, and we write $^\alpha h_1(t) = e^{\alpha t} h_1(t)$.

Then from Lemma 3.1 we get :

LEMMA 3.2. *For every real t we have $^\alpha f(t) = \overset{+\infty}{\underset{n=0}{\Sigma}} {}^\alpha\mu_T * {}^\alpha h_1(t)$*

Now we are going to study some properties of T and $ln(a_1 a_2 \cdots a_T)$ under $^\alpha\mathbb{P}$. For that purpose we consider the new random variables $g_i'(i \geq 1)$ defined by $g_i' = (a_i^{-1}, b_i a_i^{-1})$. Under $^\alpha\mathbb{P}$, there random variables are i.i.d with law $^\alpha\mu'$. We have :
$$g_n' g_{n-1}' \cdots g_1' = ((a_1 a_2 \cdots a_n)^{-1}, R_n(a_1 \cdots a_n)^{-1}),$$
hence for $T' = Inf\{n \; ; \; g_n' g_{n-1}' \cdots g_1' \in G_+\}$ we have $T' = T$. It follows that T can be interpreted as the entrance time in $\mathbb{R}_+ =]0, \infty[$ of the affine random walk on $\mathbb{R}$ defined by $^\alpha\mu'$, starting from 0. We denote by $^\alpha Q$ the Markov kernel of this affine random walk, and for $p \in \mathbb{R}$ we write $p_n = g_n' g_{n-1}' \cdots g_1' p$.

LEMMA 3.3. *1) There exists a unique probability measure $^\alpha\nu'$ on $\mathbb{R}$ such that $^\alpha Q(^\alpha\nu') = {}^\alpha\nu'$. The probability $^\alpha\nu'$ has no atoms.*

2) If $^\alpha\nu'(]0, +\infty]) > 0$ then $0 < {}^\alpha\mathbb{E}(T') < \infty$.

Now we complete the proof of Theorem 1.1 using the above Lemmas.
For assertion 3, there are two cases.
<u>First case</u> $^\alpha\nu'(]0, +\infty]) > 0$.
Then by Lemma 3.3 and the observation before Lemma 3.3, $^\alpha\mathbb{E}(T) = {}^\alpha\mathbb{E}(T') < \infty$, $^\alpha\mu_T(\mathbb{R}) = 1$. By Wald's lemma (see [**7**]), since $T' < \infty$ $^\alpha\mathbb{P} - a.e$:
$$^\alpha\mathbb{E}\{ln(a_1 a_2 \cdots a_T) = {}^\alpha\mathbb{E}(ln(a_1))^\alpha\mathbb{E}(T)$$
where $^\alpha\mathbb{E}(ln(a_1)) = \mathbb{E}(a_1^\alpha ln(a_1))$ is finite and positive, hence $^\alpha\mathbb{E}(S_T)$ is finite and positive.

Assume $c_+ = 0$, hence $\lim_{t\to\infty} {}^\alpha f(t) = 0$. Then, if we denote by ${}^\alpha h_{1,L}$ $(L > 0)$ the function $t \to {}^\alpha h_1(t) 1_{[-L,L]}(t)$, we have using Lemma 3.2 and Proposition 4.1 below : for every $L > 0$,

$$0 = \lim_{t\to+\infty} \frac{1}{t} \int_0^t \overset{+\infty}{\underset{n=0}{\Sigma}} {}^\alpha \mu_T^n * {}^\alpha h_{1,L}(s)ds = \frac{1}{{}^\alpha \mathbb{E}(\ell n(a_1))^\alpha \mathbb{E}(T)} \int_{-L}^L {}^\alpha h_1(s)ds,$$

hence $0 = \int_{\mathbb{R}} {}^\alpha h_1(s)ds$. Since ${}^\alpha h_1$ and h_1 are non negative we get $h_1 = 0$ a.e, hence Lemma 3.1 implies $f(t) = 0$, $dt - a.e.$

We conclude that for almost every real s :
$$f(s) = \mathbb{P}(R > e^s) = 0,$$
and so $\mathbb{P}(R \le 0) = 1$, hence $supp\nu \subset]-\infty, 0]$. It follows that $supp\mu$ preserves an interval $(-\infty, v_0)$ with $v_0 \le 0$.

<u>Second case</u> ${}^\alpha \nu'(]0, +\infty[) = 0$.

Denote by $v_0 \le 0$ the upper bound of the support of the probability ${}^\alpha \nu'$. Then by the stationarity property of ${}^\alpha \nu'$ we can write that for every $n \ge 1$:
$$ {}^\alpha \mathbb{P}\{g'_n g'_{n-1} \cdots g'_1 v_0 \le v_0\} = 1, \quad 1 = \mathbb{E}(a_1^\alpha \cdots a_n^\alpha 1_{\{v_0 + R_n \le a_1 \cdots a_n v_0\}}),$$
which implies that for every integer $n \ge 1$, $\mathbb{P}(R_n \le -v_0) = 1$ since $\mathbb{E}(a_1^\alpha \cdots a_n^\alpha) = 1$. Since R_n converges $\mathbb{P} - a.e$ to R we have $\mathbb{P}(R \le -v_0) = 1$ hence $c_+ = 0$.

In conclusion we see that $c_+ = 0$ if and only if the upper bound of $supp\nu$ is finite i.e if $supp\mu$ preserves an interval $]-\infty, v_0]$.

In order to show assertion 4 we will distinguish the 2 cases $c_+ > 0$ and $c_- = 0$, $c_+ > 0$, and $c_- > 0$. We observe that $supp\nu$ is invariant under $supp\mu$ and condition $\int \ell n(a(g))d\mu(g) < 0$ implies that for some $g \in (supp\mu)^2$ we have $0 < a(g) < 1$. Also the complement of $supp\nu$ is invariant under $(supp\mu)^{-1}$. We denote by T_μ the closed subsemigroup of G generated by $supp\mu$, and by $\Delta \subset \mathbb{R}$ the closure of the set of attractive fixed points of the elements of T_μ. We observe that $T_\mu\Delta \subset \Delta$. Since for any $x \in \Delta$ the law of $g_n \cdots g_1 x$ is supported by Δ and converges to ν, we obtain that $\Delta \supset supp\nu$. Since the attractive fixed points of T_μ belong to $supp\nu$, we conclude that $\Delta = supp\nu$. Then, for any open interval $I = [a, b] \subset \mathbb{R}$, $n < 0$, $g^n(I)$ is an interval of length $a^n(g)(b - a)$ which converges to $+\infty, -\infty$ or $\mathbb{R}$, depending of the relative positions of I and the fixed point x_0 of g. If $c_+ > 0$ and $c_- = 0$, then from above $supp\mu$ preserves the interval $[\tau, \infty[$ with $\tau = Inf(supp\nu)$. Since $\Delta = supp\nu$ we can choose $g \in (supp\mu)^2$ such that its fixed point $x_0 \in supp\nu$ is arbitrary close to τ, and in particular $\tau \le x_0 < a$. If $I \subset]\tau, \infty[$ satisfies $\nu(I) = 0$ then $\nu(g^n(I)) = 0$ for $n < 0$; since the length of the interval $g^n(I)$ is $a^n(g)(b - a)$ and $\lim_{n\to-\infty} a^n(g) = \infty$ this contradicts $c_+ > 0$.

If $c_+ > 0$, $c_- > 0$ the same argument is valid for any interval I with $\nu(I) = 0$.
$\square$

We now give the proofs of the above lemmas.

Proof of Lemma 3.1

1) Denote $R^n = \overset{+\infty}{\underset{k=n}{\Sigma}} a_{n+1} \cdots a_k b_{k+1}$. Under $\mathbb{P}$ the law of R^n is ν and moreover R^n is independant of the random variables $g_i (1 \le i \le n)$.

The formula $R = R_n + a_1 \cdots a_n R^n$ gives $R - b_1 = a_1 R^1$, hence :
$$\mathbb{P}\{R - b_1 > e^t\} = \mathbb{P}\{R^1 > e^t a_1^{-1}\} = \mu * f(t), \quad f(t) = \mu * f(t) + f_1(t)$$
We have also from above if $T < \infty$: $R = R_T + a_1 \cdots a_T R^T$

$$\{R > e^t\} = \{R_T + a_1 a_2 \cdots a_\tau R^T > e^t, \ T < \infty\}$$
$$= \{R^T > e^{t - \ell n(a_1 a_2 \cdots a_T)} \ ; \ T < \infty\} U \{(a_1 \cdots a_T)^{-1}(e^t - R_T)$$
$$< R^T \leq e^{t - \ell n(a_1 a_2 \cdots a_T)} \ ; \ T < \infty\}$$

Using the independence of $a_1 \cdots a_T$ and R^T we have
$$f(t) = \mathbb{P}\{R > e^t\} = \mathbb{P}\{R > e^t, \ T < \infty\} = \mu_T * f(t) + h_1(t)$$
where $h_1(t) = \mathbb{E}(1_{\{T < \infty\}} \nu] e^{t - \ell n(a_1 \cdots a_T)} - R_T, e^{t - \ell n(a_1 \cdots a_T)}]$,
2) It follows :
$$f = \sum_{k=0}^{n} \bar{\mu}^k * f_1 + \bar{\mu}^{n+1} * f$$
where $\bar{\mu}^{n+1} * f(t) = \mathbb{P}\{R > e^t (a_1 \cdots a_{n+1})^{-1}\}$. The condition $\mathbb{E}(\ell n(a_1)) < 0$ implies the $\mathbb{P} - a.e$ convergence of $(a_1 \cdots a_{n+1})^{-1}$ to ∞, hence $\lim_{n \to \infty} \bar{\mu}^{n+1} * f(t) = 0$. The first part of the formula follows.

From above we deduce that for every integer n and $t \in \mathbb{R}$.
$$f(t) = \sum_{j=0}^{n} \mu_T^j * h_1(t) + \mu_T^{n+1} * f(t).$$
We now prove that $\qquad \lim_{n \to +\infty} \mu_T^{n+1} * f_1(t) = 0.$

There are two cases

Case 1) $\mathbb{P}(T < \infty) < 1$

We have
$$0 \leq \mu_T^{n+1} * f(t) \leq (\mathbb{P}(T < \infty))^n$$
hence $\lim_{n \to \infty} \mu_T^{n+1} * f(t) = 0.$

Case 2) $\mathbb{P}(T < \infty) = 1$

Define the shift θ on Ω by $\theta(\omega) = (g_{i+1}(\omega), i \geq 1)$ where $\omega = (g_i(\omega), i \geq 1)$ and consider the sequence $(T_n(\omega))_{n \geq 1}$ of random times defined $\mathbb{P} - a.e$ by $T_{n+1} = T_0 \theta^{T_n}$, $T_1 = T$. Under $\mathbb{P}$ the sequence of random variables $[(T_1, S_{T_1}), \cdots, (T_{n+1} - T_n, S_{T_{n+1}} - S_{T_n})]$, is i.i.d and the law of S_{T_n} is μ_T^n. Because $\mathbb{E}(\ell n(a_1)) < 0$, we have $\mathbb{P} - a.e \ \lim_{n \to \infty} S_n = -\infty$ and moreover $\lim_{n \to \infty} T_n = \infty$ hence $\mathbb{P} - a.e, \ \lim_{n \to \infty} S_{T_n} = -\infty$.
We have that
$$\mu_T^{n+1} * f(t) = \mathbb{E}(f(t - S_{T_{n+1}})),$$
and $\lim_{t \to \infty} f(t) = 0$. So, using Lebesgue's theorem, we can conclude that $\lim_{n \to \infty} \mu_T^{n+1} * f(t) = 0 \quad \square$

Proof of Lemma 3.2

Lemma 3.2 is a direct consequence of the formula $^\alpha \mu_T^n * {}^\alpha h(t) = e^{\alpha t}(\mu_T^n * h_1(t))$, Lemma 3.1 Part 2, and the fact that h_1 is non negative. $\quad \square$

Proof of Lemma 3.3

The definition of $^\alpha \mu'$ and the condition $H_\mu(3)$ imply $\int |\ell n(a(g))| \ ^\alpha \mu'(dg) < \infty$. The strict convexity of the function $\ell n \int a^s(g) \mu(dg)(s > 0)$ gives $\int \ell n(a(g)) \ ^\alpha \mu'(dg) < 0$.

It follows $\int |\ell n|b(g)| \ ^\alpha \mu'(dg) < \infty$.

As observed above, the existence and uniqueness of $^\alpha \nu'$ follows.

If x_0 is a fixed point of $supp^\alpha \mu'$ then for any $(a, b) \in supp\mu$:
$a^{-1} x_0 + ba^{-1} = x_0$, i.e $x_0(a - 1) = b$.

This implies that $-x_0$ is a fixed point of $supp\mu$, which contradicts $H_\mu(4)$. Hence, as it well known (see [**5**]), $^\alpha\nu'$ has no atom.

In order to show $^\alpha\mathbb{E}(T) < \infty$ we consider the space $^\alpha\Omega^\# = \mathbb{R} \times G^{\mathbb{Z}}$ and the extended bilateral shift defined by $^\alpha\theta(p,\omega) = (p_1, \theta\omega)$ where $p_1 = g_1'(p)$ and θ is the bilateral shift on $G^{\mathbb{Z}}$. We endow $^\alpha\Omega^\#$ with the Markov measure $\kappa^\#$ associated with the $^\alpha Q$-invariant probability $^\alpha\nu'$. Clearly $\kappa^\#$ is $^\alpha\theta$-invariant and ergodic. Also we consider the fibered bilateral Markov chain (p_n, V_n) on $\mathbb{R} \times \mathbb{R}^*$ where $V_n = p^{-1}p_n(a_1a_2\cdots a_n) = p^{-1}(p + R_n)$. Let $\tau(p,\omega)$ be the first "ladder epoch" of (p_n, V_n) (see [**7**]), i.e $\tau = Inf\{n \geq 1 \ ; \ V_n > 1\}$, hence $p^{-1}p_\tau > 0$ and $\tau = T$ if $p > 0$. We observe that the conditions in $H_\mu(3)$ implies $\int |p|^\varepsilon \ ^\alpha\nu'(dp) < \infty$ for $0 < \varepsilon < \alpha$, hence $\limsup\limits_{|n|\to\infty} \frac{1}{|n|} \ell n|p_n| \leq 0$. Since $^\alpha\mathbb{E}(\ell n(a_1)) > 0$ the ergodic theorem gives $\limsup\limits_{n\to\infty} V_n = \infty$, $\lim\limits_{n\to-\infty} |V_n| = 0$, in particular τ is finite $\kappa^\# - a.e.$ Since $^\alpha\nu'(\mathbb{R}_+) > 0$ we can consider the Markov kernel $^\alpha Q_+$ induced by $^\alpha Q$ on $\mathbb{R}_+$; the normalized restriction $^\alpha\nu'_+$ of $^\alpha\nu'$ to $\mathbb{R}_+$ is $^\alpha Q_+$-invariant and ergodic. We denote by $^\alpha\Omega^\#_+$ the subset of $^\alpha\Omega^\#$ defined by the conditions $p > 0, p_n > 0$ infinitely often for $n = n_k > 0$ and $n = n_{-k} < 0$. Since $^\alpha\nu'(\mathbb{R}_+) > 0$, $^\alpha\Omega^\#_+$ has positive $\kappa^\#$-measure and we denote by $\kappa^\#_+$ the normalized restriction of $\kappa^\#$ to $^\alpha\Omega^\#_+$; then $\kappa^\#_+$ is invariant and ergodic under the corresponding induced shift $^\alpha\theta_+$. From above we know that $\lim\limits_{k\to\infty} V_{n_{-k}} = 0$, $\kappa^\#_+ - a.e$ hence the time $\tau_+(\omega) = n_{-j} \ (j \geq 0)$, of the last strict maximum of $V_{n_{-k}}$ is finite $\kappa^\#_+ - a.e.$ We define $^\alpha\Omega^\#_0 = \{(p,\omega) \in {}^\alpha\Omega^\#_+, \sup\limits_{k>0} V_{n_{-k}} < 1\} = \{\tau_+ = 0\}$. Then we have $^\alpha\kappa^\#_+ ({}^\alpha\Omega^\#_0) > 0$ since, by $^\alpha\theta_+$-invariance of $\kappa^\#_+$:

$$1 = \kappa^\#_+\{\tau_+ > -\infty\} = \sum_{n\geq 0} \kappa^\#_+\{\tau_+ = -n\} \leq \sum_{n\geq 0} \kappa^\#_+\{V_{-n} > \sup_{n_{-k}<-n} V_{n_{-k}}\} =$$

$$\sum_{n\geq 0} \kappa^\#_+\{0 > \sup_{k>0} V_{n_{-k}}\} \leq \infty \kappa^\#_+({}^\alpha\Omega^\#_0).$$

On the other hand, the definition of τ shows that for $(p,\omega) \in {}^\alpha\Omega^\#_0$, $\tau(p,\omega) = \tau(\omega)$ is the first return time of $^\alpha\theta^k(p,\omega)$ to $^\alpha\Omega^\#_0$, so that $^\alpha\theta^\tau$ is the transformation of $^\alpha\Omega^\#_0$ induced by $^\alpha\theta$ on $^\alpha\Omega^\#_0$. Then Kac's theorem (see [**24**]) implies that $^\alpha\theta^\tau$ is ergodic with respect to the normalized restriction $\kappa^\#_0$ of $\kappa^\#_+$ to $^\alpha\Omega^\#_0$ and $^\alpha\mathbb{E}_0(\tau) = \int \tau(\omega) \kappa^\#_0(d\omega) < \infty$. Also we denote by $^\alpha\nu^\tau_+$ the push forward of $\kappa^\#_+$ to $\mathbb{R}_+$ under the map $\omega \to p_0(\omega)$. Since the stopped kernel $^\alpha Q^\tau_+$ and the map $^\alpha\theta^\tau$ commute with p_0, the measure κ^τ_+ is $^\alpha Q^\tau_+$-invariant, ergodic and absolutely continuous with respect to $^\alpha\nu_+$ with $^\alpha\mathbb{E}_0(\tau) < \infty$.

We write $^\alpha\Omega^\#_+ = (\mathbb{R}_+ \times G^{-(\mathbb{N}\cup\{0\})}) \times \Omega$ and we denote by $^\alpha\overline{\Omega}^\#_0$ the projection of $^\alpha\Omega^\#_0$ on $\mathbb{R} \times G^{-(\mathbb{N}\cup\{0\})}$ so that $^\alpha\Omega^\#_0 = {}^\alpha\overline{\Omega}^\#_0 \times \Omega$. Since for $p > 0$, $\tau(p,\omega) = T(\omega)$ is a function on Ω and the projections of (p,ω) on $\mathbb{R} \times G^{-(\mathbb{N}\cup\{0\})}$ and Ω are independant with respect to $\kappa^\#_+$ we have $^\alpha\mathbb{E}(T) = {}^\alpha\mathbb{E}_0(\tau) < \infty$. $\square$

REMARK 3.4. A different proof of $^\alpha\mathbb{E}(T) < \infty$ uses the interpretation of $T = T'$ as hitting time of the open set $\mathbb{R}_+$ by the Markov chain with kernel $^\alpha Q$ starting from 0. Since $\int a^\delta(g) \ ^\alpha\mu'(dg) < \infty$, $\int |b^\delta(g)| \ ^\alpha\mu'(dg) < \infty$ for $0 < \delta < \alpha$, the operator defined by $^\alpha Q$ on a space of Hölder functions on $\mathbb{R}$ (as in [**13**]) has a spectral gap. This implies $^\alpha\mathbb{E}(T') < \infty$.

The proof given above extends to the multidimensional case under a stronger moment condition (see [12]).

4. Appendix

4.1. A weak renewal theorem.

PROPOSITION 4.1. *Let* $(Z_n)_{n\geq 1}$ *a sequence of independant, identically distributed real random variables on* $\mathbb{R}$ *with law* η. *Assume that* $\int |z| \eta(dz) < +\infty$ *and that* $\gamma = \int z \eta(dz) > 0$.
Let ψ *a bounded non negative Borel function which is supported on* $[-a, a]$.
Then the potential $U\psi = \sum\limits_{n=0}^{+\infty} \eta^n * \psi$ *is a bounded function and we have*

$$\lim_{t\to+\infty} \frac{1}{t} \int_0^t U\psi(s)ds = \frac{1}{\gamma} \int_{\mathbb{R}} \psi(t)dt$$

PROOF. If $\Sigma_n = \sum\limits_{i=1}^{n} Z_i$, we have : $U\psi(s) = \sum\limits_{n=0}^{+\infty} E[\psi(s - \Sigma_n)]$.

Because $\gamma = \int z\eta(dz) > 0$ the random walk on $\mathbb{R}$ with law η is transient and using the maximum principle we have that $\sup\limits_{s\in\mathbb{R}} U\psi(s) < +\infty$.

For $\varepsilon > 0$, $t > 0$ denote
$$n_1(t) = [\frac{1}{\gamma}\varepsilon t] = n_1, \qquad n_2(t) = [\frac{1}{\gamma}(1-\varepsilon)t] = n_2,$$
$$U_n\psi = \sum_0^{n-1} \eta^k * \psi, \qquad U^n\psi = \sum_{n+1}^{\infty} \eta^k * \psi, \qquad U_n^m\psi = \sum_n^m \eta^k * \psi.$$
Then we have
$$I(t) = \frac{1}{t}\int_0^t U\psi(s)ds = \sum_1^3 I_k(t) - I_4(t)$$
where
$$I_1(t) = \frac{1}{t}\int_{\mathbb{R}} U_{n_1}^{n_2}\psi(s)ds, \qquad I_2(t) = \frac{1}{t}\int_{\mathbb{R}} U_{n_1}\psi(s)ds,$$
$$I_3(t) = \frac{1}{t}\int_{\mathbb{R}} U^{n_2}\psi(s)ds, \qquad I_4(t) = \frac{1}{t}\int_{\mathbb{R}[0,t]} U_{n_1}^{n_2}\psi(s)ds.$$

We have
$$I_1(t) = \frac{n_2-n_1+1}{t}(\int_{\mathbb{R}} \psi(s)ds) \text{ hence } \lim_{t\to+\infty} I_1(t) = \frac{(12\varepsilon)}{\gamma} \int \psi(s)ds,$$
$$0 \leq I_4(t) \leq (\frac{(n_2-n_1+1)}{t} \sup_{s\in\mathbb{R}} |\psi(s)|) \sup_{n_1\leq n\leq n_2} [\mathbb{P}(\Sigma_n \leq a) + \mathbb{P}(t - \Sigma_n \leq a)].$$
By the law of large numbers we know that $\mathbb{P} - a.e$, $\lim\limits_{n\to+\infty} \frac{\Sigma_n}{n} = \gamma > 0$, hence :
$$\lim_{t\to+\infty} \sup_{n_1\leq n\leq n_2} (\mathbb{P}\{\Sigma_n \leq a\} + P\{t - \Sigma_n \leq a\}) = 0.$$
Hence $\lim\limits_{t\to+\infty} I_4(t) = 0$ and : $0 \leq \frac{I_2(t)}{t} \leq \frac{\varepsilon}{\gamma} \times \int \psi(s)ds$.

Consider now $I_3(t)$ and denote for $n \in \mathbb{N}$, $s > 0$: $\rho_n^s = Inf\{k \geq n ; |V_n - s| \leq a\}$.
We use the interpretation of $U^n\psi$ as the expected number of visits to ψ after time $n : U^n\psi(x) \leq (U\psi)\mathbb{P}\{\rho_n^s < \infty\}$ with $n\frac{[(1+\varepsilon)t]}{\gamma} = n_2$, hence :
$$I_3(t) \leq |U\psi|\mathbb{P}\{\Sigma_k \leq t + a \text{ for some } k \geq \frac{(1+\varepsilon)t}{\gamma}\}.$$
Since $\frac{\Sigma_n}{n}$ converges to $\gamma > 0$, $\mathbb{P} - a.e$, we get $\lim\limits_{t\to\infty} I_3(t) = 0$.

Since ε is arbitrary we get finally : $\displaystyle \lim_{t\to\infty} I(t) = \frac{1}{\gamma}\int \psi(s)ds.$ □

4.2. Analytic Proof of Theorem 1.1.

We prove now a version of Theorem 1.1, using a stronger hypothesis on μ which allows to use well known results in analytical number theory. We recall that if $m(z) = \int_0^\infty x^z \rho(dx)$ with $Rez = s \geq 0$, is the Mellin transform of a positive measure ρ on $\mathbb{R}_+$ and $m(z)$ is meromorphic in a domain $U \supset \{0 < Rez \leq \alpha\}$ with a unique (possible) simple pole at $z = \alpha$, then one has :
$$\lim_{t\to\infty} t^\alpha \rho(t,\infty) = \alpha^{-1}A \text{ with } A = \lim_{s\to\alpha_-}(\alpha - s)m(s) \geq 0.$$
This follows from Wiener-Ikehara theorem (see [**23**] p 233). We recall also the following lemma of E. Landau. If $\alpha \in \mathbb{R}_+$ is the convergence abcissa of $m(s)$, then $m(s)$ cannot be extended holomorphically to a neighbourhood of α (see [**23**] p 58). Then we have the following proposition :

PROPOSITION 4.2. *Assume that the hypothesis of Theorem 1 is valid and, for some $\delta > 0$ and $\int(|a(g)| + |b(g)|)^{\alpha+\delta}\mu(dg) < \infty$. Then there exists a domain U with $\{Rez \in]0, \alpha]\} \subset U \subset \{Rez > 0\}$ such that the Mellin transform $m(z) = \int_0^\infty x^z \nu(dx)$ of ν is meromorphic in U with unique (possible) simple pole at $z = \alpha$. If $\lim_{s\to\alpha_-}(\alpha - s)m(s) = A$ then one has $\lim_{t\to\infty} t^\alpha \nu(t,\infty) = \alpha^{-1}A = c_+$. If μ has compact support and $supp\mu$ do not preserve a half-line of the form $]-\infty, k]$, then $c_+ > 0$.*

PROOF. We give the proof in case $supp\mu \subset \{g \in G, a(g) > 0\}$. The hypothesis of Theorem 1 implies $\int_0^\infty x^s \nu(dx) = \mathbb{E}(R_+^s) < \infty$ if $s \in]0, \alpha[$. As in the proof of Lemma 1, we start from the equation $R - b_1 = a_1 R^1$, hence $(R - b_1)_+ = a_1 R_+^1$. Then for any $z \in \{Rez \in]0, \alpha[\}$, using independance of a_1 and R^1 :
$$\mathbb{E}((R - b_1)_+^z) = \mathbb{E}(a_1^z)\ \mathbb{E}(R_+^z).$$
If $k(z) = \mathbb{E}(|a_1|^z)$, $m(z) = \mathbb{E}(R_+^z)$, $h(z) = \mathbb{E}(R_+^z - (R - b_1)_+^z)$, we get :
$$(E)\quad (1 - k(z))m(z) = h(z).$$
The functions $\mathbb{E}(|a_1|^z)$ and $\mathbb{E}(|b_1|^z)$ are holomorphic in the domain $\{Rez \in]0, \alpha+\delta[\}$. We denote $\varepsilon(z) = 1_{[0,1]}(s)$, $\varepsilon'(z) = |z|1_{[1,\infty[}(s)$ and we write, using the mean value theorem :
$$|R_+^z - (R - b_1)_+^z| \leq \varepsilon(z)|b_1|^s + \varepsilon'(z)(|b_1|(R_+ + |b_1|))^{s-1}|.$$
Using Hölder's inequality and the moment condition, we get :
$$|h(z)| \leq \mathbb{E}(|R_+^z - (R - b_1)_+^z|) < \infty \quad \text{for } 0 < s < \alpha + \delta.$$
On the other hand, since $k(\alpha) = 1$ and $\bar{\mu}$ is non-arithmetic we get $|k(\alpha + it)| < 1$, if $t \neq 0$. Also, equation (E) implies that $m(z) = h(z)(1 - k(z))^{-1}$ is meromorphic in a domain U with $\{0 < Rez \leq \alpha\} \subset U \subset \{0 < Rez < \alpha + \delta\}$ and a unique simple pole (possible) at $z = \alpha$. Then the hypothesis of Wiener-Ikehara theorem is satisfied by the Mellin transform $m(z) = \int_0^\infty x^z \nu(dx)$.
   Hence :
$$\lim_{t\to\infty} t^\alpha \nu(t,\infty) = \alpha^{-1}A = \alpha^{-1}\lim_{s\to\alpha_-}(\alpha - s)m(s).$$

If $A = 0$, since $k(s) > 1$ for $s \in]\alpha, \alpha + \delta[$, $m(z)$ is holomorphic in a neighbourhood of $]0, \alpha + \delta[$ hence Landau's lemma implies that $m(z)$ is holomorphic for $0 < Re z < \alpha + \delta$. If μ has compact support then $k(z)$ is holomorphic for $Re z > 0$; since $k(s) \neq 1$ for $s \neq \alpha$ the function $(1 - k(z))^{-1}$ is holomorphic in a neighbourhood of $\mathbb{R}_+$. Then, as above, Landau's lemma and equation (E) imply that the Mellin transform $m(z)$ is holomorphic for $Re z > 0$.

Now we use the relation $k(s)^{1/s} m(s)^{1/s} = \mathbb{E}((R - b_1)_+^s)^{1/s} \leq \mathbb{E}(R_+^s)^{1/s} + \mathbb{E}(|b_1|^s)^{1/s}$ $(s > 1)$. Hence $(k(s)^{1/s} - 1) m(s)^{1/s} \leq B < \infty$ with $B = sup\{|b(g)| \; ; \; g \in supp\mu\}$.

Also $\lim_{s \to \infty} k(s)^{1/s} = K \leq sup\{a(g) \; ; \; g \in supp\mu\}$. It follows : $m(s)^{1/s} \leq B(K - 1)^{-1} < \infty$. Hence $\sup_{\omega \in \Omega} R_+(\omega) \leq B(K - 1)^{-1} < \infty$ $\mathbb{P} - a.e.$ This contradict the fact that $supp\nu \cap \mathbb{R}_+$ is unbounded. $\qquad\square$

REMARK 4.3. The above simple argument showing $c_+ > 0$ was used in ([**5**], [**11**]). However under the weaker natural hypothesis of Theorem 1, the use of the ergodic Lemma 3 above seems to be necessary. This observation is also valid for the use of the renewal theorem (see [**10**]) instead of Wiener-Ikehara theorem.

References

[1] Martine Babillot, Philippe Bougerol, and Laure Elie, *The random difference equation $X_n = A_n X_{n-1} + B_n$ in the critical case*, Ann. Probab. **25** (1997), no. 1, 478–493, DOI 10.1214/aop/1024404297. MR1428518 (98c:60075)

[2] François Bardou, Jean-Philippe Bouchaud, Alain Aspect, and Claude Cohen-Tannoudji, *Lévy statistics and laser cooling*, Cambridge University Press, Cambridge, 2002. How rare events bring atoms to rest. MR1886520 (2003h:81269)

[3] Bojan Basrak and Johan Segers, *Regularly varying multivariate time series*, Stochastic Process. Appl. **119** (2009), no. 4, 1055–1080, DOI 10.1016/j.spa.2008.05.004. MR2508565 (2010i:60078)

[4] Sara Brofferio, Dariusz Buraczewski, and Ewa Damek, *On the invariant measure of the random difference equation $X_n = A_n X_{n-1} + B_n$ in the critical case* (English, with English and French summaries), Ann. Inst. Henri Poincaré Probab. Stat. **48** (2012), no. 2, 377–395, DOI 10.1214/10-AIHP406. MR2954260

[5] Dariusz Buraczewski, Ewa Damek, Yves Guivarc'h, Andrzej Hulanicki, and Roman Urban, *Tail-homogeneity of stationary measures for some multidimensional stochastic recursions*, Probab. Theory Related Fields **145** (2009), no. 3-4, 385–420, DOI 10.1007/s00440-008-0172-8. MR2529434 (2011a:60188)

[6] D. Dolgopyat and I. Goldsheid, *Quenched limit theorems for nearest neighbour random walks in 1D random environment*, Comm. Math. Phys. **315** (2012), no. 1, 241–277, DOI 10.1007/s00220-012-1539-3. MR2966946

[7] William Feller, *An introduction to probability theory and its applications. Vol. II.*, Second edition, John Wiley & Sons, Inc., New York-London-Sydney, 1971. MR0270403 (42 #5292)

[8] János Galambos, *The distribution of the largest coefficient in continued fraction expansions*, Quart. J. Math. Oxford Ser. (2) **23** (1972), 147–151. MR0299576 (45 #8624)

[9] B. Gnedenko, *Sur la distribution limite du terme maximum d'une série aléatoire* (French), Ann. of Math. (2) **44** (1943), 423–453. MR0008655 (5,41b)

[10] Charles M. Goldie, *Implicit renewal theory and tails of solutions of random equations*, Ann. Appl. Probab. **1** (1991), no. 1, 126–166. MR1097468 (93i:60118)

[11] Yves Guivarc'h, *Heavy tail properties of stationary solutions of multidimensional stochastic recursions*, Dynamics & stochastics, IMS Lecture Notes Monogr. Ser., vol. 48, Inst. Math. Statist., Beachwood, OH, 2006, pp. 85–99, DOI 10.1214/lnms/1196285811. MR2306191 (2008h:60261)

[12] Guivarc'h Y., Le Page E. : Homogeneity at infinity of stationary solutions of Multivariate Affine Stochastic Recursions. In "Random Matrices and Iterated Random functions". Alsmeyer G. - Löwe M. Eds. Springer Proceedings in Mathematics and Statistics 53, pp 119-135 (2013).

[13] Y. Guivarc'h and Emile Le Page, *On spectral properties of a family of transfer operators and convergence to stable laws for affine random walks*, Ergodic Theory Dynam. Systems **28** (2008), no. 2, 423–446. MR2408386 (2009b:60051)

[14] Guivarc'h Y., Le Page E. : Asymptotique des valeurs extrêmes pour les marches aléatoires affines. C.R.A.S, ser I (2013).

[15] Laurens de Haan, Sidney I. Resnick, Holger Rootzén, and Casper G. de Vries, *Extremal behaviour of solutions to a stochastic difference equation with applications to ARCH processes*, Stochastic Process. Appl. **32** (1989), no. 2, 213–224, DOI 10.1016/0304-4149(89)90076-8. MR1014450 (91g:60029)

[16] Harry Kesten, *Random difference equations and renewal theory for products of random matrices*, Acta Math. **131** (1973), 207–248. MR0440724 (55 #13595)

[17] H. Kesten, M. V. Kozlov, and F. Spitzer, *A limit law for random walk in a random environment*, Compositio Math. **30** (1975), 145–168. MR0380998 (52 #1895)

[18] D. Y. Kleinbock and G. A. Margulis, *Logarithm laws for flows on homogeneous spaces*, Invent. Math. **138** (1999), no. 3, 451–494, DOI 10.1007/s002220050350. MR1719827 (2001i:37046)

[19] M. R. Leadbetter, Georg Lindgren, and Holger Rootzén, *Extremes and related properties of random sequences and processes*, Springer Series in Statistics, Springer-Verlag, New York-Berlin, 1983. MR691492 (84h:60050)

[20] Mark Pollicott, *Limiting distributions for geodesics excursions on the modular surface*, Spectral analysis in geometry and number theory, Contemp. Math., vol. 484, Amer. Math. Soc., Providence, RI, 2009, pp. 177–185, DOI 10.1090/conm/484/09474. MR1500147 (2010c:37074)

[21] Fred Solomon, *Random walks in a random environment*, Ann. Probability **3** (1975), 1–31. MR0362503 (50 #14943)

[22] Dennis Sullivan, *Disjoint spheres, approximation by imaginary quadratic numbers, and the logarithm law for geodesics*, Acta Math. **149** (1982), no. 3-4, 215–237, DOI 10.1007/BF02392354. MR688349 (84j:58097)

[23] David Vernon Widder, *The Laplace Transform*, Princeton Mathematical Series, v. 6, Princeton University Press, Princeton, N. J., 1941. MR0005923 (3,232d)

[24] J. Wolfowitz, *Remarks on the notion of recurrence*, Bull. Amer. Math. Soc. **55** (1949), 394–395. MR0029109 (10,549e)

Contemporary Mathematics
Volume **631**, 2015
http://dx.doi.org/10.1090/conm/631/12601

Dani's work on dynamical systems on homogeneous spaces

Dave Witte Morris

To Professor S. G. Dani on his 65th birthday

ABSTRACT. We describe some of S. G. Dani's many contributions to the theory
and applications of dynamical systems on homogeneous spaces, with emphasis
on unipotent flows.

S. G. Dani has written over 100 papers. They explore a variety of topics, including:

- flows on homogeneous spaces
  - unipotent dynamics
  - applications to Number Theory
  - divergent orbits
  - bounded orbits and Schmidt's game
  - topological orbit equivalence
  - Anosov diffeomorphisms
  - entropy and other invariants
- actions of locally compact groups
  - actions of lattices
  - action of Aut G on the Lie group G
  - stabilizers of points
- convolution semigroups of probability measures on a Lie group
- finitely additive probability measures
- Borel Density Theorem
- history of Indian mathematics

Most of Dani's papers (about 60) are directly related to flows on homogeneous spaces. This survey will briefly discuss several of his important contributions in this field.

NOTATION. Let:

- $G = \mathrm{SL}(n, \mathbb{R})$ (or, for the experts, G may be any connected Lie group),
- $\{g^t\}$ be a one-parameter subgroup of G,
- $\Gamma = \mathrm{SL}(n, \mathbb{Z})$ (or, for the experts, G may be any *lattice* in G), and
- $\varphi_t(x\Gamma) = g^t x\Gamma$ for $t \in \mathbb{R}$ and $x\Gamma \in G/\Gamma$.

Then $\varphi_{s+t} = \varphi_s \circ \varphi_t$, so φ_t is a flow on the homogeneous space G/Γ.

2010 *Mathematics Subject Classification.* Primary 37A17; Secondary 11H55, 37A45.
Key words and phrases. Unipotent dynamics, homogeneous dynamics, linearization, Oppenheim conjecture.

Much is known about flows generated by a general one-parameter subgroup $\{g^t\}$. (For example, the spectrum of $\{g^t\}$, as an operator on $L^2(G/\Gamma)$, is described in [**D77**, Thm. 2.1]. A few of Dani's other results on the general case are mentioned in Section 3 below.) However, the most impressive results apply only to the flows generated by one-parameter subgroups that are "unipotent:"

DEFINITION. A one-parameter subgroup $\{u^t\}$ of $\mathrm{SL}(n, \mathbb{R})$ is said to be *unipotent* if 1 is the only eigenvalue of u^t (for every $t \in \mathbb{R}$). This is equivalent to requiring that

$$\{u^t\} \text{ is conjugate to a subgroup of } \mathbb{U}_n = \begin{bmatrix} 1 & & & \\ & 1 & & * \\ & & \ddots & \\ & 0 & & 1 \\ & & & & 1 \end{bmatrix}.$$

Work of Dani, Margulis, Ratner, and others in the 1970's and 1980's showed that if $\{u^t\}$ is unipotent, then the corresponding flow on G/Γ is surprisingly well behaved. In particular, the closure of every orbit is a very nice C^∞ submanifold of G/Γ, and every invariant probability measure is quite obvious.

Section 1 describes these (and other) fundamental results on unipotent flows, and Section 2 explains some of their important applications in Number Theory. Dani's contributions are of lasting importance in both of these areas.

ACKNOWLEDGMENTS. I thank an anonymous referee for helpful comments.

1. Unipotent flows

Orbit closure (O), equidistribution (E), and measure classification (M). If $\{g^t\}$ is a (nontrivial) group of diagonal matrices, then there exists a $\{g^t\}$-orbit in G/Γ that is very badly behaved — its closure is a fractal (cf. [**St90**, Lem. 2]). Part (**O**) of the following fundamental result tells us that unipotent flows have no such pathology. The other two parts describe additional ways in which the dynamical system is very nicely behaved.

THEOREM 1.1 (Ratner [**R91a, R91b**]). *If $\{u^t\}$ is unipotent, then the following hold:*

- (**O**) *The closure of every u^t-orbit on G/Γ is a (finite-volume, homogeneous) C^∞ submanifold.*
- (**E**) *Every u^t-orbit on G/Γ is uniformly distributed in its closure.*
- (**M**) *Every ergodic u^t-invariant probability measure on G/Γ is the natural Lebesgue measure on some (finite-volume, homogeneous) C^∞ submanifold.*

REMARK 1.2. Here is a more precise statement of each part of Theorem 1.1.

(**O**) For each $a \in G/\Gamma$, there is a closed subgroup $L = L_a$ of G, such that $\overline{\{u^t\}a} = La$. Furthermore, the closed submanifold La of G/Γ has finite volume, with respect to an L-invariant volume form on La.

(**E**) Given $a \in G/\Gamma$, let dg be the L_a-invariant volume form on $\overline{\{u^t\}a}$ that is provided by (**O**). (After multiplying dg by a scalar, we may assume the volume of $\overline{\{u^t\}a}$ is 1.) Then, for any continuous function f on G/Γ, with compact support, the average value of f along the orbit of a is equal to the average value of f on the closure of the orbit. That is, we have

$$\lim_{T \to \infty} \frac{1}{T} \int_0^T f(u^t a) \, dt = \int_{\overline{\{u^t\}a}} f \, dg.$$

(**M**) Suppose μ is a Borel measure on G/Γ, such that
- $\mu(G/\Gamma) = 1$ (this means μ is a *probability measure*),
- $\mu(u^t A) = \mu(A)$ for every measurable subset A of G/Γ (this means μ is u^t-*invariant*), and
- for every u^t-invariant, measurable subset A of G/Γ, either A has measure 0 or the complement of A has measure 0 (this means μ is *ergodic*).

Then there exists a closed subgroup $L = L_\mu$ of G, and some $a = a_\mu \in G$, such that the orbit La is closed, and μ is an L-invariant Lebesgue measure on the submanifold La. (The L-invariant measure on La is unique up to multiplication by a positive scalar, so the probability measure μ is uniquely determined by L.)

DANI'S CONTRIBUTIONS. Dani was a central figure in the activity that paved the way for Ratner's proof of Theorem 1.1. In particular:

(1) Dani's early work on actions of unipotent subgroups was one of the ingredients that inspired Raghunathan to conjecture the truth of (**O**). Raghunathan did not publish the conjecture himself — its first appearance in the literature was in a paper of Dani [**D81**, p. 358].

(2) (**M**) was conjectured by Dani [**D81**, p. 358]. This was an important insight, because the methods developed in the 1980's were able to prove this part of Theorem 1.1 first, and the other two parts are corollaries of it (cf. Section 1 below).

(3) For $G = \mathrm{SL}(2, \mathbb{R})$, (**M**) and (**E**) were proved by Furstenberg [**F73**] when G/Γ is compact, but it was Dani who tackled the noncompact case. First, he [**D78a**] proved (**M**). Then he [**D82**] proved the special case of (**E**) in which $\Gamma = \mathrm{SL}(2, \mathbb{Z})$. Subsequent joint work with Smillie [**DS84**] established (**E**) for the remaining lattices in $\mathrm{SL}(2, \mathbb{R})$.

(4) Dani [**D81, D86**] proved analogues of (**O**) and (**M**) in which u^t replaced with the larger unipotent subgroup $\mathbb{U}_n$. (See Section 1 below for more discussion of this.)

(5) Dani and Margulis [**DM90a**] proved (**O**) under the assumption that $G = \mathrm{SL}(3, \mathbb{R})$ and u^t is generic.

Furthermore, Dani [**D79, D84a**] established a very important special case of (**E**), long before the full theorem was proved. Note that if $a \in G/\Gamma$ and $\epsilon > 0$, then, since (**O**) tells us that $\overline{\{u^t\}a}$ has finite volume, there exists $C > 0$, such that the complement of the ball of radius C has volume less than ϵ. Therefore, letting
- λ be the usual Lebesgue measure on $\mathbb{R}$, and
- $d(x, y)$ be the distance from x to y in G/Γ,

(**E**) implies that

(1.3) $$\limsup_{T \to \infty} \frac{\lambda\{\, t \in [0, T] \mid d(u^t a, a) > C \,\}}{T} < \epsilon.$$

Dani proved this fundamental inequality:

THEOREM 1.4 (Dani [**D79, D84a**]). *For every $a \in G/\Gamma$ and $\epsilon > 0$, there exists $C > 0$, such that (1.3) holds.*

This is a strengthening of the following fundamental result:

COROLLARY 1.5 (Margulis [**M71, M75**]). *No u^t-orbit diverges to ∞.*
More precisely, for every $a \in G/\Gamma$, we have $d(u^t a, a) \not\to \infty$ as $t \to +\infty$.

PROOF. If $d(u^t a, a) \to \infty$, then there exists $T_0 > 0$, such that $d(u^t a, a) > C$ for all $t > T_0$. Therefore

$$\limsup_{T \to \infty} \frac{\lambda\{ t \in [0, T] \mid d(u^t a, a) > C \}}{T} \geq \lim_{T \to \infty} \frac{T - T_0}{T} = 1 > \epsilon. \qquad \square$$

Theorem 1.4 has the following important consequences:

COROLLARY 1.6 (Dani). *Suppose μ is an ergodic u^t-invariant measure on G/Γ.*
If every compact subset of G/Γ has finite measure, then all of G/Γ has finite measure. (In other words, if μ is locally finite, then μ is finite.)

COROLLARY 1.7 (Dani). *Suppose μ is an ergodic $\mathrm{SL}(n, \mathbb{Z})$-invariant measure on $\mathbb{R}^n$. If every compact set has finite measure, and every finite set has measure 0, then μ is a scalar multiple of Lebesgue measure.*

The proof of Theorem 1.4 employs an ingenious induction argument of Margulis that was appropriately modified by Dani. The best exposition of this idea is in an appendix of a paper by Dani and Margulis [**DM90b**].

For applications, it is important to have strengthened versions of Theorem 1.4 and (**E**) giving estimates that are *uniform* as the starting point of the orbit varies over a compact set. The first such theorems were proved by Dani and Margulis [**DM91, DM93**].

REMARK 1.8. Margulis [**M88**, Rem. 3.12(II)] observed that Dani's Theorem 1.4 provides a short (and rather elementary) proof of the fundamental fact that arithmetic groups are lattices. That is, if **G** is an algebraic subgroup of $\mathbf{SL}_n$ that is defined over $\mathbb{Q}$, and **G** has no characters that are defined over $\mathbb{Q}$, then the homogeneous space $\mathbf{G}(\mathbb{R})/\mathbf{G}(\mathbb{Z})$ has finite volume.

Linearization. Dani and Margulis [**DM93**, §3] developed an important method that is called "*Linearization*," because it replaces the action of u^t on G/Γ with the much simpler action of u^t by linear transformations on a vector space. It has become a crucial tool in applications of unipotent flows in Number Theory and related areas. To see the main idea (which has its roots in earlier work of Dani-Smillie [**DS84**] and Shah [**Sh91**]), consider the following proof that (**E**) is a consequence of (**O**) and (**M**):

IDEA OF PROOF OF (**E**) FROM (**O**) AND (**M**). The start of the proof is straightforward. Fix $a \in G/\Gamma$. From (**O**), we know there is a closed subgroup L of G, such that $\overline{\{u^t\}a} = La$. Furthermore, there is an L-invariant probability measure μ on La.

Assume, for simplicity, that G/Γ is compact, and let $\mathrm{Meas}(G/\Gamma)$ be the set of probability measures on G/Γ. (The Riesz Representation Theorem tells us that $\mathrm{Meas}(G/\Gamma)$ can be identified with the set of positive linear functionals on $C(G/\Gamma)$ that have norm 1, so it has a natural weak* topology.) For each $T > 0$, define $M_T \in \mathrm{Meas}(G/\Gamma)$ by

$$M_T(f) = \frac{1}{T} \int_0^T f(u^t a) \, dt.$$

We wish to show $M_T \to \mu$ as $T \to \infty$.

Since $\mathrm{Meas}(G/\Gamma)$ is a closed, convex subset of the unit ball in $C(G/\Gamma)^*$, the Banach-Alaoglu Theorem tells us that $\mathrm{Meas}(G/\Gamma)$ is compact (in the weak* topology). Therefore, in order to show $M_T \to \mu$, it suffices to show that μ is the only accumulation point of $\{M_T\}$.

Thus, given an accumulation point μ_∞ of $\{M_T\}$, we wish to show $\mu_\infty = \mu$. It is not difficult to see that μ_∞ is u^t-invariant (since M_T is nearly invariant when T is large). Assume, for simplicity, that μ_∞ is ergodic. Then (**M**) tells us there is a closed subgroup L_∞ of G, and some $a_\infty \in G$, such that the orbit $L_\infty a_\infty$ is closed, μ_∞ is L_∞-invariant, and μ_∞ is supported on $L_\infty a_\infty$.

Since $M_T \to \mu_\infty$, we know $L_\infty a_\infty \subseteq \overline{\{u^t\}a} = La$. To complete the proof, it suffices to show the opposite inclusion $La \subseteq L_\infty a_\infty$. (This suffices because the two inclusions imply $L = L_\infty$, which means that μ_∞ must be the (unique) L-invariant probability measure on La, which is μ.)

We will now use *Linearization* to show $u^t a \in L_\infty a_\infty$ for all t. (Since $\overline{\{u^t\}a} = La$, and $L_\infty a_\infty$ is closed, this implies the desired inclusion $La \subseteq L_\infty a_\infty$.) Roughly speaking, one can show that the subgroup L_∞ is Zariski closed (since the ergodicity of μ_∞ implies that the unipotent elements generate a cocompact subgroup). Therefore, Chevalley's Theorem (from the theory of Algebraic Groups) tells us there exist:

- a finite-dimensional real vector space V,
- a homomorphism $\rho \colon G \to \mathrm{SL}(V)$, and
- $\vec{v} \in V$,

such that $L_\infty = \mathrm{Stab}_G(\vec{v})$. The Euclidean distance formula tells us that $d(\vec{x}, \vec{v})^2$ is a polynomial of degree 2 (as a function of $\vec{x}$). Also, since $\rho(u^t)$ is unipotent, elementary Lie theory shows that each matrix entry of $\rho(u^t)$ is a polynomial function of t. Therefore, if we choose $g \in G$, such that $a = g\Gamma$, then

$$d_g(t) := d\big(\rho(u^t g)\vec{v}, \vec{v}\big)$$

is a polynomial function of t. Furthermore, the degree of this polynomial is bounded, independent of g.

Since $M_T \to \mu_\infty$, and μ_∞ is supported on $L_\infty a_\infty$, we know that if T is large, then for most $t \in [0, T]$, the point $u^t a$ is close to $L_\infty a_\infty$. Assuming, for simplicity, that $a_\infty = e$, so $\rho(L_\infty a_\infty)\vec{v} = \rho(L_\infty)\vec{v} = \vec{v}$, this implies there exists $\gamma \in \Gamma$, such that $d_{g\gamma}(t)$ is very small. However, a polynomial of bounded degree that is very small on a large fraction of an interval must be small on the entire interval. We conclude that $d_{g\gamma}(t)$ is small for all $t \in \mathbb{R}^+$ (and that γ is independent of t). Since constants are the only bounded polynomials, this implies that the distance from $u^t a$ to $L_\infty a_\infty$ is a constant (independent of t). Then, from the first sentence of this paragraph, we see that this distance must be 0, which means $u^t a \in L_\infty a_\infty$ for all t, as desired. $\square$

REMARK 1.9. The above argument can be modified to derive both (**O**) and (**E**) from (**M**), without needing to assume (**O**). Therefore, as was mentioned on page 133, (**M**) implies both (**O**) and (**E**).

REMARK 1.10. The above proof assumes that G/Γ is compact. To eliminate this hypothesis, one passes to the one-point compactification, replacing $\mathrm{Meas}(G/\Gamma)$ with $\mathrm{Meas}\big(G/\Gamma \cup \{\infty\}\big)$. The bulk of the argument can remain unchanged, because

Dani's Theorem 1.4 tells us that if μ_∞ is any accumulation point of $\{M_T\}$, then $\mu_\infty(\{\infty\}) = 0$, so $\mu_\infty \in \mathrm{Meas}(G/\Gamma)$.

Actions of horospherical subgroups and actions of lattices. As has already been mentioned on page 133, Dani proved the analogues of (**M**) and (**O**) in which the one-dimensional unipotent subgroup $\{u^t\}$ is replaced by a unipotent subgroup that is maximal:

THEOREM 1.11 (Dani [**D81, D86**]). *Let*

- $G = \mathrm{SL}(n, \mathbb{R})$ (*or, more generally, let G be a connected, reductive, linear Lie group*),
- $U = \mathbb{U}_n$ (*or, more generally, let U be any maximal unipotent subgroup of G*), *and*
- Γ *be a discrete subgroup of G, such that G/Γ has finite volume* (*in other words, let Γ be a lattice in G*).

Then:

- (**O′**) *The closure of every U-orbit on G/Γ is a* (*finite-volume, homogeneous*) *C^∞ submanifold.*
- (**M′**) *Every ergodic U-invariant probability measure on G/Γ is the natural Lebesgue measure on some* (*finite-volume, homogeneous*) *submanifold.*

REMARK 1.12. Although the statement of Theorem 1.11 requires the unipotent subgroup U to be maximal, Dani actually proved (**O′**) under the weaker assumption that U is "horospherical:"

For each $g \in G$, the corresponding *horospherical subgroup* is
$$U_g = \{\, u \in G \mid g^{-k} u g^k \to e \text{ as } k \to +\infty \,\}.$$

Since 1 is the only eigenvalue of the identity matrix, and similar matrices have the same eigenvalues, it is easy to see that every element of U_g is unipotent. Conversely, the maximal unipotent subgroup $\mathbb{U}_n$ is horospherical. (Namely, we have $\mathbb{U}_n = U_g$ if $g = \mathrm{diag}(\lambda_1, \lambda_2, \ldots, \lambda_n)$ is any diagonal matrix with $\lambda_1 > \lambda_2 > \cdots > \lambda_n > 0$.)

Dani's first published paper was joint work with Mrs. Dani [**DD73**], while they were students at the Tata Institute of Fundamental Research. It proved a p-adic version of the following interesting consequence of (**O′**):

COROLLARY 1.13 (Greenberg [**G63**]). *Let $G = \mathrm{SL}(n, \mathbb{R})$, and let Γ be a discrete subgroup of G, such that G/Γ is compact. Then the Γ-orbit of every nonzero vector is dense in $\mathbb{R}^n$.*

PROOF. 1.11(**O′**) tells us that, for each $g \in G$, there is a (unique) connected subgroup L_g of G, such that $\overline{\mathbb{U}_n g \Gamma} = L_g \Gamma$. Since $\mathbb{U}_n$ is horospherical (see Remark 1.12) and G/Γ is compact, one can show that $L_g = G$. This means $\mathbb{U}_n g \Gamma$ is dense in G (for all $g \in G$).

So $\Gamma g \mathbb{U}_n$ is dense in G. Since $\mathbb{U}_n$ fixes the vector $\vec{e}_n = (0, \ldots, 0, 1)$, and $G\vec{e}_n = \mathbb{R}^n \smallsetminus \{0\}$, this implies that $\Gamma g \vec{e}_n$ is dense in $\mathbb{R}^n$. This is the desired conclusion, since $g\vec{e}_n$ is an arbitrary nonzero vector in $\mathbb{R}^n$.   □

Corollary 1.13 provides information about an action of the "lattice" Γ. This particular result is a consequence of the theory of unipotent dynamics, but other theorems of Dani about lattice actions do not come from this theory. As an example of this, we mention Dani's topological analogue of a famous measure-theoretic result of Margulis [**M78**, Thm. 1.14.2]:

THEOREM 1.14 (Dani [**D84b**]). *Let*

- $G = \mathrm{SL}(n, \mathbb{R})$ *with* $n \geq 3$,
- $\Gamma = \mathrm{SL}(n, \mathbb{Z})$,
- B *be the "Borel" subgroup of all upper-triangular matrices,*
- X *be a compact, Hausdorff space on which* Γ *acts by homeomorphisms, and*
- $\phi \colon G/B \to X$ *be a continuous, surjective,* Γ*-equivariant map.*

Then B *is contained in a closed subgroup* P *of* G, *such that* X *is* Γ*-equivariantly homeomorphic to* G/P.

REMARK 1.15. To make the statement more elementary, we assume $G = \mathrm{SL}(n, \mathbb{R})$ and $\Gamma = \mathrm{SL}(n, \mathbb{Z})$ in Theorem 1.14, but Dani actually proved the natural generalization in which G is allowed to be any semisimple Lie group of real rank greater than one, and Γ is an irreducible lattice in G.

2. Applications in Number Theory

Unipotent dynamics (which was discussed in Section 1) became well known through its use in the solution of the "Oppenheim Conjecture." Before explaining this, let us look at a similar problem whose solution is much more elementary.

EXAMPLE 2.1. Suppose

$$L(x_1, x_2, \ldots, x_n) = \sum_i a_i x_i = a_1 x_1 + \cdots + a_n x_n$$

is a homogeneous polynomial of degree 1, with real coefficients. (Assume, to avoid degeneracies, that L is not identically zero.) For any $b \in \mathbb{R}$, it is easy to find a solution of the equation $L(\vec{x}) = b$.

However, a number theorist may wish to require the coordinates of $\vec{x}$ to be integers (i.e., $\vec{x} \in \mathbb{Z}^n$). Since $\mathbb{Z}^n$ is countable, most choices of b will yield an equation that does not have an integral solution, so it is natural to ask for an approximate solution: for every $\epsilon > 0$,

does there exist $\vec{x} \in \mathbb{Z}^n$, such that $|L(\vec{x}) - b| < \epsilon$?

Obviously, to say an approximate solution exists for every $b \in \mathbb{R}$ (and every $\epsilon > 0$) is the same as saying that $L(\mathbb{Z}^n)$ is dense in $\mathbb{R}$.

It is obvious that if the coefficients of L are integers, then L takes integer values at any point whose coordinates are integers. This means $L(\mathbb{Z}^n) \subseteq \mathbb{Z}$, so $L(\mathbb{Z}^n)$ is not dense in $\mathbb{R}$. More generally, if L is a scalar multiple of a polynomial with integer coefficients, then $L(\mathbb{Z}^n)$ is not dense in $\mathbb{R}$. It is less obvious (but not difficult to prove) that the converse is true:

If $L(\vec{x})$ *is a a homogeneous polynomial of degree 1, with real coefficients, and* L *is not a scalar multiple of a polynomial with integer coefficients, then* $L(\mathbb{Z}^n)$ *is dense in* $\mathbb{R}$.

Everything in the above discussion is trivial, but the problem becomes extremely difficult if polynomials of degree 1 are replaced by polynomials of degree 2. In this setting, a conjecture made by A. Oppenheim [**O29**] in 1929 was not proved until almost 60 years later. The statement of the result needs to account for the following counterexamples:

EXAMPLES 2.2. Let $Q(x_1, x_2, \ldots, x_n) = \sum_{i,j} a_{i,j} x_i x_j$ be a homogeneous polynomial of degree 2, with real coefficients. (In other words, Q is a "real quadratic form.")

(1) We say Q is *positive-definite* if $Q(\mathbb{R}^n) \subseteq \mathbb{R}^{\geq 0}$. (Similarly, Q is *negative-definite* if $Q(\mathbb{R}^n) \subseteq \mathbb{R}^{\leq 0}$.) Obviously, this implies that $Q(\mathbb{R}^n)$ is not dense in $\mathbb{R}$, so it is obvious that the smaller set $Q(\mathbb{Z}^n)$ is not dense in $\mathbb{R}$. For example, if we let $Q(x_1, x_2, \ldots, x_n) = a_1 x_1^2 + \cdots a_n x_n^2$, with each a_i positive, then Q is positive-definite, so $Q(\mathbb{Z}^n)$ is not dense in $\mathbb{R}$.

(2) Let $Q(x_1, x_2) = x_1^2 - \alpha^2 x_2^2$. It is not difficult to see that if α is badly approximable, then 0 is *not* an accumulation point of $Q(\mathbb{Z}^2)$. Obviously, then $Q(\mathbb{Z}^2)$ is not dense in $\mathbb{R}$. (Recall that, to say $\alpha \in \mathbb{R}$ is *badly approximable* means there exists $\epsilon > 0$, such that $|\alpha - (p/q)| > \epsilon/q^2$ for all $p, q \in \mathbb{Z}$. It is well known that quadratic irrationals, such as $1 + \sqrt{2}$, are always badly approximable.) This means that the "obvious" converse can fail when $n = 2$.

(3) Given a 2-variable counterexample $x_1^2 - \alpha^2 x_2^2$, it is easy to construct counterexamples in any number of variables, such as

$$Q(x_1, x_2, \ldots, x_n) = (x_1 + \cdots + x_{n-1})^2 - \alpha^2 x_n^2.$$

Note that this quadratic form has n variables, but a linear change of coordinates can transform it into a form with less than n variables. This means it is *degenerate*.

The following theorem shows that all quadratic counterexamples are of the above types.

THEOREM 2.3 (Margulis [**M87**]). *Suppose:*

- $Q(x_1, x_2, \ldots, x_n)$ *is a homogeneous polynomial of degree 2, with real coefficients,*
- Q *is not a scalar multiple of a polynomial with integer coefficients,*
- Q *is neither positive-definite nor negative-definite,*
- $n \geq 3$, *and*
- Q *is not degenerate.*

Then $Q(\mathbb{Z}^n)$ is dense in $\mathbb{R}$.

IDEA OF PROOF. We will explain how the result can be obtained as a corollary of Ratner's Theorem (1.1). (The result was originally proved directly, because Ratner's theorem was not yet available in 1987. See [**D01**] for an elementary exposition of a direct proof of this type.)

Assume, for simplicity, that $n = 3$. Given $b \in \mathbb{R}$, we wish to show there exists $\vec{m} \in \mathbb{Z}^3$, such that $Q(\vec{m}) \approx b$. Since Q is neither positive-definite nor negative-definite, we have $Q(\mathbb{R}^3) = \mathbb{R}$, so there exists $\vec{v} \in \mathbb{R}^3$, such that $Q(\vec{v}) = b$.

Let $G = \mathrm{SL}(3, \mathbb{R})$, $\Gamma = \mathrm{SL}(3, \mathbb{Z})$, and

$$H = \mathrm{SO}_3(Q) = \{\, h \in \mathrm{SL}(3, \mathbb{R}) \mid Q(h\vec{x}) = Q(\vec{x}) \text{ for all } \vec{x} \in \mathbb{R}^3 \,\}.$$

Note that H is a subgroup of G that is generated (up to finite index) by unipotent one-parameter subgroups. Therefore, Ratner's Theorem (1.1) implies that the closure $\overline{H\Gamma}$ of $H\Gamma$ is a very nice submanifold of G. More precisely, there is a closed subgroup L of G, such that $\overline{H\Gamma} = L\Gamma$, and $H \subseteq L$. However, it can be shown that H is a maximal subgroup of G, and, since $Q(\vec{x})$ does not have integer coefficients (up to a scalar multiple), that $\overline{H\Gamma} \neq H\Gamma$ (so $L \neq H$). This implies $L = G$. In other words, $H\Gamma$ is dense in G.

For convenience, let $\vec{e_1} = (1,0,0) \in \mathbb{R}^3$. Since $G = \mathrm{SL}(3,\mathbb{R})$ is transitive on the nonzero vectors in $\mathbb{R}^3$, we know there exists $g \in G$, such that $g\vec{e_1} = \vec{v}$. Then the conclusion of the preceding paragraph implies there exist $h \in H$ and $\gamma \in \Gamma = \mathrm{SL}(3,\mathbb{Z})$, such that

$$(*) \qquad\qquad h\gamma\vec{e_1} \approx \vec{v}.$$

Let $\vec{m} = \gamma\vec{e_1} \in \mathbb{Z}^3$. Then

$$\begin{aligned} Q(\vec{m}) &= Q(\gamma\vec{e_1}) && \text{(definition of } \vec{m}) \\ &= Q(h\gamma\vec{e_1}) && \text{(definition of } H) \\ &\approx Q(\vec{v}) && \text{(polynomial } Q \text{ is continuous, and } (*)) \\ &= b && \text{(definition of } \vec{v}). \end{aligned}$$ $\qquad\square$

DANI'S CONTRIBUTIONS. Joint work of Dani and Margulis made the following important improvements to this theorem:

[**DM89**] *approximation by primitive vectors:* $Q(\mathcal{P})$ is dense in $\mathbb{R}$, where

$$\mathcal{P} = \{\,(m_1, m_2, \ldots, m_n) \in \mathbb{Z}^n \mid \gcd(m_1, m_2, \ldots, m_n) = 1\,\}.$$

[**DM90a**] *simultaneous approximation:* Given two quadratic forms Q_1 and Q_2 (satisfying appropriate conditions), the set $\{\,(Q_1(\vec{m}), Q_2(\vec{m})) \mid \vec{m} \in \mathcal{P}\,\}$ is dense in $\mathbb{R}^2$.

[**DM93**] *quantitative estimates:* For any nonempty open interval $I \subset \mathbb{R}$, Theorem 2.3 shows there exists at least one $\vec{m} \in \mathbb{Z}^n$ with $Q(\vec{m}) \in I$. In fact, there exist *many* such $\vec{m}$ (of bounded norm). Namely, if λ is the Lebesgue measure on $\mathbb{R}^n$, then

$$\liminf_{C \to \infty} \frac{\#\left\{ \vec{m} \in \mathbb{Z}^n \,\middle|\, \begin{matrix} Q(\vec{m}) \in I, \\ \|\vec{m}\| < C \end{matrix} \right\}}{\lambda\left(\left\{ \vec{v} \in \mathbb{R}^n \,\middle|\, \begin{matrix} Q(\vec{v}) \in I, \\ \|\vec{v}\| < C \end{matrix} \right\}\right)} > 0.$$

Furthermore, the estimate is uniform when Q varies over any compact set of quadratic forms that all satisfy the hypotheses of Theorem 2.3.

The paper [**DM93**] has been especially influential, because it introduced *Linearization* (which was discussed in Section 1).

Dani has continued his contributions to this field of research in recent years. For example, the corollary of the following theorem is an approximation theorem for linear forms.

THEOREM 2.4 (Dani-Nogueira [**DN09**]). *Let Γ^+ be the set of matrices in $\mathrm{SL}(n,\mathbb{Z})$ that do not have any negative entries. If $\vec{a} \in \mathbb{R}^n$, such that a_1/a_2 is an irrational number that is negative, and $n \geq 3$, then $\Gamma^+\vec{a}$ is dense in $\mathbb{R}^n$.*

COROLLARY 2.5 (Dani-Nogueira). *Let $L(\vec{x}) = \sum_{i=1}^n a_i x_i$, and assume a_1/a_2 is an irrational number that is negative. Then $L(\mathcal{P}_k)$ is dense in $\mathbb{R}$, for every k, where*

$$\mathcal{P}_k = \left\{\,(m_1, \ldots, m_n) \in \mathbb{Z}^n \,\middle|\, \begin{matrix} \gcd(m_1, \ldots, m_n) = 1, \\ \text{and } m_i > k, \,\forall i \end{matrix} \right\}.$$

Dani [**D08**] also proved a simultaneous approximation theorem for certain pairs consisting of a linear form and a quadratic form.

3. Dynamics of non-unipotent homogeneous flows

This section presents a few fundamental questions that Dani worked on, but remain open.

Kolmogorov automorphisms. It is believed that almost all translations on homogeneous spaces are isomorphic to Bernoulli shifts:

DEFINITION 3.1. If $p\colon X \to \mathbb{R}^+$ is a probability distribution on a finite set $X = \{x_1, \ldots, x_n\}$, then there is a natural product measure $p^{\mathbb{Z}}$ on the infinite product $X^{\mathbb{Z}} = \{f\colon \mathbb{Z} \to X\}$, and the associated *Bernoulli shift* is the measurable map $B_{X,p}\colon X^{\mathbb{Z}} \to X^{\mathbb{Z}}$, defined by $B_{X,p}(f)(k) = f(k-1)$.

CONJECTURE 3.2. *Suppose*

- *$G = \mathrm{SL}(n, \mathbb{R})$ (or, more generally, let G be a connected, semisimple, linear Lie group with no compact factors),*
- *$\Gamma = \mathrm{SL}(n, \mathbb{Z})$ (or, more generally, let Γ be an irreducible lattice in G),*
- *$g \in G$, and*
- *$T_g\colon G/\Gamma \to G/\Gamma$ be defined by $T_g(x\Gamma) = gx\Gamma$.*

If there is an eigenvalue λ of g, such that $|\lambda| \neq 1$ (in other words, if the entropy of T_g is nonzero), then T_g is measurably isomorphic to a Bernoulli shift.

DANI'S CONTRIBUTIONS. Two results of Dani [**D76**] are the inspiration for this conjecture:

(1) The conjecture is true if the matrix g is diagonalizable over $\mathbb{C}$.
(2) In the general case, T_g has no zero-entropy quotients (if $\exists \lambda, |\lambda| \neq 1$).

A transformation with no zero-entropy quotients is called a "*Kolmogorov automorphism.*" Examples of non-Bernoulli Kolmogorov automorphisms are rare, so (2) is good evidence that the conjecture is true for all g, not just those that are diagonalizable.

Anosov diffeomorphisms. Dani has worked on the longstanding conjecture that Anosov diffeomorphisms can only be constructed on infranilmanifolds:

DEFINITION 3.3. A diffeomorphism f of a compact, connected manifold M is *Anosov* if, at every point $x \in M$, the tangent space $T_x M$ has a splitting $T_x M = \mathcal{E}^+ \oplus \mathcal{E}^-$, such that

- for $v \in \mathcal{E}^+$, $D(f^k)(v) \to 0$ exponentially fast as $k \to -\infty$, and
- for $v \in \mathcal{E}^-$, $D(f^k)(v) \to 0$ exponentially fast as $k \to +\infty$.

CONJECTURE 3.4 (from the 1960's). *If there is an Anosov diffeomorphism f on M, then some finite cover of M is a nilmanifold. (This means the cover is a homogeneous space G/Γ, where G is a nilpotent Lie group, and Γ is a discrete subgroup of G.) Furthermore, lifting f to the finite cover yields an affine map on the nilmanifold.*

DANI'S CONTRIBUTIONS.

[**D78b**] Dani constructed many nilmanifolds that have Anosov diffeomorphisms. Much more recently, joint work with M. Mainkar [**DMa05**] constructed examples of every sufficiently large dimension.

[**D80**]   Dani proved Conjecture 3.4 under the assumption that M is a double-coset
space $K \backslash G / H$ (and certain additional technical conditions are satisfied).

Divergent trajectories. We know, from Corollary 1.5, that if u^t is unipotent,
then no u^t-orbit diverges to ∞. On the other hand, orbits of a diagonal subgroup
can diverge to ∞.

EXAMPLE 3.5. Let

$$G = \mathrm{SL}(2, \mathbb{R}), \quad a^t = \begin{bmatrix} e^{-t} & 0 \\ 0 & e^t \end{bmatrix}, \quad \Gamma = \mathrm{SL}(2, \mathbb{Z}), \quad \text{and} \quad \vec{e}_1 = \begin{bmatrix} 1 \\ 0 \end{bmatrix}.$$

Then $a^t \vec{e}_1 \to \vec{0}$ as $t \to \infty$, so $a^t \Gamma \to \infty$ in G/Γ.

The divergent orbits of the diagonal matrices in $\mathrm{SL}(2, \mathbb{R}) / \mathrm{SL}(2, \mathbb{Z})$ are well
known, and quite easy to describe. Dani vastly generalized this, by proving that
all divergent orbits are obvious in a much wider setting. Here is a special case of
his result:

THEOREM 3.6 (Dani [**D85**]). *Let*

- $G = \mathrm{SO}(1, n)$ (*or, more generally, let G be a connected, almost simple
  algebraic $\mathbb{Q}$-subgroup of $\mathrm{SL}_{n+1}(\mathbb{R})$, with $\mathrm{rank}_{\mathbb{Q}} G = 1$*),
- $\Gamma = G \cap \mathrm{SL}_{n+1}(\mathbb{Z})$,
- $\{a^t\}$ *be a one-parameter subgroup of G that is diagonalizable over $\mathbb{R}$, and*
- $g \in G$.

Then $a^t g \Gamma$ diverges to ∞ in G/Γ if and only if there exist

- *a continuous homomorphism $\rho \colon G \to \mathrm{SL}(\ell, \mathbb{R})$, for some ℓ, such that
  $\rho(\Gamma) \subseteq \mathrm{SL}(\ell, \mathbb{Z})$, and*
- *a nonzero vector $\vec{v} \in \mathbb{Z}^\ell$,*

such that $\rho(a^t g) \vec{v} \to \vec{0}$ as $t \to \infty$.

Surprisingly, he was also able to exhibit many cases in which there are divergent
orbits that do not come from the construction in Theorem 3.6:

THEOREM 3.7 (Dani [**D85**]). *Let*

- $G = \mathrm{SL}(n, \mathbb{R})$, *with $n \geq 3$ (or, more generally, let G be the $\mathbb{R}$-points of a
  connected, almost simple algebraic $\mathbb{Q}$-group with $\mathrm{rank}_{\mathbb{Q}} G = \mathrm{rank}_{\mathbb{R}} G \geq 2$*),
- $\Gamma = \mathrm{SL}(n, \mathbb{Z})$ (*or, in general, let G be the $\mathbb{Z}$-points of G*), *and*
- $\{a^t\}$ *be a (nontrivial) one-parameter subgroup of G that is diagonalizable
  over $\mathbb{R}$.*

*Then there are a^t-orbits in G/Γ that diverge to ∞, but do not correspond to a
continuous homomorphism $\rho \colon G \to \mathrm{SL}(\ell, \mathbb{R})$, as in Theorem 3.6.*

It remains an open problem to determine the set of divergent orbits in G/Γ in
cases where Theorem 3.6 does not apply.

Bounded orbits and Schmidt's game. Dani also pioneered the study of
orbits at the opposite extreme from the divergent trajectories discussed in Section 3:
an orbit in G/Γ is *bounded* if its closure is compact.

It is well known that if $\{a^t\}$ is any nontrivial one-parameter group of diagonal
matrices in $\mathrm{SL}(n, \mathbb{R})$, then almost every a^t-orbit is dense in $\mathrm{SL}(n, \mathbb{R}) / \mathrm{SL}(n, \mathbb{Z})$.
Since bounded orbits cannot be dense, this implies that the set $\mathcal{B}$ of bounded orbits
has measure 0, which obviously means that $\mathcal{B}$ is very small. However, Dani showed
that, from the viewpoint of Hausdorff dimension, this set can be as large as possible:

THEOREM 3.8 (Dani [**D89**, Thm. 4.2]). *There is a nontrivial one-parameter group* $\{a^t\}$ *of diagonal matrices in* $\mathrm{SL}(n,\mathbb{R})$, *such that the Hausdorff dimension of the set of bounded orbits in* $\mathrm{SL}(n,\mathbb{R})/\mathrm{SL}(n,\mathbb{Z})$ *is equal to* $n^2 - 1$, *the dimension of* $\mathrm{SL}(n,\mathbb{R})$.

The proof uses a method, now called "Schmidt's game," which had been introduced into the theory of Diophantine Approximation by W. M. Schmidt [**Sc66**]. We have two players, A and B. In the setting of Theorem 3.8, Player B starts the game, by choosing any closed ball B_0 in the space $X = G/\Gamma$, and the game proceeds inductively. Given ball B_{i-1}, Player A chooses a closed ball A_i that is contained in B_{i-1}, and whose radius is exactly one-half of the radius of B_{i-1}. Then Player B chooses a ball B_i that is contained in A_i, and whose radius is exactly one-half of the radius of A_i. We say $\mathcal{B}$ is a *winning set* if Player A has a strategy to guarantee that the unique point in $\bigcap_{i=0}^{\infty} B_i$ belongs to $\mathcal{B}$. (More precisely, $\mathcal{B}$ is a "$(1/2, 1/2)$-winning set," since we have specified that the radius of each ball is one-half of the radius of the preceding ball.)

Results like Theorem 3.8 can be proved by showing that the set of bounded orbits is a winning set. It is obvious that a winning set must be dense in G/Γ. (Otherwise, Player B could win by choosing B_0 to be disjoint from $\mathcal{B}$.) A more careful argument shows that the Hausdorff dimension of a winning set must be equal to the dimension of G/Γ.

Dani's introduction of this method into the study of dynamical systems was very influential; Schmidt's game has now been used to prove that numerous other sets of exceptional orbits are large. For example, it can be used to study the set of orbits whose closure does not contain a certain point in the space.

References

[DD73]   J. S. Dani and S. G. Dani, *Discrete groups with dense orbits*, J. Indian Math. Soc. (N.S.) **37** (1973), 183–195 (1974). MR0360932 (50 #13379)

[D76]   S. G. Dani, *Kolmogorov automorphisms on homogeneous spaces*, Amer. J. Math. **98** (1976), no. 1, 119–163. MR0419728 (54 #7746)

[D77]   S. G. Dani, *Spectrum of an affine transformation*, Duke Math. J. **44** (1977), no. 1, 129–155. MR0444835 (56 #3182)

[D78a]   S. G. Dani, *Invariant measures of horospherical flows on noncompact homogeneous spaces*, Invent. Math. **47** (1978), no. 2, 101–138. MR0578655 (58 #28260)

[D78b]   S. G. Dani, *Nilmanifolds with Anosov automorphism*, J. London Math. Soc. (2) **18** (1978), no. 3, 553–559, DOI 10.1112/jlms/s2-18.3.553. MR518242 (80k:58082)

[D79]   S. G. Dani, *On invariant measures, minimal sets and a lemma of Margulis*, Invent. Math. **51** (1979), no. 3, 239–260, DOI 10.1007/BF01389917. MR530631 (80d:58039)

[D80]   S. G. Dani, *On affine automorphisms with a hyperbolic fixed point*, Topology **19** (1980), no. 4, 351–365, DOI 10.1016/0040-9383(80)90019-1. MR584560 (81m:58059)

[D81]   S. G. Dani, *Invariant measures and minimal sets of horospherical flows*, Invent. Math. **64** (1981), no. 2, 357–385, DOI 10.1007/BF01389173. MR629475 (83c:22009)

[D82]   S. G. Dani, *On uniformly distributed orbits of certain horocycle flows*, Ergodic Theory Dynamical Systems **2** (1982), no. 2, 139–158 (1983). MR693971 (84g:58068)

[D84a]   S. G. Dani, *On orbits of unipotent flows on homogeneous spaces*, Ergodic Theory Dynam. Systems **4** (1984), no. 1, 25–34, DOI 10.1017/S0143385700002248. MR758891 (86b:58068)

[D84b]   S. G. Dani, *Continuous equivariant images of lattice-actions on boundaries*, Ann. of Math. (2) **119** (1984), no. 1, 111–119, DOI 10.2307/2006965. MR736562 (85i:22009)

[D85]   S. G. Dani, *Divergent trajectories of flows on homogeneous spaces and Diophantine approximation*, J. Reine Angew. Math. **359** (1985), 55–89, DOI 10.1515/crll.1985.359.55. MR794799 (87g:58110a)

[D86]   S. G. Dani, *Orbits of horospherical flows*, Duke Math. J. **53** (1986), no. 1, 177–188, DOI
         10.1215/S0012-7094-86-05312-3. MR835804 (87i:22026)
[D89]   S. G. Dani, *On badly approximable numbers, Schmidt games and bounded orbits of
         flows*, Number theory and dynamical systems (York, 1987), London Math. Soc. Lec-
         ture Note Ser., vol. 134, Cambridge Univ. Press, Cambridge, 1989, pp. 69–86, DOI
         10.1017/CBO9780511661983.006. MR1043706 (91d:58200)
[D01]   Shrikrishna G. Dani, *On the Oppenheim conjecture on values of quadratic forms*, Essays
         on geometry and related topics, Vol. 1, 2, Monogr. Enseign. Math., vol. 38, Enseignement
         Math., Geneva, 2001, pp. 257–270. MR1929329 (2003m:11100)
[D08]   Shrikrishna G. Dani, *Simultaneous Diophantine approximation with quadratic and
         linear forms*, J. Mod. Dyn. **2** (2008), no. 1, 129–138, DOI 10.3934/jmd.2008.2.129.
         MR2366232 (2009m:11103)
[DMa05] S. G. Dani and Meera G. Mainkar, *Anosov automorphisms on compact nilmanifolds
         associated with graphs*, Trans. Amer. Math. Soc. **357** (2005), no. 6, 2235–2251, DOI
         10.1090/S0002-9947-04-03518-4. MR2140439 (2006j:22005)
[DM89]   S. G. Dani and G. A. Margulis, *Values of quadratic forms at primitive integral points*,
         Invent. Math. **98** (1989), no. 2, 405–424, DOI 10.1007/BF01388860. MR1016271
         (90k:22013b)
[DM90a] S. G. Dani and G. A. Margulis, *Orbit closures of generic unipotent flows on homogeneous
         spaces of* SL(3, **R**), Math. Ann. **286** (1990), no. 1-3, 101–128, DOI 10.1007/BF01453567.
         MR1032925 (91k:22026)
[DM90b] S. G. Dani and G. A. Margulis, *Values of quadratic forms at integral points: an elemen-
         tary approach*, Enseign. Math. (2) **36** (1990), no. 1-2, 143–174. MR1071418 (91k:11053)
[DM91]   S. G. Dani and G. A. Margulis, *Asymptotic behaviour of trajectories of unipotent flows
         on homogeneous spaces*, Proc. Indian Acad. Sci. Math. Sci. **101** (1991), no. 1, 1–17, DOI
         10.1007/BF02872005. MR1101994 (92g:22027)
[DM93]   S. G. Dani and G. A. Margulis, *Limit distributions of orbits of unipotent flows and
         values of quadratic forms*, I. M. Gel'fand Seminar, Adv. Soviet Math., vol. 16, Amer.
         Math. Soc., Providence, RI, 1993, pp. 91–137. MR1237827 (95b:22024)
[DN09]   S. G. Dani and Arnaldo Nogueira, *On* SL(n, ℤ)$_+$*-orbits on* ℝ^n *and positive integral
         solutions of linear inequalities*, J. Number Theory **129** (2009), no. 10, 2526–2529, DOI
         10.1016/j.jnt.2008.12.010. MR2541029 (2010g:11111)
[DS84]   S. G. Dani and John Smillie, *Uniform distribution of horocycle orbits for Fuchsian
         groups*, Duke Math. J. **51** (1984), no. 1, 185–194, DOI 10.1215/S0012-7094-84-05110-X.
         MR744294 (85f:58093)
[F73]   Harry Furstenberg, *The unique ergodicity of the horocycle flow*, Recent advances in
         topological dynamics (Proc. Conf., Yale Univ., New Haven, Conn., 1972; in honor of
         Gustav Arnold Hedlund), Springer, Berlin, 1973, pp. 95–115. Lecture Notes in Math.,
         Vol. 318. MR0393339 (52 #14149)
[G63]   L. Greenberg, Discrete groups with dense orbits, in L. Auslander, L. Green, and F. Hahn,
         Flows on Homogeneous Spaces. Princeton University Press, Princeton, N. J., 1963,
         pp. 85–103. MR0167569
[M71]   G. A. Margulis, *The action of unipotent groups in a lattice space* (Russian), Mat. Sb.
         (N.S.) **86(128)** (1971), 552–556. MR0291352 (45 #445)
[M75]   G. A. Margulis, *On the action of unipotent groups in the space of lattices*, Lie groups
         and their representations (Proc. Summer School, Bolyai, János Math. Soc., Budapest,
         1971), Halsted, New York, 1975, pp. 365–370. MR0470140 (57 #9907)
[M78]   G. A. Margulis, *Factor groups of discrete subgroups and measure theory* (Russian),
         Funktsional. Anal. i Prilozhen. **12** (1978), no. 4, 64–76. MR515630 (80k:22005)
[M87]   Gregori Aleksandrovitch Margulis, *Formes quadratriques indéfinies et flots unipotents
         sur les espaces homogènes* (English, with French summary), C. R. Acad. Sci. Paris Sér.
         I Math. **304** (1987), no. 10, 249–253. MR882782 (88f:11027)
[M88]   G. A. Margulis, *Lie groups and ergodic theory*, Algebra—some current trends (Varna,
         1986), Lecture Notes in Math., vol. 1352, Springer, Berlin, 1988, pp. 130–146, DOI
         10.1007/BFb0082022. MR981823 (91a:22009)
[O29]   A. Oppenheim: The minima of indefinite quaternary quadratic forms, *Proc. Nat. Acad.
         Sci. U.S.A.* 15 (1929) 724–727. zbMATH: JFM 55.0722.01

[R91a]   Marina Ratner, *On Raghunathan's measure conjecture*, Ann. of Math. (2) **134** (1991),
          no. 3, 545–607, DOI 10.2307/2944357. MR1135878 (93a:22009)
[R91b]   Marina Ratner, *Raghunathan's topological conjecture and distributions of unipotent
          flows*, Duke Math. J. **63** (1991), no. 1, 235–280, DOI 10.1215/S0012-7094-91-06311-8.
          MR1106945 (93f:22012)
[Sc66]   Wolfgang M. Schmidt, *On badly approximable numbers and certain games*, Trans. Amer.
          Math. Soc. **123** (1966), 178–199. MR0195595 (33 #3793)
[Sh91]   Nimish A. Shah, *Uniformly distributed orbits of certain flows on homogeneous spaces*,
          Math. Ann. **289** (1991), no. 2, 315–334, DOI 10.1007/BF01446574.   MR1092178
          (93d:22010)
[St90]   A. N. Starkov, *The structure of orbits of homogeneous flows and the Raghunathan
          conjecture* (Russian), Uspekhi Mat. Nauk **45** (1990), no. 2(272), 219–220, DOI
          10.1070/RM1990v045n02ABEH002338; English transl., Russian Math. Surveys **45**
          (1990), no. 2, 227–228. MR1069361 (91i:22017)

DEPARTMENT OF MATHEMATICS AND COMPUTER SCIENCE, UNIVERSITY OF LETHBRIDGE, LETH-
BRIDGE, ALBERTA, T1K 3M4, CANADA
    E-mail address: Dave.Morris@uleth.ca
    URL: http://people.uleth.ca/~dave.morris/

Contemporary Mathematics
Volume **631**, 2015
http://dx.doi.org/10.1090/conm/631/12602

Calculus of generalized Riesz products

e. H. el Abdalaoui and M. G. Nadkarni

ABSTRACT. In this paper we discuss generalized Riesz products bringing into consideration H^p theory, the notion of Mahler measure, and the zeros of polynomials appearing in the generalized Riesz product. Formula for Radon-Nikodym derivative between two generalized Riesz product is established under suitable conditions. This is then used to formulate a Dichotomy theorem and prove a conditional version of it. A discussion involving flat polynomials is given.

1. Introduction

Generalized Riesz products considered in this paper are the ones defined in [**17**],[**21**] where one of the aims was to describe the spectrum of a non-singular rank one map as a generalized Riesz product. Generalized Riesz product has remained only in the state of definition although much deep work has appeared over last two decades on special generalized Riesz products arising in the spectral study of measure preserving rank one maps of ergodic theory.

The purpose of this paper is to set forth some basic facts of generalized Riesz products and bring into play Hardy class theory to discuss some of the problems arising in the subject. It is surprising that one can garner so much information simply from the fact that L^2 norm of the trigonometric polynomials appearing in a generalized Riesz product is one. These facts are discussed in section 2 and 3 of the paper. Section 4 gives a formula for Radon-Nikodym derivative of two generalized Riesz products. In section 5 we discuss a conditional dichotomy result. Connection with flat polynomials is discussed in section 6 and a result on zeros of polynomials under consideration is given in section 7.

In the rest of this section we give some background material.

2010 *Mathematics Subject Classification.* Primary 37A05, 37A30, 37A40; Secondary 42A05, 42A55.

Key words and phrases. Simple spectrum, simple Lebesgue spectrum, Banach problem, singular spectrum, Mahler measure, rank one maps, Generalized Riesz products, Hardy spaces, outer functions, inner functions, flat polynomials, ultraflat polynomials, Littelwood problem, zeros of polynomials.

Riesz Products. Consider a trigonometric series

$$\sum_{n=-\infty}^{+\infty} a_n z^n, \ z \in S^1,$$

where $S^1 = \{z \ : \ |z| = 1\}$, the circle group. If we ignore those terms whose coefficients are zero, then we can write the trigonometric series as

$$\sum_{k=-\infty}^{+\infty} a_{n_k} z^{n_k}, \ z \in S^1.$$

Now if $|\frac{n_{k+1}}{n_k}| > q, \ \forall k$, for some $q > 1$, then the series is said to be lacunary. The convergence questions for a lacunary trigonometric are answered by:

THEOREM 1.1 ([**27**, p.203]). *A lacunary trigonometric series converges on a set of positive Lebesgue measure if and only if its coefficients form an ℓ^2 sequence. If the coefficients are not square summable then the lacunary trigonometric series is not a Fourier series (of any $L^1(S^1, dz)$ function).*

Next we need the notion of dissociated polynomials. Consider the following two products:

$$(1+z)(1+z) = 1 + z + z + z^2 = 1 + 2z + z^2,$$
$$(1+z)(1+z^2) = 1 + z + z^2 + z^3.$$

In the first case we group terms with the same power of z, while in the second case all the powers of z in the formal expansion are distinct. In the second case we say that the polynomials $1+z$ and $1+z^2$ are dissociated. More generally we say that a set of trigonometric polynomials is dissociated if in the formal expansion of product of any finitely many of them, the powers of z in the non-zero terms are all distinct. (see section 2 for a detailed definition).

Now consider the infinite Product due to F. Riesz: ([**27**, p.208])

$$\prod_{k=1}^{+\infty} \Big(1 + a_k \cos(n_k x)\Big), \ -1 \le a_k \le 1, \ \frac{n_{k+1}}{n_k} \ge 3.$$

Each term of this product is non-negative and integrates to 1 with respect to the normalized Lebesgue measure on the circle group. We rewrite this product as

$$\prod_{k=1}^{+\infty} \Big(1 + \frac{a_k}{2}\big(z^{n_k} + z^{-n_k}\big)\Big).$$

Because of the lacunary nature of the n_k's, the polynomials

$$1 + \frac{a_k}{2}\big(z^{n_k} + z^{-n_k}\big), \ k = 1, 2, \cdots$$

are dissociated. If we expand the finite product $\prod_{k=1}^{L}\Big(1 + \frac{a_k}{2}\big(z^{n_k} + z^{-n_k}\big)\Big)$, we get a finite sum of the type

$$1 + \sum_{\substack{k \ne 0 \\ k=-M}}^{M} \gamma_k z^{m_k},$$

and for the infinite product we get the series

$$\sum_{k=-\infty}^{+\infty} \gamma_k z^{m_k},$$

both sums being formal expansions of the corresponding products. Since the finite products are non-negative and integrate to 1, they are probability densities and the corresponding probability measures converge weakly to a probability measure, say μ, whose Fourier-Stieltjes series is the formal expansion of the infinite product. The main theorem about Riesz products is

THEOREM 1.2 ([**27**, p.209]). *The Riesz product* $\prod_{k=1}^{+\infty}\left(1 + \frac{a_k}{2}\left(z^{n_k} + z^{-n_k}\right)\right)$, $-1 \le a_k \le 1$, $\frac{n_{k+1}}{n_k} \ge 3, \forall k$. *represents a continuous measure* μ *on* S^1 *which is absolutely continuous or singular with respect to the Lebesgue measure on* S^1 *according as the sequence* $a_k, k = 1, 2, \cdots$ *is in* ℓ^2 *or not. The finite products*

$$\prod_{k=1}^{L}\left(1 + \frac{a_k}{2}\left(z^{n_k} + z^{-n_k}\right)\right), \quad L = 1, 2, \cdots$$

converge to $\frac{d\mu}{dz}$ *a.e. (dz) as* $L \longrightarrow +\infty$.

We will improve this theorem using Hardy class theory later.

The above account about Riesz products is based on parts of chapter V of Zygmund [**27**]. The original four page paper of F. Riesz appeared nearly 95 years ago in 1918 [**26**], and has led to much further work. The aim of the paper was to give a continuous function of bounded variation whose Fourier-Stieltjes coefficients do not tend to zero.

Connection with Ergodic Theory. In [**20**] F. Ledrappier showed that Riesz products appear as maximal spectral type of a class of measure preserving transformation. We will assume here that the reader is familiar with the stacking method of constructing measure preserving transformations. The construction is recalled in full detail and generality in section 6. Consider a sequence stacks Σ_n, $n = 1, 2, \cdots$ of pairwise disjoint intervals, beginning with the unit interval as Σ_1. Each stack comes equipped with the usual linear maps among its element. Let h_n be the height of Σ_n, $n = 1, 2, \cdots$. For each $n \ge 2$, the stack Σ_n is obtained from Σ_{n-1} by dividing Σ_{n-1} into two equal parts and adding a finite number, say a_{n-1}, of spacers on the left subcolumn. The top of the left subcolumn (after adding spacers) is mapped linearly onto the bottom of the right subcolumn, and the resulting new stack is Σ_n. If T is the measure preserving transformation given by this system of stacks, then, as shown by Ledrappier, the associated unitary operator U_T has simple spectrum whose maximal spectral type,(except possibly for some discrete part), is given by the weak$*$ limit of probability measures

$$\mu_L = \prod_{n=1}^{L}\left|\frac{1}{\sqrt{2}}\left(1 + z^{h_n+a_n}\right)\right|^2 dz = \prod_{n=1}^{L}\left(1 + \frac{1}{2}\left(z^{h_n+a_n} + z^{-h_n-a_n}\right)\right)dz, L = 1, 2, \cdots$$

as $L \longrightarrow +\infty$. We write this measure formally as

(A) $$\mu = \prod_{n=1}^{+\infty} \left| \frac{1}{\sqrt{2}} \left(1 + z^{h_n + a_n} \right) \right|^2.$$

More generally, consider a rank one measure preserving transformation made of a sequence of stacks Σ_n, $n = 1, 2, \cdots$ with h_n as the height of Σ_n, and Σ_1 being the unit interval. For $n \geq 2$, Σ_n is obtained from Σ_{n-1} by dividing Σ_{n-1} into m_{n-1} equal parts, and placing a certain number $a_j^{(n-1)}$ of spacers on the jth subcolumn, $1 \leq j \leq m_{n-1} - 1$. Let T denote the resulting measure preserving transformation and U_T the associated unitary operator. Here again the operator U_T has simple spectrum and the maximal spectral type (except possibly some discrete part) is given by the weak$*$ limit of the probability measures [17],[21],[19]:

$$\mu_n = \prod_{k=1}^{n} \frac{1}{m_k} \left| \left(1 + \sum_{j=1}^{m_k-1} z^{jh_k + \sum_{i=1}^{j} a_i^{(k)}} \right) \right|^2 dz, \ n = 1, 2, 3, \cdots$$

We denote this weak$*$ limit μ by the infinite product:

(B) $$\mu = \prod_{k=1}^{+\infty} \frac{1}{m_k} \left| \left(1 + \sum_{j=1}^{m_k-1} z^{jh_k + \sum_{i=1}^{j} a_i^{(k)}} \right)^2 \right|$$

Ornstein [24] has constructed a family of rank one measure preserving maps which are mixing. Bourgain [10] has shown that almost all of these rank one map have singular spectrum. It is not known if there exists a rank one measure preserving map whose maximal spectral type has a part which is absolutely continuous with respect to the Lebesgue measure on S^1. This question is naturally related to a question of Banach (in The Scottish Book) which asks if there is a measure preserving transformation T on the Real line (with Lebesgue measure) which admits a function f such that $f \circ T^n$, $n = 1, 2, \cdots$ are orthogonal and span $L^2(\mathbb{R})$.

Let Σ_n, $n = 1, 2, \cdots$ and T be as above, and let ϕ be a function of absolute value one which is constant on interval of Σ_n except the top piece, $n - 1, 2, \cdots$. It is known that the unitary operator $V_\phi = \phi \cdot U_T$ also has multiplicity one and its maximal spectral type is continuous whenever ϕ is not a coboundary. It is given by the weak$*$ limit of a sequence of probability measures given by:

$$\mu_L = \prod_{k=1}^{L} \left| P_k(z) \right|^2 dz,$$

where P_k's are polynomials of the type:

$$P_k(z) = c_0^{(k)} + \sum_{j=1}^{m_k-1} c_j^{(k)} z^{jh_k + \sum_{i=1}^{j} a_i^{(k)}}, \quad \sum_{j=0}^{m_k-1} |c_j^{(k)}|^2 = 1.$$

The constants $c_j^{(k)}$ are determined by the m_k's and the function ϕ. We may write this weak$*$ limit as

(C) $$\mu_\phi = \prod_{k=1}^{+\infty} \left| P_k(z) \right|^2.$$

Note that in all the products (A), (B), and (C) there is no lacunarity condition imposed on the powers of z from outside. The gap between two consecutive nonzero terms of the polynomials are determined by parameters of construction, and need not be lacunary. In the rest of this paper we will, for most part, dispense with the dynamical origin of the measures of the type (A), (B) and (C) and discuss a larger class of measures, called generalized Riesz product, which include these measures.

2. Generalized Riesz Products and their Weak Dichotomy

In this section we introduce generalized Riesz products and derive a weak dichotomy result about infinite product of polynomials associated to it. This also yields conditions for absolute continuity of the generalized Riesz product.

DEFINITION 2.1. Let $P_1, P_2, \cdots$, be a sequence of trigonometric polynomials such that

(i) for any finite sequence $i_1 < i_2 < \cdots < i_k$ of natural numbers

$$\int_{S^1} \left| (P_{i_1} P_{i_2} \cdots P_{i_k})(z) \right|^2 dz = 1,$$

where S^1 denotes the circle group and dz the normalized Lebesgue measure on S^1,

(ii) for any infinite sequence $i_1 < i_2 < \cdots$ of natural numbers the weak* limit of the measures $| (P_{i_1} P_{i_2} \cdots P_{i_k})(z) |^2 \, dz, k = 1, 2, \cdots$ as $k \to \infty$ exists,

then the measure μ given by the weak* limit of $| (P_1 P_2 \cdots P_k)(z) |^2 \, dz$ as $k \to \infty$ is called generalized Riesz product of the polynomials $| P_1 |^2, | P_2 |^2, \cdots$ and denoted by

(1.1) $$\mu = \prod_{j=1}^{\infty} |P_j|^2$$

For an increasing sequence $k_1 < k_2 < \cdots$ of natural numbers the product $\prod_{j=1}^{\infty} |P_{k_j}|^2$ makes sense as the weak* limit of probability measures $|(P_{k_1} P_{k_2} \cdots P_{k_n})(z)|^2 dz$ as $n \to \infty$. It depends on the sequence $k_1 < k_2 \cdots$, and called a subproduct of the given generalized Riesz product.

Since the object under consideration is the generalized Riesz product $\prod_{j=1}^{\infty} |P_j|^2$, without loss of generality we assume that the polynomials $P_j, j = 1, 2, \cdots$ are analytic with positive constant term. Their domain of definition will mainly be the circle group, but with option to look at then as functions on the complex plane. Since $\int_{S^1} |P_j|^2(z) dz = 1$, the sum of the squares of the absolute values of coefficients of P_j is one, and so each coefficient of P_j is at most one in absolute value. Let $a_0^{(j)}$ denote the constant term of P_j, which is positive by assumption. The sequence of products $\prod_{j=1}^{n} a_0^{(j)}, n = 1, 2, \cdots$ is non-increasing, and so has a limit which is either zero or some positive constant which can be at most 1. The case when this constant is one is obviously the trivial case when each P_j is the constant 1.

Consider the sequence of polynomials $S_n \stackrel{\text{def}}{=} \prod_{j=1}^{n} P_j, n = 1, 2, \cdots$ (without the absolute value squared). For each n, let $b_0^{(n)} \stackrel{\text{def}}{=} \prod_{j=1}^{n} a_0^{(j)}$ denote the constant term of S_n. Write $b = \lim_{n \to \infty} b_0^{(n)}$. We have the following weak dichotomy theorem for generalized Riesz products.

THEOREM 2.2. *If $b = 0$, the sequence of polynomials $S_n = \prod_{i=1}^{n} P_i$, $i = 1, 2, \cdots$ converges to zero weakly in $L^2(S^1, dz)$. If b is positive it converges in $L^1(S^1, dz)$ (and in H^1) norm to a non-zero function f which is also in H^2 with H^2 norm at most 1, $\log(|f|)$ has finite integral.*

PROOF. Assume that $b = 0$. We show that the sequence $S_n, n = 1, 2, \cdots$ has zero weak limit as functions in $L^2(S^1, dz)$. Assume that a subsequence $S_{k_n}, n = 1, 2, \cdots$ converges weakly to $f \in L^2(S^1, dz)$. We show that f is the zero function. By choosing a further subsequence if necessary we can assume without any loss that the constant term of $\frac{S_{k_{n+1}}}{S_{k_n}}, n = 1, 2, \cdots$ goes to zero as $n \to \infty$. Since $b = 0$, the zeroth Fourier coefficient of f is zero. Since each S_n is an analytic trigonometric polynomial, the negative Fourier coefficients of f are all zero. Assume now that for $0 \leq j < l$, $b_j = \int_{S^1} z^{-j} f(z) dz = 0$. Then, given ϵ, for large enough m, $\left| \int_{S^1} z^{-j} S_{k_m} dz \right| < \epsilon$, for $0 \leq j < l$, and moreover the constant term of $\frac{S_{k_{m+1}}}{S_{k_m}}$ is less than ϵ. For $n > m$,

$$\prod_{j=1}^{k_n} P_i = \prod_{j=1}^{k_m} P_j \prod_{j=k_m+1}^{k_n} P_j.$$

Since P_j's are one sided trigonometric polynomials, it is easy to see from this that $\left| \int_{S^1} z^{-l} S_{k_n}(z) dz \right| \leq (l+1)\epsilon$. Since this holds for all $n > m$ we see that $\int_{S^1} z^{-l} f(z) dz = 0$. Induction completes the proof.

Assume now that $b > 0$. Then $a_0^{(n)}$ as well as $\prod_{j=m+1}^{n} a_0^{(j)}, m < n$, converge to 1 as $m, n \to \infty$. Since $L^2(S^1, dz)$ norm of all the finite products is one,

$$(2.1) \qquad P_n, \frac{S_n}{S_m} \to 1 \text{ in } L^2(S^1, dz), \quad m < n, \quad \text{as } m, n \to \infty.$$

Moreover by Cauchy-Schwarz inequality

$$\left\| S_n - S_m \right\|_1 = \left\| S_m \left(1 - \prod_{m+1}^{n} P_i \right) \right\|_1 \leq \left\| S_m \right\|_2 \left\| 1 - \prod_{j=m+1}^{n} P_j \right\|_2$$

$$\to 0 \text{ as } m, n \to \infty.$$

Thus the sequence of analytic polynomials $S_n, n = 1, 2, \cdots$ converges in $L^1(S^1, dz)$ to a function which we denote by f, and view it also as a function in the Hardy class H^1. A subsequence of $S_n, n = 1, 2, \cdots$ converges to f a.e (with respect to the Lebesgue measure of S^1), whence, over the same subsequence $S_n^2, |S_n|^2, n = 1, 2, \cdots$ converge to f^2 and $|f|^2$, respectively. Since $\|S_n\|_2 = 1, n = 1, 2, \cdots$, by Fatou's lemma we conclude that f is square integrable with $L^2(S^1, dz)$ norm at most 1. Thus f is in H^2, and $\log |f|$ has finite integral.
                                                                                  $\square$

We do not know if the $L^2(S^1, dz)$ norm of f is 1, equivalently, if $S_n, n = 1, 2, \cdots$ converges to f in $L^2(S^1, dz)$. We give some sufficient conditions under which this holds. Let $S_n = \sum_{j=0}^{m_n} b_j^{(n)} z^j$, where m_n is the degree of the trigonometric polynomial S_n. Now

$$b_j^{(n)} = \int_{S^1} z^{-j} S_n(z) dz \longrightarrow \int_{S^1} z^{-j} f(z) dz \stackrel{\text{def}}{=} b_j.$$

The series $\sum_{j=0}^{\infty} b_j z^j$ is the Fourier series of f and we call this series the formal expansion of $\prod_{j=1}^{\infty} P_j$. Since b is positive, the infinite product $\prod_{j=n+1}^{\infty} a_0^{(j)}$ is also positive, so the infinite product $\prod_{j=n+1}^{\infty} P_j$ has a formal expansion which we denote by $\sum_{j=0}^{\infty} c_j^{(n)} z^j$. Note that $c_0^{(n)} = \prod_{j=n+1}^{\infty} a_0^{(j)} \longrightarrow 1$ as $n \longrightarrow \infty$, as a result $\sum_{j=1}^{\infty} |c_j^{(n)}|^2 \le 1 - (c_0^{(n)})^2 \longrightarrow 0$ as $n \to \infty$.

At this point, let us recall the following important notion in the Riesz product theory.

DEFINITION 2.3. Finitely many trigonometric polynomials $q_0, q_1, \cdots, q_n$, $q_j = \sum_{i=-N_j}^{N_j} d_i^{(j)} z^i, j = 0, 1, 2, \cdots, n$ are said to be dissociated if in their product $q_0(z) q_1(z) \cdots q_n(z)$, (when expanded formally, i.e., without grouping terms or canceling identical terms with opposite signs), the powers $i_0 + i_1 + \cdots + i_n$ of z in non-zero terms

$$d_{i_0}^{(0)} d_{i_1}^{(1)} \cdots d_{i_n}^{(n)} z^{i_0 + i_1 + \cdots + i_n}$$

are all distinct.

A sequence $q_0, q_1, \cdots$, of trigonometric polynomials is said to be dissociated if for each n the polynomials $q_0, q_1, \cdots, q_n$ are dissociated.

Suppose now that the polynomials $P_1, P_2, \cdots$ (without the squares) appearing in generalized Riesz product (1.1) are dissociated. Then, whenever $b_j^{(n)}$ is a non-zero coefficient in the expansion of S_n, $b_j^{(l)} = b_j^{(n)} \frac{b_0^{(l)}}{b_0^{(n)}}$ for all $l \ge n$. Thus, if the polynomials $P_j, j = 1, 2, \cdots$ are dissociated, then we see on letting $l \to \infty$ that $b_j = b_j^{(n)} c_0^{(n)}$, provided $b_j^{(n)} \ne 0$. We therefore have for any n

$$\sum_{j=0}^{m_n} |b_j|^2 \ge \sum_{j=1}^{m_n} |b_j^{(n)}|^2 (c_0^{(n)})^2 = (c_0^{(n)})^2 \longrightarrow 1$$

as $n \to \infty$. Thus f has $L^2(S_1, dz)$ norm 1. We have proved:

THEOREM 2.4. If the polynomials $P_n, n = 1, 2, \cdots$ are dissociated and b is positive then the partial products $S_n, n = 1, 2, \cdots$ converge in H^2 to a non-zero function f and the generalized product $\prod_{j=1}^{\infty} |P_j|^2$ is the measure $|f|^2 dz$. Further, $\int_{S^1} \log(|f(z)|) dz$ is finite.

If we replace the condition that the polynomials $P_n, n = 1, 2, \cdots$ are dissociated by the condition that coefficients of the polynomials $P_n, n = 1, 2, \cdots$ are all non-negative, then it is easy to verify that for $0 \le k \le m_n$,

$$b_k \ge c_0^{(n)} b_k^{(n)} + b_0^{(n)} c_k^{(n)},$$

whence

$$\sum_{k=0}^{\infty} |b_k|^2 \ge (\sum_{k=0}^{m_n} |b_k^{(n)}|^2) |c_0^{(n)}|^2 = 1 \cdot |c_0^{(n)}|^2 \to 1,$$

as $n \to \infty$. Thus, if the coefficients of all the $P_n, n = 1, 2, \cdots$ are non-negative, and if $b = b_0 > 0$, we necessarily have convergence of $S_n, n = 1, 2, \cdots$ in H^2.

We continue with the assumption that b is positive, but no more assume that the polynomials $P_n, n = 1, 2, \cdots$ are dissociated or have non-negative coefficients. Fix n, and let $1 \leq j \leq m_n$, then

$$\sum_{j=0}^{\infty} b_j z^j = \left(\sum_{i=0}^{m_n} b_i^{(n)} z^i \right) \left(\sum_{k=0}^{\infty} c_k^{(n)} z^k \right).$$

This gives any $j \geq 0$,

$$b_j = b_j^{(n)} c_0^{(n)} + \sum_{i=0}^{j-1} b_i^{(n)} c_{j-i}^{(n)}.$$

Hence, for any $j \geq 1$,

$$\mid b_j - b_j^{(n)} c_0^{(n)} \mid^2 \leq \left(\sum_{i=0}^{j-1} \mid b_i^{(n)} \mid^2 \right) \left(\sum_{i=0}^{j-1} \mid c_{j-i}^{(n)} \mid^2 \right) \leq \sum_{j=1}^{\infty} \mid c_j^{(n)} \mid^2$$

$$\leq 1 - (c_0^{(n)})^2 \longrightarrow 0 \text{ as } n \to \infty$$

Assume now that $m_n(1 - c_0^{(n)}) \longrightarrow 0$ as $n \to \infty$ Then $\sum_{j=0}^{m_n} \mid b_j - b_j^{(n)} c_0^{(n)} \mid^2 \to 0$ as $n \to \infty$. Another use of the assumption that $m_n(1 - c_0^{(n)}) \longrightarrow 0$ as $n \to 0$ allows us to conclude that $\sum_{j=0}^{m_n} |b_j - b_j^{(n)}|^2 \longrightarrow 0$ as $n \to \infty$. Since $\sum_{j=1}^{m_n} |b_j^{(n)}|^2 = 1$, we conclude that $\sum_{j=1}^{\infty} |b_j|^2 = 1$, so that $L^2(S^1, dz)$ norm of f is one and $S_n, n = 1, 2, \cdots$ converges to f in H^2. We have proved:

THEOREM 2.5. *If b is positive and $m_n(1 - c_0^{(n)}) \longrightarrow 0$ as $n \to \infty$ then $S_n \to f$ in H^2 and $|f|^2 dz$ is the generalized Riesz product $\prod_{j=1}^{\infty} |P_j|^2$. Moreover $\log(|f|)$ has finite integral.*

Our calculus can be interpreted as follows. Put

$$B = (b_j)_{j=0}^{+\infty}, \quad B^{(n)} = (b_0^{(n)}, b_1^{(n)}, \cdots, b_{m_n}^{(n)}, 0, 0, \cdots) \text{ and } C^{(n)} = (c_j^{(n)})_{j=0}^{+\infty},$$

then

$$\widehat{B}(z) = \sum_{j=0}^{\infty} b_j z^j = \left(\sum_{i=0}^{m_n} b_i^{(n)} z^i \right) \left(\sum_{k=0}^{\infty} c_k^{(n)} z^k \right) = \widehat{B^{(n)}}(z) \widehat{C^{(n)}}(z) = B^{(n)} * C^{(n)}(z).$$

Therefore

$$\left\| B - B^{(n)} \right\|_2 = \left\| B^{(n)} * C^{(n)} - B^{(n)} \right\|_2 = \left\| B^{(n)} * \left(C^{(n)} - (1, 0, 0, \cdots) \right) \right\|_2.$$

Hence

$$\left\| B - B^{(n)} \right\|_2 \leq \left\| B^{(n)} \right\|_1 \left\| C^{(n)} - (1, 0, 0, \cdots) \right\|_2.$$

It follows under our condition $(m_n(1 - c_0^{(n)}) \longrightarrow 0$ as $n \to \infty)$ that

$$\left\| B^{(n)} \right\|_1 \left\| C^{(n)} - (1, 0, 0, \cdots) \right\|_2 \longrightarrow 0.$$

COROLLARY 2.6. *If $b > 0$ then there is always a subproduct $\prod_{k=1}^{\infty} P_{n_k}$ for which the condition of the above theorem is satisfied, so that if $b > 0$ holds, then a subproduct $\prod_{k=1}^{\infty} | P_k(z) |^2$ has the same null sets as Lebesgue measure.*

PROOF. Put $k_1 = 1$ and $P_{k_1} = P_1$. Let m_1 be the degree of P_{k_1}. Since $b > 0$, $c_0^{(n)} \to 1$ as $n \to \infty$. Choose $k_2 > k_1$ such that $m_1(1 - c_0^{(k_2)}) \le \frac{1}{2}$. Consider $P_{k_1} \cdot P_{k_2+1}$. Suppose its degree is m_2. Choose $k_3 > k_2$ such that $m_2(1 - c_0^{(k_3)}) \le \frac{1}{4}$. Assume that we have chosen $k_1 < k_2 < \cdots < k_{l-1}$ such that for any $i, 1 \le i \le l-2$ if m_i is the degree of $P_{k_1} P_{k_2+1} \cdots P_{k_i+1}$, then

$$m_i(1 - c_0^{(k_{(i+1)})}) \le \frac{1}{2^i}.$$

Choose $k_l > k_{l-1}$ such that

$$m_{l-1}(1 - c_0^{(k_l)}) \le \frac{1}{2^{l-1}}.$$

Thus we have inductively chosen a sequence $k_1 < k_2 < k_3 < \cdots < k_i < \cdots$. Write $J_1 = P_{k_1}, J_2 = P_{k_2+1}, \cdots, J_n = P_{k_n+1}, \cdots$ and $R = \prod_{i=0}^{\infty} | J_i(z) |^2$. If γ_n denotes the constant term of $\prod_{i=n+1}^{\infty} J_i$, then it is easy to see that $\gamma_n > c_0^{(k_n)}$ so that $p_n(1 - \gamma_{n+1}) \le \frac{1}{2^n}$, where p_n is the degree of $\prod_{i=1}^{n} J_i$. By the theorem above the Riesz product $R = \prod_{i=1}^{\infty} | J_i(z) |^2$ is equivalent to the Lebesgue measure. This completes the proof of the corollary. □

Assume that b is positive. Consider $L(z) = \prod_{j=m+1}^{n} |P_j(z)|^2$. If $d_k(m,n) = d_k$ is the coefficient of z^k in $\prod_{j=m+1}^{n} P_j$, then for $k > 0$, the coefficient of z^k in $L(z)$ is in absolute value

$$\left| \sum_{j \ge k} d_j \overline{d_{j-k}} \right| \le \left(\sum_{j \ge k} | d_j |^2 \right)^{\frac{1}{2}} \le (1 - d_0^2)^{\frac{1}{2}}.$$

Under the assumption that b is positive we can make this coefficient (which depends on m and n) as small as we please by choosing m large. We have proved:

THEOREM 2.7. If b is positive, the generalized Riesz products $\mu_n \overset{def}{=} \prod_{n+1}^{\infty} | P_i |^2$, $n = 1, 2, \cdots$ converge weakly to the Lebesgue measure on S^1 as $n \to \infty$.

We do not know if the conclusion of Theorem 2.7 always holds when b is zero, but such generalized Riesz products form an important class of measures and will be discussed in section 5.

REMARK 2.8. The weak dichotomy theorem (Theorem 2.2) is rather weak in the sense that no information can be garnered about μ, such as absolute continuity or singularity, when b is zero. Consider the classical Riesz product

$$\mu = \prod_{j=1}^{\infty} | \cos(\theta_j) + \sin(\theta_j)z^{n_j} |^2, \ \frac{n_{j+1}}{n_j} \ge 3, \ 0 < \theta_j < \frac{\pi}{2}, \ j = 1, 2, \cdots$$

It is known to be absolutely continuous if $\sum_{j=1}^{\infty} \cos^2(\theta_j) \sin^2(\theta_j)$ is finite and singular otherwise. Clearly the condition for absolute continuity is satisfied with $\cos(\theta_j) = \frac{1}{j}, j = 1, 2, \cdots$ and also with $\cos(\theta_j) = \sqrt{1 - \left(\frac{1}{j}\right)^2}$, $j = 1, 2, \cdots$. In the first case the product of the constant terms is zero, while in the second case it is positive. This defect is rectified if we replace the polynomials P_j with their outer parts, as discussed in the next section.

3. Outer Polynomials and Mahler Measure

Let

(1)
$$\mu = \prod_{j=1}^{\infty} \mid P_j(z) \mid^2$$

be a generalized Riesz product. Let μ_a denote the part of μ absolutely continuous with respect to dz. We write $\frac{d\mu}{dz}$, to mean $\frac{d\mu_a}{dz}$. In this section we use the classical prediction theoretic ideas to evaluate $\exp\left(\int_{S^1} \log\left(\frac{d\mu}{dz}\right) dz\right)$ a quantity which we call the Mahler measure of μ (denoted by $M(\mu)$) with respect to the Lebesgue measure.

We will prove:

THEOREM 3.1.
$$\int_{S^1} \log \frac{d\mu_a}{dz} dz = \lim_{n\to\infty} \int_{S^1} \log \prod_{j=1}^{n} \mid P_j(z) \mid^2 dz.$$

Note that the theorem is false if we drop the log on both sides of the equation, for then the right hand side is always one, while the left hand side is zero for μ singular to Lebesgue measure. For the proof we begin by recalling Beurling's inner and outer factors for the case of polynomials and the expression for one step 'prediction error', namely the quantity:

$$\inf_{q\in\mathcal{Q}} \int_{S^1} \mid 1 - q(z) \mid^2 \mid P(z) \mid^2 dz,$$

where $P(z)$ is an analytic trigonometric polynomial with $L^2(S^1, dz)$ norm 1 and non-zero constant term. $\mathcal{Q}$ is the class of all analytic trigonometric polynomials with zero constant term. To this end we have to bring into consideration the zeros of polynomials $P_j, j = 1, 2, \cdots$. Consider the kth polynomial of the generalized Riesz product product $\prod_{k=1}^{\infty} |P_k|^2$. Suppressing the index k, it is of the type:

$$P(z) = a_0 + a_1 z + \cdots + a_m z^m.$$

assuming that it is of degree m. Let
$$A = \{a : P(a) = 0, \mid a \mid < 1\},$$
$$B = \{b : P(b) = 0, \mid b \mid = 1\},$$
$$C = \{c : P(c) = 0, \mid c \mid > 1\}.$$

Then
$$P(z) = a_m \prod_{a\in A}(z-a) \prod_{b\in B}(z-b) \prod_{c\in C}(z-c)$$
$$= \prod_{a\in A}\frac{(z-a)}{(1-\bar{a}z)} a_m \prod_{a\in A}(1-\bar{a}z) \prod_{b\in B}(z-b) \prod_{c\in C}(z-c).$$

Write
$$I(z) = \bar{\gamma} \prod_{a\in A}\frac{(z-a)}{(1-\bar{a}z)},$$
$$O(z) = \gamma a_m \prod_{a\in A}(1-\bar{a}z) \prod_{b\in B}(z-b) \prod_{c\in C}(z-c).$$

where γ is a constant of absolute value 1 such that the constant term of $O(z)$ is positive, while $\overline{\gamma}$ is the complex conjugate of γ. We have,

$$P(z) = I(z)O(z).$$

Note that for $z \in S^1$, $| I(z) | = 1$, $| P(z) | = | O(z) |$. The function $O(z)$ is non-vanishing inside the unit disc. The functions I and O are Beurling's inner and outer parts of the polynomial P. Note that, since constant term of P is non-zero, the degree of O is same as that of P and that $O(0) = $ constant term of $O \geq P(0) = a_0$. Recall that outer functions in H^2 are precisely those functions f in H^2 for which the functions $z^n f, n \geq 0$ span H^2 in the closed linear sense. Hence, if f is an outer function in H^2, then the closed linear span of $\{z^n f, n \geq 1\}$ is zH^2. The orthogonal projection of $O(z)$ on zH^2 is $O(z) - O(0)$ where $O(0)$ is the constant term of $O(z)$ which we denote by α. Note that

$$|\alpha| = \left| a_m \prod_{b \in B} b \prod_{c \in C} c \right| = |a_m| \prod_{c \in C} |c|.$$

We have

$$|\alpha|^2 = \int_{S^1} |\alpha|^2 \, dz = \int_{S^1} \left| (O(z) - (O(z) - \alpha) \right|^2 dz$$

$$= \int_{S^1} \left| 1 - \frac{(O(z) - \alpha)}{O(z)} \right|^2 |O(z)|^2 dz$$

$$= \inf_{q \in \mathcal{Q}} \int_{S^1} \left| 1 - q(z) \right|^2 |O(z)|^2 dz,$$

$$= \inf_{q \in \mathcal{Q}} \int_{S^1} \left| 1 - q(z) \right|^2 |P(z)|^2 dz$$

where the infimum is taken over the class $\mathcal{Q}$ of all analytic trigonometric polynomials q with zero constant term. Thus $\frac{O(z) - \alpha}{O(z)}$ is the orthogonal projection of the constant function 1 on the closed linear span of $\{z^n, n \geq 1\}$ in $L^2(S^1, |P(z)|^2 dz)$.

LEMMA 3.2. *If λ is a probability measure on S^1 such that $d\nu = | P(z) |^2 \, d\lambda$ is again a probability measure then*

$$| \alpha |^2 \geq \inf_{q \in \mathcal{Q}} \int_{S^1} | 1 - q(z) |^2 \, d\nu.$$

PROOF. If $O(z)$ has no zeros on the unit circle then $\dfrac{O(z) - \alpha}{O(z)}$ is analytic on the closed unit disk. The partial sums of the power series of this function converge to it uniformly on the unit circle. Let $q_k, k = 1, 2, \cdots$ be the sequence of these partial sums. Then

$$| \alpha |^2 = \int_{S^1} \left| 1 - \frac{O(z) - \alpha}{O(z)} \right|^2 |O(z)|^2 d\lambda$$

$$= \int_{S^1} \left| 1 - \frac{O(z) - \alpha}{O(z)} \right|^2 d\nu$$

$$= \lim_{k \to \infty} \int_{S^1} \left|1 - q_k\right|^2 d\nu \geq \inf_{q \in \mathcal{Q}} \int_{S^1} \left|1 - q(z)\right|^2 d\nu$$

This conclusion remains valid even if $O(z)$ has zeros on the circle but the proof is slightly different. For fixed r, $0 \leq r < 1$, the function $\frac{O(z) - \alpha}{O(rz)}$ is analytic on the closed unit disk, so the partial sums of its power series converge to it uniformly on S^1. Now for any fixed real θ, $\frac{z - e^{i\theta}}{rz - e^{i\theta}}$ remains bounded by 2 for $z \in S^1$ and $0 \leq r < 1$, and converges to 1 as $r \to 1$, for $z \neq e^{i\theta}$. Therefore $z \neq \theta$, $\frac{O(z)}{O(rz)} \to 1$ boundedly as $r \to 1$, whence

$$\left|1 - \frac{O(z) - \alpha}{O(rz)}\right|^2 \left|O(z)\right|^2 \to |\alpha|^2$$

boundedly as $r \to 1$. It is easy to see from this that

$$\inf_{q \in \mathcal{Q}} \int_{S^1} \left|1 - q\right|^2 d\nu \leq \lim_{r \to 1} \int_{S^1} \left|1 - \frac{O(z) - \alpha}{O(rz)}\right|^2 d\nu = |\alpha|^2 .$$

This proves the lemma. $\square$

Consider now the polynomials P_k, $k = 1, 2, \cdots$ and the associated finite products $\prod_{k=1}^n P_k$, $n = 1, 2, 3, \cdots$. Let $A_k, B_k, C_k, I_k, O_k, \alpha_0^{(k)}$ have the obvious meaning: they are for P_k what A, B, C, I, O, α are for P. Note that the inner and outer parts of $\prod_{k=1}^n P_k$ are $\prod_{k=1}^n I_k$ and $\prod_{k=1}^n O_k$ respectively and the constant term of the outer part is $\prod_{k=1}^n \alpha_0^{(k)}$. Note that $\prod_{k=1}^{\infty} \alpha_0^{(k)} \overset{\text{def}}{=} \beta \geq \prod_{k=1}^{\infty} a_0^{(k)} = b$, so if b is positive, then so is β. On the other hand, the positivity of β does not in general imply positivity of b as shown by the case of classical Riesz product (see remark 6.8 below).

To prove Theorem 3.1 We apply the lemma above to

$$P = \prod_{k=1}^n P_k(z)$$

and

$$\lambda = \prod_{k=n+1}^{\infty} |P_k(z)|^2,$$

Note that

$$d\nu = \left(\prod_{k=1}^n |P_k(z)|^2\right) d\lambda = d\mu.$$

We see that for any n

$$\inf_{q \in \mathcal{Q}} \int_{S^1} \left|1 - q\right|^2 d\mu \leq \prod_{k=1}^n |\alpha_0^{(k)}|^2,$$

whence

$$\inf_{q \in \mathcal{Q}} \int_{S^1} \left|1 - q\right|^2 d\mu \leq \prod_{k=1}^{\infty} |\alpha_0^{(k)}|^2 .$$

To prove the above inequality in the reverse direction we note that by Szegö's theorem as generalized by Kolmogorov and Krein (K. H. Hoffman [16]) that

$$\exp\left\{\int_{S^1}\log\left(\frac{d\mu_a}{dz}\right)dz\right\}=\inf_{q\in\mathcal{Q}}\int_{S^1}\mid 1-q(z)\mid^2 d\mu.$$

Denote this infimum by l. Then, given $\epsilon > 0$, there is a polynomial $q_0 \in \mathcal{Q}$ such that

$$l\le\int_{S^1}\mid 1-q_0\mid^2 d\mu < l+\epsilon,$$

whence for large enough n

$$\int_{S^1}\left|1-q_0\right|^2\prod_{k=1}^{n}|P_k|^2\,dz < l+\epsilon.$$

Since

$$\left|\prod_{k=1}^{n}\alpha_0^{(k)}\right|^2=\inf_{q\in\mathcal{Q}}\int_{S^1}\mid 1-q\mid^2\prod_{k=1}^{n}\mid P_k(z)\mid^2\,dz,$$

we see that $\mid\alpha_0^{(1)}\cdot\alpha_0^{(2)}\cdots\alpha_0^{(n)}\mid^2\le l+\epsilon$. Since ϵ is arbitrary positive real number, and $\mid\alpha_0^{(k)}\mid< 1$ for all k, we have

$$\prod_{k=1}^{\infty}\mid\alpha_0^{(k)}\mid^2\le l.$$

We also note that

$$\prod_{j=1}^{n}\mid\alpha_0^{(j)}\mid=\exp\left\{\int_{S^1}\log(\mid\prod_{j=1}^{n}P_j(z)\mid)dz\right\}$$

Thus we have proved (see [22]):

$$\prod_{k=1}^{\infty}\mid\alpha_0^{(k)}\mid^2=\exp\left\{\int_{S^1}\log\left(\frac{d\mu_a}{dz}\right)dz\right\}$$

$$=\lim_{n\to\infty}\exp\left\{\int_{S^1}\log(\prod_{j=1}^{n}\mid P_j(z)\mid^2)dz\right\}.$$

which is indeed theorem 3.1 with some additional information.

COROLLARY 3.3. If each $P_i, i = 1, 2, \cdots$ is outer, then $\log\left(\frac{d\mu}{dz}\right)$ has finite integral if and only if β is positive.

REMARK 3.4.   i) For the trigonometric polynomial $P(z) = \cos\theta_j + \sin\theta_j z^{n_j}$ appearing in the classical Riesz product of remark 2.8, its outer part has the constant term $\max\{\cos\theta_j, \sin\theta_j\}$ and the condition $\sum_{j=1}^{\infty}\cos^2\theta_j\sin^2\theta_j < \infty$ is equivalent to the condition $\prod_{j=1}^{\infty}\max\{\cos\theta_j, \sin\theta_j\}$ is positive. The additional information we have now is that in case μ is absolutely continuous with respect to dz, $\log\frac{d\mu}{dz}$ has finite integral.

ii)   Using a deep result of Bourgain [**10**], the first author has shown recently [**3**] that if the cutting parameter $m_n, n = 1, 2, \cdots$ of a rank one transformation T satisfies $\frac{m_n}{n^\beta} < K$ for all n for some constant K and some $\beta \in (0, 1]$, then the Mahler measure of the spectrum of T is zero.

iii) It is to be noted that if each P_n is outer and the product $\prod_{k=1}^\infty P_k(0)$ is non-zero, then formal expansion f of $\prod_{k=1}^\infty P_k(z)$ is an outer function. Indeed the Mahler measure of $\mid f \mid^2$ is $\mid f(0) \mid^2$, so f can not admit a non-trivial inner factor. Also the measure $1 \cdot dz$ can be expressed as a generalized Riesz product only by choosing each $P_n = 1$, for if any of the P_n is not the constant equal to 1, then its normalized outer part will have constant term less than one, which will force the Mahler measure of $1 \cdot dz$ to be less than 1, which is false. It is not known if the measure $cdz + d\delta_1, c, d > 0, c + d = 1$ can be expressed as a generalized Riesz product, where δ_1 denotes the Dirac measure at 1.

iv)   Let μ be a probability measure on S^1, and let q be a natural number. We contract the measure to the arc $A = \{z : z = \exp\{i\theta\}, 0 \le \theta < \frac{2\pi}{q}\}$, namely we consider the measure ν_1 supported on this arc given by $\nu_1(B) = \mu(z^q : z \in B), B \subset A$. We write similarly $\nu_j(C) = \nu_1(\exp\{-\frac{2\pi ij}{q}\}C), C \subset \exp\{\frac{2\pi ij}{q}\}A$. Let $\Pi_q(\mu) = \frac{1}{q}\sum_{j=1}^q \nu_j$. It can be verified that if $\mu = \mid P(z) \mid^2 dz$, then

$$\Pi_q(\mu) = \mid P(z^q) \mid^2 dz,$$

from which we conclude that if $\mu = \Pi_{k=1}^\infty \mid P_k(z) \mid^2$ then

$$\Pi_q(\mu) = \Pi_{k=1}^\infty \mid P_k(z^q) \mid^2 .$$

We see immediately that the Mahler measure of a generalized Riesz product is invariant under the application of Π_q for any q.

4. A Formula for Radon Nikodym Derivative

Consider two generalized Riesz products μ and ν based on polynomials $P_j, j = 1, 2, \cdots$ and $Q_j, j = 1, 2, \cdots$ where ν is continuous except for a possible mass at 1. Under suitable assumptions we prove the formula:

$$\frac{d\mu}{d\nu} = \lim_{n\to\infty} \frac{\prod_{j=1}^n \mid P_j \mid^2}{\prod_{j=1}^n \mid Q_j \mid^2},$$

in the sense of $L^1(S^1, \nu)$ convergence.

Let σ and τ be two measures on the circle. Then, by Lebesgue decomposition of σ with respect to τ, we have

$$\sigma = \frac{d\sigma}{d\tau}d\tau + \sigma_s,$$

where σ_s is singular to τ and $\frac{d\sigma}{d\tau}$ is the Radon-Nikodym derivative. In the case of two Riesz products $\mu = \prod_{j=1}^\infty \mid P_j \mid^2$ and $\nu = \prod_{j=1}^\infty \mid Q_j \mid^2$, we are able to see that their affinities, namely the ratios $\frac{\prod_{j=1}^n |P_j|}{\prod_{j=1}^n |Q_j|}, n = 1, 2, \cdots$, converge in L^1 to $\sqrt{\frac{d\mu}{d\nu}}$, assuming that ν has no point masses except possibly at 1. This result extends a theorem of G. Brown and W. Moran [**8**]. Let δ_1 denote the unit mass at one. We have (see [**3**])

THEOREM 4.1. *Let* $\mu = \prod_{j=0}^{\infty} | P_j |^2$, $\nu = \prod_{j=0}^{\infty} | Q_j |^2$ *be two generalized Riesz products. Let*

$$\mu_n = \prod_{j=n+1}^{\infty} | P_j |^2, \ \nu_n = \prod_{j=n+1}^{\infty} | Q_j |^2$$

Assume that

(1) $\nu = \nu' + b\delta_1$, ν' *is continuous measure,* $0 \le b < 1$.
(2) $\prod_{j=0}^{n} | P_j |^2 \, d\nu_n \longrightarrow \mu$ *weakly as* $n \longrightarrow \infty$
(3) $\prod_{j=0}^{n} | Q_j |^2 \, d\mu_n \longrightarrow \nu$ *weakly as* $n \longrightarrow \infty$

Then the finite products $R_n = \prod_{k=1}^{n} \left| \frac{P_k(z)}{Q_k(z)} \right|, n = 1, 2, \cdots$ *converge in* $L^1(S^1, \nu)$ *to* $\sqrt{\frac{d\mu}{d\nu}}$.

To prove this we need the following proposition.

PROPOSITION 4.2. *The sequence* $\prod_{j=0}^{n} \left| \frac{P_j(z)}{Q_j(z)} \right|, n = 1, 2, \cdots$ *converges weakly in* $L^2(S^1, \nu)$ *to* $\sqrt{\frac{d\mu}{d\nu}}$.

PROOF. Put $f = \sqrt{\frac{d\mu}{d\nu}}$ and let n be a positive integer. Now

$$\int_{S^1} R_n^2 d\nu = \int_{S^1} \prod_{j=1}^{n} | P_j |^2 d\nu_n \to \int_{S^1} d\mu = 1$$

by assumption (2). Hence $\int_{S^1} R_n^2 d\nu, n = 1, 2, \cdots$ remain bounded. Thus, the weak closure of $R_n(z), n = 1, 2, \cdots$ in $L^2(S^1, \nu)$ is not empty.

We show that this weak closure has only one point, namely, $\sqrt{\frac{d\mu}{d\nu}}$. Indeed, let g be a weak subsequential limit, say, of $R_{n_j}(z), j = 1, 2, \cdots$. Then, for any continuous positive function h, we have, by judicious applications of Cauchy-Schwarz inequality,

$$\left(\int_{S^1} f(z)h(z)d\nu(z) \right)^2 = \left(\int_{S^1} h(z)R_{n_j}(z)\frac{1}{R_{n_j}}\sqrt{\frac{d\mu}{d\nu}}d\nu(z) \right)^2$$

$$\le \left(\int_{S^1} h(z)R_{n_j}(z)d\nu(z) \right)\left(\int_{S_1} h(z)R_{n_j}(z)\frac{1}{R_{n_j}^2(z)}\frac{d\mu}{d\nu}d\nu(z) \right)$$

$$\le \left(\int_{S^1} h(z)R_{n_j}(z)d\nu(z) \right)\left(\int_{S^1} h(z)\frac{1}{R_{n_j}(z)}d\mu \right)$$

$$\le \int_{S^1} h(z)R_{n_j}(z)d\nu(z)\left(\int_{S^1} h(z)d\mu \right)^{\frac{1}{2}}\left(\int_{S^1} h(z)\frac{d\mu}{R_{n_j}^2(z)} \right)^{\frac{1}{2}}$$

$$\le \left(\int_{S^1} h(z)R_{n_j}(z)d\nu(z) \right)\left(\int_{S^1} h(z)d\mu \right)^{\frac{1}{2}}\left(\int_{S^1} h(z) \mid \prod_{k=1}^{n_j} Q_k \mid^2 d\mu_{n_j} \right)^{\frac{1}{2}}$$

Letting $j \to +\infty$, from our assumption (3), we get

$$(2) \qquad \left(\int_{S^1} fh d\nu \right)^2 \le \left(\int_{S^1} hg d\nu \right) \left(\int_{S^1} h d\mu \right)^{\frac{1}{2}} \left(\int_{S^1} h d\nu \right)^{\frac{1}{2}}$$

But, since the space of continuous functions is dense in $L^2(\mu + \nu)$, we deduce from (2) that, for any Borel set B,

$$\left(\int_B f d\nu \right)^2 \le \left(\int_B g d\nu \right) \left(\int_B d\mu \right)^{\frac{1}{2}} \left(\int_B d\nu \right)^{\frac{1}{2}}.$$

By taking a Borel set E such that $\mu_s(E) = 0$ and $\nu(E) = 1$, we thus get, for any $B \subset E$,

$$\left(\int_B f d\nu \right)^2 \le \left(\int_B g d\nu \right) \left(\int_B f^2(z) d\nu \right)^{\frac{1}{2}} \left(\int_B d\nu \right)^{\frac{1}{2}}.$$

It follows from Martingale convergence theorem that:

$$f(z) \le g(z) \text{ for almost all } z \text{ with respect to } \nu.$$

Indeed, let $\mathcal{P}_n = \{A_{n,1}, A_{n,2} \cdots, A_{n,k_n}\}$, $n = 1, 2, \cdots$, be a refining sequence of finite partitions of E into Borel sets such that such that they tend to the partition of singletons. If $\{x\} = \cap_{n=1}^\infty A_{n,j_n}$,

$$\left(\frac{1}{\mu(A_{n,j_n})} \int_{A_{n,j_n}} f d\nu \right)^2 \le$$

$$\left(\frac{1}{\mu(A_{n,j_n})} \int_{A_{n,j_n}} g d\nu \right) \left(\frac{1}{\mu(A_{n,j_n})} \int_{A_{n,j_n}} f^2(z) d\nu \right)^{\frac{1}{2}} \left(\frac{1}{\mu(A_{n,j_n})} \int_{A_{n,j_n}} d\nu \right)^{\frac{1}{2}}.$$

Letting $n \to \infty$ we have, by Martingale convergence theorem as applied to the theory of derivatives, for a.e $x \in E$ w.r.t. ν,

$$(f(x))^2 \le g(x) f(x), \text{ whence } f(x) \le g(x)$$

For the converse note that for any continuous positive function h we have

$$\int_{S^1} gh d\nu = \lim_{j \to +\infty} \int_{S^1} h(z) R_{n_j} d\nu$$

$$\le \lim_{j \to \infty} \left(\int_{S^1} h R_{n_j}^2 d\nu \right)^{\frac{1}{2}} \left(\int_{S^1} h d\nu \right)^{\frac{1}{2}}$$

$$\le \left(\int_{S^1} h d\mu \right)^{\frac{1}{2}} \left(\int_{S^1} h d\nu \right)^{\frac{1}{2}}.$$

As before we deduce that $g(z) \le f(z)$ for almost all z with respect to ν. Consequently, we have proved that $g = f$ for almost all z with respect to ν and this completes the proof of the proposition. $\square$

Proof of Theorem 4.1. We will show that $\beta_n \stackrel{\text{def}}{=} \int_{S^1} \mid R_n - f \mid d\nu \to 0$ as $n \to \infty$, where $f = \sqrt{\frac{d\mu}{d\nu}}$. Now,

$$\frac{d\mu}{d\nu} = R_n^2(z) \frac{d\mu_n}{d\nu_n} \text{ and } \sqrt{\frac{d\mu}{d\nu}} = R_n(z) \sqrt{\frac{d\mu_n}{d\nu_n}}$$

Put

$$f_n^2 = \frac{d\mu_n}{d\nu_n},$$

Then,

$$\int_{S^1} f_n^2 d\nu = \int_{S^1} \prod_{k=1}^{n} \mid Q_k \mid^2 d\mu_n \to \int_{S^1} d\nu = 1,$$

by assumption (3). The functions $f_n, n = 1, 2, \cdots$ are therefore bounded in $L^2(S^1, \nu)$. Hence, there exists a subsequence $f_{n_j} = \sqrt{\frac{d\mu_{n_j}}{d\nu_{n_j}}}, j = 1, 2, \cdots$ which converges weakly to some $L^2(S^1, \nu)$-function ϕ. We show that $0 \leq \phi \leq 1$ a.e (ν). For any continuous positive function h, we have

$$\left(\int_{S^1} h f_{n_j} d\nu \right)^2 \leq \left(\int_{S^1} h d\nu \right) \left(\int_{S^1} h f_{n_j}^2 d\nu \right)$$

$$\leq \left(\int_{S^1} h d\nu \right) \left(\int_{S^1} h \frac{d\mu_{n_j}}{d\nu_{n_j}} d\nu \right).$$

Hence, by letting j go to infinity combined with our assumption (3), we deduce that

$$\int_{S^1} h(z)\phi(z) d\nu \leq \int_{S^1} h(z) d\nu.$$

Since this hold for all continuous positive functions h, we conclude that $0 \leq \phi \leq 1$ for almost all z with respect to ν. Thus any subsequential limit of the sequence $f_n, n = 1, 2, \cdots$ assumes values between 0 and 1. Now, for any subsequence $n_j, j = 1, 2, \cdots$ over which $f_{n_j}, j = 1, 2, \cdots$ has a weak limit , from our assumption (2) combined with Cauchy-Schwarz inequality, we have

$$\left(\int_{S^1} |R_{n_j} - f| d\nu \right)^2 = \left(\int_{S^1} |R_{n_j} - R_{n_j} f_{n_j}| d\nu \right)^2$$

$$= \left(\int_{S^1} R_{n_j} |1 - f_{n_j}| d\nu \right)^2$$

$$\leq \left(\int_{S^1} R_{n_j} |1 - f_{n_j}|^2 d\nu \right) \left(\int_{S^1} R_{n_j} d\nu \right)$$

$$\leq \left(\int_{S^1} R_{n_j} d\nu - 2 \int_{S^1} R_{n_j} f_{n_j} d\nu + \int_{S^1} R_{n_j} (f_{n_j})^2 d\nu \right) \left(\int_{S^1} R_{n_j} d\nu \right)$$

$$\leq \left(\int_{S^1} R_{n_j} d\nu - 2 \int_{S^1} f d\nu + \int_{S^1} R_{n_j} f_{n_j} . f_{n_j} d\nu \right) \left(\int_{S^1} R_{n_j} d\nu \right)$$

$$\leq \left(\int_{S^1} R_{n_j} d\nu - 2 \int_{S^1} f d\nu + \int_{S^1} f . f_{n_j} d\nu \right) \left(\int_{S^1} R_{n_j} d\nu \right)$$

Hence, letting j go to infinity,

$$\left(\lim_{j \to \infty} \int_{S^1} | R_{n_j} - f | d\nu \right)^2$$

$$\leq \int_{S^1} f d\nu - 2 \int_{S^1} f d\nu + \int_{S^1} f . \phi d\nu$$

$$\leq \int_{S^1} (\phi(z) - 1) f(z) d\nu(z).$$

$$\leq 0,$$

and this implies that $R_{n_j}, j = 1, 2, \cdots$ converges to f in $L^1(S^1, \nu)$ and the proof of the theorem is achieved. $\square$

REMARK 4.3. Notice that $\int_{S^1} \frac{d\mu}{d\nu} d\nu = 1$, implies the convergence of $\prod_{j=0}^{N} |R_j|$ to $\sqrt{\frac{d\mu}{dz}}$ in $L^2(d\nu)$, by virtue of the classical results on "when weak convergence implies strong convergence".

We further have [21]

COROLLARY 4.4. Two generalized Riesz products

$$\mu = \prod_{j=1}^{\infty} |P_j|^2, \qquad \nu = \prod_{j=1}^{\infty} |Q_j|^2$$

satisfying the conditions of Theorem 4.1 are mutually singular if and only if

$$\int_{S^1} \prod_{j=0}^{n} \left| \frac{P_j}{Q_j} \right| d\nu \to 0 \text{ as } n \to \infty.$$

5. A Conditional Strong Dichotomy and Other Discussion

An important class of generalized Riesz products is the one arising in the study of rank one transformations of ergodic theory [21]. Indeed much of the recent work on generalized Riesz products (including the present contribution) is motivated by or focussed on the question whether these generalized Riesz products are always singular to Lebesgue measure. For in the contrary case, a counter example to this belief, will in all probability yield an affirmative answer to an old problem of Banach as to whether there exists a measure preserving transformation on an atom free measure space with simple Lebesgue spectrum.

The kth polynomial in the generalized Riesz product arising in the study of measure preserving rank one transformation is of the type

$$P_k(z) = \frac{1}{\sqrt{m_k}}(1 + \sum_{j=1}^{m_k-1} z^{jh_{k-1}+a_1^{(k)}+\cdots+a_j^{(k)}}),$$

$$h_{k-1} = m_{k-2}h_{k-2} + \sum_{j=1}^{m_{k-2}} a_j^{(k)}, h_0 = 1,$$

where m_k's are the cutting parameter and $a_j^{(k)}$'s are the spacers of the rank one transformation under consideration. It is easy to see that the partial products $\prod_{j=1}^k P_j, j = 1, 2, \cdots$ converge weakly to zero in $L^2(S^1, dz)$. These generalized Riesz products $\prod_{j=1}^\infty |P_j|^2$ have the property that the sequence of their tails $\prod_{j=n+1}^\infty |P_j|^2$, $n = 1, 2, \cdots$ converges weakly to the Lebesgue measure. In the rest of this section we will assume that the generalized Riesz products have this additional property, although it is not assumed that they arise from rank one transformations as above.

DEFINITION 5.1. A generalized Riesz product $\mu = \prod_{j=1}^\infty |P_j|^2$ is said to be of class (L) if for each sequence $k_1 < k_2 < \cdots$ of natural numbers, the tail measures $\prod_{j=n+1}^\infty | P_{k_j} |^2, n = 1, 2, \cdots$ converge weakly to Lebesgue measure.

PROPOSITION 5.2. If the generalized Riesz product $\mu = \prod_{j=1}^\infty | P_j |^2$ is of class (L) then the partial products $\prod_{j=1}^n | P_j |, n = 1, 2, \cdots$ converge in $L^1(S^1, dz)$ to $\sqrt{\frac{d\mu}{dz}}$, and the convergence is almost everywhere (w.r.t dz) over a subsequence.

PROOF. In Theorem 4.1 we put $Q_j(z) = 1$ for all j, so that ν is the Lebesgue measure on S^1. The first conclusion follows from theorem 4.1. The second conclusion follows since L^1 convergence implies convergence a.e over a subsequence. □

The following formula follows immediately from this:

COROLLARY 5.3. Let a generalized Riesz product μ be of class (L). Let $\mathcal{K}_1, \mathcal{K}_2$ be two disjoint subsets of natural numbers and let $\mathcal{K}_0$ be their union. Let μ_1, μ_2 and μ_0 be the generalized Riesz subproducts of μ over $\mathcal{K}_1, \mathcal{K}_2$, and $\mathcal{K}_0$ respectively. Then we have:

(1) $$\frac{d\mu_0}{dz} = \frac{d\mu_1}{dz}\frac{d\mu_2}{dz},$$

where equality is a.e. with respect to the measure dz.

Let $\mu = \prod_{j=1}^\infty | P_j |^2$ be a generalized Riesz product of class (L). We assume that the polynomials $P_j, j = 1, 2, \cdots$ are outer. Write $S_n = \prod_{j=1}^n P_j$, and let $\phi_n = \frac{S_n}{|S_n|}$, a function of absolute value one. The functions $\phi_n, n = 1, 2, \cdots$ admit weak* limits as functions in $L^\infty(S^1, dz)$. By Theorem 4.1 if β is positive then there is a unique nowhere vanishing weak star limit $\frac{f}{|f|}$ which is indeed also a limit in $L^1(S^1, dz)$. On the other hand consider the simplest classical Riesz product given by

$$\mu = \prod_{j=1}^\infty \frac{1}{\sqrt{2}}\left|1 + z^{n_j}\right|^2, \frac{n_j}{n_{j-1}} \geq 3,$$

which is singular to the Lebesgue measure on S^1. Since $1 + e^{it} = \mid 1 + e^{it} \mid e^{i\frac{t}{2}}$, we see that

$$\phi_k(e^{it}) = e^{i(\sum_{j=1}^{k} n_j)\frac{t}{2}} \longrightarrow 0$$

in the weak$*$ topology as $k \to \infty$, by virtue of the Riemann-Lebesgue lemma. However the following conditional strong dichotomy holds.

THEOREM 5.4. *If the functions* $\phi_n, n = 1, 2, \cdots$ *admit a weak star limit* ϕ *in* $L^\infty(S^1, dz)$ *which is non-vanishing a.e.* (dz) *on the set* $\{z : \frac{d\mu}{dz} > 0\}$, *then* μ *is either singular to Lebesgue measure, or its absolutely continuous part has positive Mahler measure.*

PROOF. Let $f = \sqrt{\frac{d\mu}{dz}}$. Fix an integer k, then

$$\left| \int_{S^1} z^k S_n(z) - z^k \phi_n(z) f(z) dz \right| = \left| \int_{S^1} z^k \phi_n(z)(|S_n(z)| - f(z)) dz \right|$$

$$\leq \int_{S^1} ||S_n(z)| - f(z)| dz \to 0 \text{ as } n \to \infty$$

by Theorem 4.1. On the other hand, by assumption, since $f \in L^1(S^1, dz)$,

$$\int_{S^1} z^k \phi_n(z) f(z) dz \to \int_{S^1} z^k \phi(z) f(z) dz$$

Now for $k < 0$, $\int_{S^1} z^k S_n(z) dz = 0$, so for $k < 0$,

$$\int_{S^1} z^k \phi(z) f(z) dz = 0$$

By F and M Riesz theorem ϕf is either the zero function or a non-zero function in H^1. In the first case f is the zero function a.e dz, since ϕ is assumed to be non-vanishing a.e. (dz) on the set where f is positive. In the second case $\mid \phi f \mid$ has an integrable log, which implies that f has an integrable log. Thus f is either the zero function or has an integrable log. $\square$

REMARK 5.5. The proof of Theorem 5.4 in fact shows that any weak limit ϕ of ϕ_n's either never vanishes or vanishes on the set where f does not vanish. Suppose S_n has degree m_n and let $z_1, z_2, \cdots, z_{m_n}$ be the zeros of S_n, counting multiplicity. Since S_n is outer, $\mid z_j \mid \geq 1, j = 1, 2, \cdots, m_n$, whence $\mid \frac{1}{z_j} \mid \leq 1, j = 1, 2, \cdots, m_n$, so the function $(1 - \frac{z}{z_j}), j = 1, 2, \cdots, m_n$ has continuous arguments except when z_j has absolute value 1, in which case z_j is the only point where the argument is not defined, and a continuous argument can be defined at all other points. Thus the polynomials $B_n \stackrel{def}{=} \prod_{j=1}^{m_n}(1 - \frac{z}{z_j})$ admits a continuous argument, denoted by A_n, except at points z_j with $\mid z_j \mid = 1$. If $\lim_{n\to\infty} A_n(z)$ exists at almost every point where $\frac{d\mu}{dz}$ is positive, then it is clear that $\phi_n, n = 1, 2, \cdots$ admit a weak limit not vanishing a.e. on the set $\{z : \frac{d\mu}{dz} > 0\}$. Theorem 5.4 is a soft version in the context of generalized Riesz product of similar results in the context of lacunary series (see Theorem 1.1 and Theorem 6.4 in [**27**, T1, p.202]).).

View the functions $S_n(z), n = 1, 2, \cdots$ as outer analytic functions on the open unit disk. From weak dichotomy theorem 1.1, we immediately see that $S_n, n = 1, 2, \cdots$ converge uniformly on every compact subset of the open unit disk to a function which is non-zero and in H^1 if β is positive and the identically zero function if β is zero. We have, using notation from H^p theory, with $0 \le r < 1$,

$$\lim_{r \to 1} \frac{1}{2\pi} \int_0^{2\pi} \left| S_n(re^{i\theta}) \right| d\theta = \frac{1}{2\pi} \int_0^{2\pi} \left| S_n(e^{i\theta}) \right| d\theta$$

We prefer to write this in our notation as

$$\lim_{r \to 1} \int_{S^1} |S_n(rz)| \, dz = \int_{S^1} |S_n(z)| \, dz$$

Letting $n \to \infty$ we get

$$\lim_{n \to \infty} \left(\lim_{r \to 1} \int_{S^1} | S_n(rz) | \, dz \right) = \lim_{n \to \infty} \left(\int_{S^1} | S_n(z) | \, dz \right) = \int \sqrt{\frac{d\mu}{dz}} \, dz.$$

However, in general one can not interchange the order of taking limits and write this as

$$= \lim_{r \to 1} \lim_{n \to \infty} \int_{S^1} | S_n(rz) | \, dz,$$

for that would immediately establish the singularity of μ with respect to the Lebesgue measure when $\beta = 0$, which is false in general, see remark 6.7.

6. Non-Singular Rank One Maps and Flat Polynomials

In this section we will discuss generalized Riesz product in connection with spectral questions about non-singular and measure preserving rank one transformations. We will give necessary and sufficient conditions under which a generalized Riesz product is the maximal spectral type (up to possibly a discrete component) of a unitary operator associated with a rank one non-singular transformation and certain functions of absolute value one. We will expose in more detail and generality, possibly adding new perspective, the known connection of these questions (see [13], [14]) with problems about flat polynomials. Proposition 5.2 is particularly useful in this discussion.

Non-Singular Rank One Maps. Let T be a non-singular rank one transformation obtained by cutting and stacking [15]. This is done as follows. Let $\Omega_0 = \Omega_{0,0}$ denote the unit interval. At stage one of the construction we divide Ω_0 into m_1 pairwise disjoint intervals, $\Omega_{0,1}, \Omega_{1,1} \cdots, \Omega_{m_1-1,1}$, of lengths $p_{0,1}, p_{1,1}, \cdots, p_{m_1-1,1}$, respectively, each $p_{i,j}$ being positive. Obviously $\sum_{j=0}^{m_1-1} p_{j,1} = 1$. For each $j, 0 \le j \le m_1 - 2$, we stack $a_{j,1} \ge 0$ pairwise disjoint intervals of length $p_{j,1}$ on $\Omega_{j,1}$. Each interval is mapped linearly onto the one above it, except that $a_{j,1}$-th spacer is mapped linearly onto $\Omega_{j+1,1}, 0 \le j \le m_1 - 2$. We thus get a stack of certain height h_1, together with a map T which is defined on all intervals of the stack except the interval at the top of the stack. Note that if $p_{j,1} \ne p_{j+1,1}$ for some j, T_1 will not be measure preserving.This completes the first stage of the construction.

At the k-th stage we divide the stack obtained at the the $(k-1)$-th stage in the ratios

$$p_{0,k}, p_{1,k}, \cdots, p_{m_k-1,k}, \sum_{i=0}^{m_k-1} p_{i,k} = 1,$$

where each $p_{i,k}$ is positive. The spacers are added in the usual manner by which we mean that the spacers stacked above the j-th column are all of the same length which is the length of the top piece of the j-th column. The extension of T to the spacers is done linearly as usual. Note that the top of the spacers above the j-th column is mapped linearly onto the bottom of the $(j+1)$-th column, so that if $p_{j,k} \neq p_{j+1,k}$, T will not be measure preserving. Note that at the k-th stage the measure is defined only on the algebra Γ_k generated by the levels of the k-th stack, except the top piece. The resulting T, after all the stages of the construction are completed, is defined on the space X consisting increasing union of stack intervals. Let ν denote the Lebesgue measure defined on the σ-algebra Γ generated by $\cup_{n=1}^{\infty} \Gamma_n$. Note that $\prod_{j=1}^{k} p_{m_j-1,j}$ is the measure of the top piece of the column of height h_k at the end of kth stage of construction. We require that this goes to 0 as as $k \to 0$. This ensures that T is eventually defined for almost every point of Ω_0. Note that T is non-singular (see remark below), ergodic with respect to ν, and $\frac{d\nu \circ T}{d\nu}$ is constant on all but the top layer of every stack. If no spacers are added at every stage of the construction, the resulting transformation will be called non-singular odometer.

REMARK 6.1. The transformation T is non-singular in the sense that $m(T^{-1}(A))$ $=0$ whenever $m(A) = 0$. For each j, let $p_{l_j,j} = \max\{\{p_{i,j} : 0 \leq i \leq m_j - 1\}$.

Consider the case when $\prod_{j=1}^{\infty} p_{l_j,j} > 0$. Then $\sum_{j=0}^{\infty} (1 - p_{l_j,j}) < \infty$. Now $\lambda_k \stackrel{\text{def}}{=} \prod_{j=1}^{k} p_{l_j,j}$ is the length of the largest of subinterval of $[0,1)$ which appears as a level after the kth stage of construction. For each k, let I_k denote this level. Then $I_{k+1} \subset I_k$, the length of $I_{k+1} = \lambda_{k+1} = \lambda_k \times p_{l_{k+1},k+1}$ and $W \stackrel{\text{def}}{=} \cap_{k=1}^{\infty} I_k$ has positive length $= \prod_{j=1}^{\infty} p_{l_j,j}$. The Lebesgue measure of $(I_k - I_{k+1})$ is $\lambda_k - \lambda_{k+1}$. Write $L_k = (\cup_{j=-a_k}^{b_k} T^j (I_k - I_{k+1})) \cap [0,1)$, where $T^{a_k} \Omega_{0,k} = I_k$, and $b_k = h_k - 1 - a_k$. Then the Lebesgue measure of L_k is $1 - p_{l_{k+1},k+1}$, so by Borel-Canterlli Lemma the Lebesgue measure of $L \stackrel{\text{def}}{=} \limsup_{k \to \infty} L_k = \cap_{k=1}^{\infty} \cup_{j=k}^{\infty} L_j$ is zero. Now if $x \in [0,1) - L$, then x is in at most finitely many L_js. This means that either $x \in W$ or for some fixed $y \in W$ and for some fixed integer $n(x)$, $x = T^{n(x)} y$. Thus we see that when $\prod_{j=1}^{\infty} p_{l_j,j}$ is non-zero, T induces a dissipative transformation on $[0,1)$ which implies that T itself is dissipative in this case. There are two subcases: (i) if $l_j = 0$ for all j bigger than a fixed integer $N > 0$, then T is non-invertible and dissipative; W is the required wandering set which admits only finitely many negative iterates, but admits all positive iterates; (ii) in case $l_j \neq 0$ for infinitely many j, then T is invertible and dissipative, W admits pairwise disjoint iterates over all integers. Note that measure v is defined on the σ-algebra generated by levels of the stacks, and we really have a discrete measure space, and ergodicity holds.

In case $\prod_{j=1}^{\infty} p_{l_j,j} = 0$, then T is defined on an atomfree measure space and the ergodicity of T follows from the usual Lebesgue density argument.

Unitary Operators U_T and V_ϕ. Let ϕ be a function on X of absolute value 1 which is constant on all but the top layer of every stack. On $L^2(X, \Gamma, \nu)$ define

$$(U_T f)(x) = (\frac{d\nu \circ T}{d\nu}(x))^{1/2} f(Tx), f \in L^2(X, \Gamma, \nu)$$

$$(V_\phi f)(x) = (Vf)(x) = \phi(x) \cdot (U_T f)(x), f \in L^2(X, \Gamma, \nu).$$

U_T, and V are unitary operators, except when T is non-invertible, in which case U_T, V_ϕ are isometries isomorphic to the shift on l^2. The following argument, which is for the case when V is unitary, can be modified suitably to cover the case when it is an isometry.

$$(U_T^n f)(x) = (\frac{d\nu \circ T^n}{d\nu}(x))^{1/2} f(T^n x),$$

$$(V^n f)(x) = \prod_{j=0}^{n-1} \phi(T^j(x))(\frac{d\nu \circ T^n}{d\nu}(x))^{1/2} f(T^n x).$$

It is known [21] that the V has simple spectrum whose maximal spectral type (except possibly for some discrete part) is given by the generalized Riesz product

$$\prod_{j=1}^{\infty} p_{0,j} \mid P_j(z) \mid^2,$$

where

$$P_j(z) = 1 + c_{1,j}(\frac{p_{1,j}}{p_{0,j}})^{1/2} z^{-R_{1,j}} + \cdots + c_{m_j-1,j}(\frac{p_{m_j-1,j}}{p_{0,j}})^{1/2} z^{-R_{m_j-1,j}}$$

The constants $c_{i,j}, 1 \le i \le m_j - 1, j = 1, 2, \cdots$, are of absolute value 1. They are determined by ϕ. The exponent $R_{i,j}, 1 \le i \le m_j - 1, j = 1, 2, \cdots$, is the i-th return time of a point in $\Omega_{0,j}$ into $\Omega_{0,j-1}$. It is equal to

$$R_{i,j} = ih_{j-1} + a_{0,j} + a_{1,j} + \cdots + a_{i-1,j}, 1 \le i \le m_j - 1$$

We give the steps involved in proving this as it will allow us to make some needed observations. Write $Tf = f \circ T$. We have

$$(V^{-n} f)(\cdot) = T^{-n} \circ ((\prod_{j=0}^{n-1} \phi(T^j(\cdot)))^{-1}(\frac{d(\nu \circ T^n)}{d\nu}(\cdot))^{-1/2} f(\cdot))$$

$$= (\prod_{j=0}^{n-1} \phi(T^{j-n}(\cdot)))^{-1}(\frac{d\nu \circ T^n}{d\nu}(T^{-n}(\cdot)))^{-1/2} f(T^{-n}(\cdot)),$$

whence

$$(T^{-n} f)(\cdot) = \prod_{j=0}^{n-1} \phi(T^{j-n}(\cdot))(\frac{d(\nu \circ T^n)}{d\nu})^{1/2}(T^{-n}(\cdot))(V^{-n} f)(\cdot)$$

Let $\Omega_{0,k-1}$ denote the base of the stack of height h_{k-1} after $(k-1)$-th stage of construction. Let $\Omega_{0,k}, \Omega_{1,k}, \cdots, \Omega_{m_k-1,k}$ be the partition of $\Omega_{0,k-1}$ during the k-th stage of construction, and let $a_{i,k}$ denote the number of spacers put on the column with base $\Omega_{j,k}, 0 \le j < m_k - 1$. We have

$$\Omega_{0,k-1} = \cup_{j=0}^{m_k-1} T^{jh_{k-1}+\sum_{i=0}^{j-1} a_{i,k}}(\Omega_{0,k})$$

$$= \cup_{j=0}^{m_k-1} T^{R_{j,k}}(\Omega_{0,k})$$

where

$$R_{j,k} = jh_{k-1} + \sum_{i=0}^{j-1} a_{i,k}$$

= the j–th return time of a point in $\Omega_{0,k}$ into $\Omega_{0,k-1}$.

$$1_{\Omega_{0,k-1}} = \sum_{j=0}^{m_k-1} 1_{\Omega_{0,k}} \circ T^{-R_{j,k}},$$

$$= \sum_{j=0}^{m_k-1} c_{j,k} \left(\frac{d\nu \circ T^{R_{j,k}}}{d\nu} (T^{-R_{j,k}}) \right)^{1/2} (\cdot)(V^{-R_{j,k}} 1_{\Omega_{0,k}})(\cdot)$$

where $c_{j,k} = \prod_{j=0}^{R_{j,k}-1} \phi(T^{j-R_{j,k}}(\cdot))$, a constant of absolute value one. Note that the constants $c_{j,k}$'s can be preassigned and ϕ can be so defined that the above relation holds for all (j,k). We now observe that for $x \notin T^{R_{j,k}} \Omega_{0,k}$,

$$V^{-R_{j,k}} 1_{\Omega_{0,k}}(x) = 0,$$

and that for $x \in T^{R_{j,k}} \Omega_{0,k}$,

$$\frac{d\nu \circ T^{R_{j,k}}}{d\nu} (T^{-R_{j,k}}(x)) = \frac{p_{j,k}}{p_{0,k}}.$$

We thus have

$$1_{\Omega_{0,k-1}} = \sum_{j=0}^{m_k-1} c_{j,k} \left(\frac{p_{j,k}}{p_{0,k}} \right)^{1/2} (V^{-R_{j,k}} 1_{\Omega_{0,k}})(\cdot)$$

Let us normalize $1_{\Omega_{0,k}}$ and write

$$f_k = \left(\frac{1}{m(\Omega_{0,k})} \right)^{1/2} 1_{\Omega_{0,k}} = \left(\frac{1}{\prod_{j=1}^k p_{0,j}} \right)^{1/2} 1_{\Omega_{0,k}}$$

$$f_{k-1} = (p_{0,k})^{1/2} \left(1 + c_{1,k} \left(\frac{p_{1,k}}{p_{0,k}} \right)^{1/2} V^{-R_{1,k}} + \cdots + c_{m_k-1,k} \left(\frac{p_{m_k-1,k}}{p_{0,k}} \right)^{1/2} V^{-R_{m_k-1,k}} \right) f_k$$

Now $m(\Omega_{0,0}) = 1$ so $f_0 = 1_{\Omega_{0,0}}$. We have by iteration

$$f_0 = \left(\prod_{j=1}^k P_j(V) \right) f_k,$$

where

$$P_j(z) = (p_{0,j})^{1/2} \left(1 + c_{1,j} \left(\frac{p_{1,j}}{p_{0,j}} \right)^{1/2} z^{-R_{1,j}} + \cdots + c_{m_j-1,j} \left(\frac{p_{m_j-1,j}}{p_{0,j}} \right)^{1/2} z^{-R_{m_j-1,j}} \right)$$

Let $V^n = \int_{S^1} z^{-n} dE, n \in \mathbb{Z}$, be the spectral resolution of the unitary group $V^n, n \in \mathbb{Z}$, and let

$$(V^n f_k, f_k) = \int_{S^1} z^{-n} (E(dz) f_k, f_k) = \int_{S^1} z^{-n} d\sigma_k$$

where $\sigma_k(\cdot) = (E(\cdot) f_k, f_k)$

We therefore have for all integers l

$$(V^l f_0, f_0) = \int_{S^1} z^{-l} d\sigma_0 = \int_{S^1} z^{-l} \prod_{j=0}^k | P_j(z) |^2 \, d\sigma_k,$$

whence

$$d\sigma_0 = \prod_{j=1}^{k} |P_j(z)|^2 \, d\sigma_k$$

Now we will show, as in the measure preserving case [**21**], that σ_0 is the generalized Riesz product:

$$\sigma_0 = \prod_{j=1}^{\infty} |P_j(z)|^2 .$$

Let N_k denote the the set of integers consisting of zero together with the entry times of a point in $\Omega_{0,k}$ into $\Omega_{0,0}$ which are less than the height h_k of the kth stack. We have

$$f_0 = (\prod_{j=1}^{k} P_j(V))f_k = Q_k(V)f_k,$$

where

$$Q_k(z) = \prod_{j=1}^{k} P_j(z) \overset{\text{def}}{=} \sum_{j=0}^{h_k-1} q_j(k)z^j = \sum_{j \in N_k} q_j(k)z^j.$$

an expansion of the product of dissociated polynomials $P_1, P_2, \cdots, P_k$. Note that
(i) $|q_j| \leq 1$
(ii) $|q_r| \leq \prod_{j=1}^{k-1} p_{m_{j-1},j} \to 0$ as $k \to \infty$, for $h_k - h_{k-1} < r < h_k$

$$\int_{S^1} z^n |Q_k|^2 \, dz = \sum_{r-s+n=0} q_r(k)\overline{q_s(k)}$$

where $r, s \in N_k$.

Now fix $n \in \mathbb{Z}$ and let k be so large that the first return time for any $x \in \Omega_{0,k}$ back to $\Omega_{0,k}$ is bigger than $|n|$, i.e, k is so large that $h_k \geq |n|$. We actually choose k so large that $|n| < \frac{h_{k-1}}{2}$. If $r, s \in N_k$ then $r - s + n$ can never exceed or equal the second return of an $x \in \Omega_{0,k}$ back to $\Omega_{0,k}$ (under T or T^{-1}). Moreover there can be at most n^2 pairs (r, s) with $r, s \in N_k$ with $T^{r+n-s}\Omega_{0,k} \cap \Omega_{0,k} \neq \emptyset$. For suppose $n > 0$ and $T^{r+n-s}\Omega_{0,k} \cap \Omega_{0,k} \neq \emptyset$ and $r + n - s \neq 0$, $r, s \in N_k$. Then $r + n - s = u$ where u is the first return time of a point $x \in \Omega_{0,k}$ back to $\Omega_{0,k} \geq h_k$. $s = r + n - u$. Since n, r, s are less than h_k, $h_k \leq u$ and $s \geq 0$, we have $0 \leq s < n$ and $n - s + r = u > h_k$, so $r \geq h_k - (n - s) \geq h_k - n$. Thus there can be at most n^2 pairs (r, s) with $r, s \in N_k$ such that $T^{n+r-s}\Omega_{0,k} \cap \Omega_{0,k} \neq \emptyset$. Thus if $T^{n+r-s}\Omega_{0,k} \cap \Omega_{0,k} \neq \emptyset, r, s \in N_k$ then $n + r - s = 0$ except for at most n^2 pairs (r, s), $r, s \in N_k$. This in turn implies that $(V^{n+r-s}1_{\Omega_{0,k}}, 1_{\Omega_{0,k}}) = 0$ except when $n + r - s = 0$ and at most n^2 other pair (r,s), $r, s \in N_k$.

$$(V^n f_0, f_0) = (V^n Q_k(V)f_k, Q_k(V)f_k) = (V^n |Q_k|^2 (V)f_k, f_k)$$

$$= \sum_{n+r-s=0, r,s\in N_k} q_r\overline{q_s}(V^{n+r-s}f_k, f_k) + \sum_{1}$$

$$\sum_{r-s+n} q_r(k)\overline{q_s(k)} + \sum_{1}$$

where $\sum_1$ is a sum of at most n^2 terms of the type

$$q_r\overline{q_s}(V^{n+r-s}f_k, f_k), n + r - s \neq 0$$

Now $h_k - h_{k-1} < h_k - n \le r \le h_k - 1$, so that $\mid q_r(k) \mid \le \prod_{j=1}^{k-1} p_{m_j-1,j} \to 0$ as $k \to \infty$. Clearly then the sum $\sum_1$ goes to zero as $k \to \infty$ and the claim is proved.

REMARK 6.2. Let $p_{l_j,j}$ be as in remark 6.1. Note that if $\prod_{j=1}^{\infty} p_{l_j,j} > 0$ then T is dissipative, so V_ϕ has Lebesgue spectrum. For the subcase when, in addition, $l_j = 0$ for all but finitely many j, the Mahler measure of σ_0 is positive. If T is non-dissipative and $\sum_{j=-\infty}^{\infty} p_{m_j-1,j} p_{0,j} = \infty$ then one can adapt the method of I. Klemes and K. Reinhold [19] to show that σ_0 is singular to Lebesgue measure.

If $\prod_{j=1}^{\infty} p_{l_j,j} = 0$ and $V_\phi = U_T$, then it can be verified that $\sum_{k=1}^{\infty} \mid \hat{\sigma}_0(k) \mid^2 \ge \sum_{j=1}^{\infty} (1 - p_{l_j,j}) = \infty$. In addition if $m_k \le K < \infty$ for all k, then σ_0 is singular to Lebesgue measure.

Generalized Riesz Products of Dynamical Origin. Consider now the polynomials appearing in the above generalized Riesz product.

$$P_j(z) = (p_{0,j})^{1/2} (1 + c_{1,j} (\frac{p_{1,j}}{p_{0,j}})^{1/2} z^{-R_{1,j}} + \cdots + c_{m_j-1,j} (\frac{p_{m_j-1,j}}{p_{0,j}})^{1/2} z^{-R_{m_j-1,j}})$$

The exponent $R_{i,j}, 1 \le i \le m_j - 1, j = 1, 2, \cdots$, is the i-th return time of a point in $\Omega_{0,j}$ into $\Omega_{0,j-1}$. Also

$$R_{i,j} = ih_{j-1} + a_{0,j} + a_{1,j} + \cdots + a_{i-1,j}, 1 \le i \le m_j - 1$$

where h_{j-1} is the height of the tower after $(j-1)$-th stage of the construction is complete, and $a_{k,j}$ is the number of spacers on the k-th column, $0 \le k \le m_j - 2$. We observe that

(1) $h_1 = R_{m_1-1,1} + 1$,

(2) $R_{1,j} \ge h_{j-1} > R_{m_j-1,j-1}$,

(3) $R_{i+1,j} - R_{i,j} \ge h_{j-1}$.

These properties (1), (2), (3) of the powers $R_{i,j}, 1 \le i \le m_j - 1, j = 1, 2, \cdots$ indeed characterize generalized Riesz products which arise from nonsingular rank one transformations (together with a ϕ) in the above fashion. More precisely consider a generalized Riesz product

$$\prod_{j=1}^{\infty} \mid Q_j(z) \mid^2 .$$

where

$$Q_j(z) = \sum_{i=0}^{n_j} b_{i,j} z^{r_{i,j}}, \ b_{i,j} \ne 0, \ \sum_{i=0}^{n_j} \mid b_{i,j} \mid^2 = 1, \ \prod_{j=1}^{\infty} \mid b_{n_j,j} \mid = 0.$$

Define inductively:

$$h_0 = 1, h_1 = r_{n_1,1} + h_0, \cdots, h_j = r_{n_j,j} + h_{j-1}, j \ge 2$$

Note that $h_j > r_{n_j,j}$.

PROPOSITION 6.3. *Assume that for each* $j = 1, 2, \cdots$,

$$r_{1,j} \geq h_{j-1}, \; r_{i+1,j} - r_{i,j} \geq h_{j-1}$$

Then $r_{i,j}, h_j$, *satisfy* (1), (2) *and* (3). *The generalized product* $\prod_{j=1}^{\infty} | Q |^2$ *describes the maximal spectral type (up to possibly a discrete part) of a suitable* V_ϕ.

PROOF. That the $r_{i,j}, h_j$ satisfy (1), (2), (3) is obvious. The needed non-singular T is given by cutting parameters $p_{i,j} = | b_{i,j} |^2, i = 0, 1, \cdots, n_j, j = 1, 2, \cdots$, and spacers $a_{i-1,j} = r_{i,j} - r_{i-1,j} - h_{j-1}, 1 \leq i \leq n_j, j = 1, 2, \cdots$. The needed ϕ (which need not be unique) is given by constants $\frac{b_{i,j}}{|b_{i,j}|}, 0 \leq i \leq n_j, j = 1, 2, \cdots$. This proves the proposition.

$\square$

DEFINITION 6.4. A generalized Riesz product $\mu = \prod_{j=1}^{\infty} | Q_j(z) |^2$, where $Q_j(z) = \sum_{i=0}^{n_j} b_{i,j} z^{r_{i,j}}, b_{i,j} \neq 0, \sum_{i=0}^{n_j} | b_{i,j} |^2 = 1, \prod_{j=1}^{\infty} | b_{n_j,j} | = 0$, is said to be of dynamical origin if with

$$h_0 = 1, h_1 = r_{n_1,1} + h_0, \cdots, h_j = r_{n_j,j} + h_{j-1}, j \geq 2$$

it is true that for $j = 1, 2, \cdots$,

$$r_{1,j} \geq h_{j-1}, \; r_{i+1,j} - r_{i,j} \geq h_{j-1}$$

If, in addition, the coefficients $b_{i,j}$ are all positive, then we say that μ is of purely dynamical origin.

Flat Polynomials and Generalized Riesz Products.

LEMMA 6.5. *Given a sequence* $P_n = \sum_{j=0}^{m_n} a_{j,n} z^j, , n = 1, 2, \cdots$ *of analytic trigonometric polynomials in* $L^2(S^1, dz)$ *with non-zero constant terms and* $L^2(S^1, dz)$ *norm* 1, $\prod_{n=1}^{\infty} | a_{m_n,n} | = 0$, *there exist a sequence of positive integers* $N_1, N_2, \cdots$ *such that*

$$\prod_{j=1}^{\infty} | P_j(z^{N_j}) |^2$$

is a generalized Riesz product of dynamical origin.

PROOF. For each $j \geq 1$, let

$$P_j = \sum_{i=0}^{n_j} b_{i,j} z^{r_{i,j}}, b_{i,j} \neq 0, \; b_{0,j} > 0, \; \sum_{i=1}^{n_j} | b_{i,j} |^2 = 1.$$

Let $N_1 = 1$ and $h_1 = H_1 = r_{n_1,1} + 1$. Choose $N_2 \geq 2H_1 > 2r_{n_1,1}$. Then

$$N_2 \cdot r_{1,2} > h_1, N_2(r_{i+1,2} - r_{i,2}) > h_1.$$

Since $N_2 > 2r_{n_1,1}$ the polynomials $| P_1(z^{N_1}) |^2$ and $| P_2(z^{N_2}) |^2$ are dissociated. Consider now $P_1(z^{N_1})P_2(z^{N_2})$. Write $H_2 = N_1 r_{n_1,1} + N_2 r_{n_2,2} + h_1 > N_2 r_{n_2,2} + h_1 \stackrel{\text{def}}{=} h_2$. Choose $N_3 \geq 2H_2$. Then

$$N_3 \cdot r_{1,3} \geq h_2, N_3(r_{i+1,3} - r_{i,3}) > h_2.$$

Since $N_3 \geq 2H_2 > 2(N_1 r_{n_1,1} + N_2 r_{n_2,2})$ the polynomial $| P_3(z^{N_3}) |^2$ is dissociated from $| P_1(z^{N_1})P_2(z^{N_2}) |^2$. Proceeding thus we get $N_j, j = 1, 2, \cdots$ and polynomials $Q_j(z) = P_j(z^{N_j}), j = 1, 2, \cdots$ such that

(i) $\| Q_j \|_2 = 1$ (since $\| P_j \|_2 = 1$ and the map $z \to z^{N_j}$ is measure preserving.) (ii) the polynomials $| Q_j |^2, j = 1, 2, \cdots$ are dissociated, (iii) for each $j \geq 1$,

$$h_{j-1} < N_j r_{1,j}, \ h_{j-1} < N_j(r_{i+1,j} - r_{i,j})$$

Since the polynomials $Q_j, j = 1, 2, \cdots$ have $L^2(S^1, dz)$ norm 1 and their absolute squares are dissociated, the generalized Riesz product $\prod_{j=1}^{\infty} | P(z^{N_j}) |^2$ is well defined. Moreover, (iii) shows that the conditions for it to arise from a non-singular rank one T and a ϕ in the above fashion are satisfied. The lemma follows. $\quad\square$

An immediate application of this Lemma is the following:

THEOREM 6.6. *Let* $P_j = \sum_{i=0}^{m_j} a_{i,j} z^i, j = 1, 2, \cdots$ *be a sequence of analytic trigonometric polynomials of* $L^2(S^1, dz)$ *norm 1 satisfying for all* j, $| a_{m_j,j} | < \lambda < 1$, *and such that* $| P_j(z) | \to 1$ *a.e.* (dz) *as* $j \to \infty$. *Then there exists a subsequence* $P_{j_k}, k = 1, 2, \cdots$ *and natural numbers* $N_1 < N_2 < \cdots$ *such that the product* $\mu = \prod_{k=1}^{\infty} | P_{j_k}(z^{N_k}) |^2$ *is a generalized Riesz product of dynamical origin with* $\frac{d\mu}{dz} > 0$ *a.e.* (dz).

PROOF. Since $| P_j(z) | \to 1$ as $j \to \infty$ a.e. (dz), by Egorov's theorem we can extract a subsequence $P_{j_k}, k = 1, 2, \cdots$ such that the sets

$$E_k \stackrel{\text{def}}{=} \{z : | (1 - | P_{j_l}(z) |) | < \frac{1}{2^k} \ \forall \ l \geq k\}$$

increase to S^1 (except for a dz null set), and $\sum_{k=1}^{\infty}(1 - dz(E_k)) < \infty$. Write $Q_k = P_{j_k}$. Then for $z \in E_k$, $| (1 - | Q_k(z) |) | < \frac{1}{2^k}$. Note that $\prod_{k=1}^{\infty} | a_{m_{j_k},j_k} | = 0$. By lemma 6.5 we can choose $N_1, N_2, \cdots$ such that

$$\prod_{k=1}^{\infty} | Q_k(z^{N_k}) |^2$$

is a generalized Riesz product of dynamical origin. We show that $\lim_{L \to \infty} \prod_{k=1}^{L} | Q_k(z^{N_k}) |$ is nonzero a.e. (dz), which will imply, by proposition 5.2., that $\frac{d\mu}{dz} > 0$ a.e (dz).

Now the maps $S_k : z \to z^k, k = 1, 2, \cdots$ preserve the measure (dz), and since $\sum_{k=1}^{\infty} dz(S^1 - E_k) < \infty$ we have $\sum_{k=1}^{\infty} dz(S_{N_k}^{-1}(S^1 - E_k)) < \infty$. Let $F_k = S_{N_k}^{-1}(S^1 - E_k)$ and $F = \limsup_{k \to \infty} F_k = \cap_{k=1}^{\infty} \cup_{l=k}^{\infty} F_l$. Then $dz(F) = 0$, and if $z \notin F$, $z \notin S_{N_k}^{-1}(S^1 - E_k)$ hold for all but finitely many k, which in turn implies that $S_{N_k} z \in E_k$ for all but finitely many k. Thus, if $z \notin F$, then $| (1 - | Q_k(z^{N_k}) |) | < \frac{1}{2^k}$ for all but finitely many k. Also the set of points z for which some finite product $\prod_{k=1}^{L} | Q_k(z^{N_k}) |$ vanishes is countable. Clearly $\lim_{L \to \infty} \prod_{k=1}^{L} | Q_k(z^{N_k}) |$ is nonzero a.e. (dz) and the theorem is proved.

$\quad\square$

COROLLARY 6.7. (i) If $P_k, k = 1, 2, \cdots$ are as in the above theorem and if $\limsup_{k \to \infty} M(P_k) = 1$, then we can choose $P_{j_k}, k = 1, 2, \cdots$ and $N_1, N_2, \cdots$, in such a way that $M(\mu)$ is positive.
(ii) If $P_k, k = 1, 2, \cdots$ are as in the above theorem and if $\liminf_{k \to \infty} M(P_k) < 1$ then we can choose $P_{j_k}, k = 1, 2, \cdots$ and $N_1, N_2, \cdots$ in such a way that $M(\mu) = 0$, and $\frac{d\mu}{dz} > 0$ a.e. (dz).

REMARK 6.8. Now it is easy to construct polynomials $P_k, k = 1, 2, \cdots$ satisfying the hypothesis of part (ii) of the above corollary, so one can obtain generalized Riesz products μ with zero Mahler measure and $\frac{d\mu}{dz}$ positive a.e (dz). Thus the interchange of order of limits suggested in remark 5.5. is therefore not valid without some additional conditions.

Let (U) denote the class of all unimodular polynomials, i.e., polynomials of the type

$$\left\{ \sum_{j=0}^{n} a_j z^j : \mid a_j \mid = 1, 0 \leq j \leq n, n \geq 1 \right\}.$$

Note that $\|\frac{1}{\sqrt{n+1}} P\|_2 = 1$ for any polynomial in the class (U) of degree n. A question of Littlewood, answered in the affirmative by J-P. Kahane [18], [25], asks if there is a sequence $P_j, j = 1, 2, \cdots$ of polynomials in the class (U) such that $\frac{1}{\sqrt{d_j+1}} |P_j|$, $j = 1, 2, \cdots$ converges to the constant function 1 uniformly on S^1, where d_j is the degree of the polynomial $P_j, j = 1, 2, \cdots$. Littlewood problem remains open if we require that the coefficients of P_j be either -1 or 1, for all j. Let $P_j, j = 1, 2, \cdots$ be a sequence Kahane polynomials. Clearly then $\int_{S^1} \log |P_j(z)| dz \to 0$ as $j \to \infty$, which in turn implies that the Mahler measure of P_j converges to 1 as $j \to \infty$. Thus the sequence of Kahane polynomials satisfies the conditions of theorem 6.6 and we see that Kahane polynomials give rise to generalized Riesz products of Dynamical origin absolutely continuous with respect to the Lebesgue measure and with positive Mahler measure.

A sequence of polynomials $P_k, k = 1, 2, \cdots$ in the class U is said to be ultraflat if $\frac{|P_k|}{\|P_k\|_2}, k = 1, 2, \cdots$ converge uniformly to the constant function 1.

As mentioned above it is not known if there is a sequence of ultra flat polynomial with coefficients $+1$ and -1. However, M. Guenais [13] has shown that there is a sequence $P_k, k = 1, 2, \cdots$ of polynomials in U with coefficients in $\{-1, 1\}$ such that $\frac{|P_k|}{\|P_k\|_2} \to 1$ a.e.(dz) if and only if there is a measure preserving general odometer action T and a ϕ taking values in $\{-1, 1\}$ such that V_ϕ has Lebesgue spectrum. Here ϕ has to be of the special kind described above, namely, it is constant on all but the top level of the stacks associated to T.

A finite sequence $(e_0, e_2, \cdots, e_{n-1})$ of $+1$ and -1 is said to be a Barker sequence if for all $k, 0 < k \leq n - 1$, the aperiodic correlation

$$\sum_{j=0}^{n-k} e_j e_{j+k}$$

does not exceed 1 in absolute value. It is known that there are only finitely many Barker sequences of odd length, and there are no Barker sequences of odd length greater than 13. For more information on Barker sequences and their significance in Radar signal processing theory we refer the reader to [14], [6], [7]. It is not known if there are infinitely many Barker sequences, and it is conjectured that there are only finitely many Barker sequences. P. Borwein and M. Mossinghoff[6] have shown that if $e_0, e_1, \cdots, e_{n-1}$ is a Barker sequence of length n and if

$$P(z) = \frac{\sum_{j=0}^{n-1} e_j z^j}{\sqrt{n}},$$

then the Mahler measure $M(P)$ of P is $> 1 - \frac{1}{n}$. This immediately implies, in the light of the result of M. Guenais, or by the corollary above the following theorem.

THEOREM 6.9. *If there are infinitely many Barker sequences then there is a generalized Riesz product $\prod_{j=1}^{\infty} \mid P_j \mid^2$ of dynamical origin with measure preserving T, with positive Mahler measure, and such that coefficients of each P_j are real and equal in absolute value. The measure preserving T can be chosen to be an odometer action.*

One can ask the question if there is a sequence $P_k, k = 1, 2, \cdots$ of polynomials with coefficients in $\{-1, 0, 1\}$ such that $\frac{|P_k|}{\|P_k\|_2} \to 1$ a.e. (dz) as k $\to \infty$. Theorem 6.5. at once implies that this is possible if and only if there is a generalized Riesz product $\mu = \prod_{j=1}^{\infty} \mid Q_j \mid^2$, $\frac{d\mu}{dz} > 0$ a.e. (dz), of dynamical origin and such that for each j the non-zero coefficients of Q_j are real and equal in absolute value.

Consider the class of (B) of all polynomials of the type

$$P(z) = \frac{1}{\sqrt{m+1}}(1 + z^{n_1} + z^{n_2} + \cdots + z^{n_m}),$$

where $0 < n_1 < n_2 < \cdots < n_m$. Since $L^2(S^1, dz)$ norm of such a P is one, its $L^1(S^1, dz)$-norm is at most one. J. Bourgain[10] has raised the question if the supremum of the $L^1(S^1, dz)$-norms of elements in (B) can be 1, see [10]. We have the following result due to M. Guenais, proved here more generally than in [13].

PROPOSITION 6.10. *Let $\mu = \prod_{k=1}^{\infty} \mid P(z) \mid^2$ be a generalized Riesz product of class (L). If $\sum_{k=1}^{\infty}(1- \| P_n \|_1^2)^{1/2}$ is finite then $\frac{d\mu}{dz} > 0$ on a set of positive Lebesgue measure in S^1.*

PROOF. Write $v_k^2 = 1- \| P_k \|_1^2$. Then $\sum_{k=1}^{\infty} v_k < \infty$, equivalently $\prod_{k=1}^{\infty} \| P_k \|_1 > 0$. For all functions $f, g \in L^2(S^1, dz)$, Cauchy-Schwarz inequality gives

$$\mid (\| f \cdot g \|_1 - \| f \|_1 \| g \|_1) \mid \leq (\| f \|_2^2 - \| f \|_1^2)^{1/2}(\| g \|_2^2 - \| g \|_1^2)^{1/2}.$$

Fix an integer $n_0 > 1$ and let $k > n_0$. Then

$$\mid (\| \prod_{j=n_0}^{k} P_j \|_1 - \| \prod_{j=n_0}^{k-1} P_j \|_1 \| P_k \|_1) \mid$$

$$\leq (\| \prod_{j=n_0}^{k-1} P_j \|_2^2 - \| \prod_{j=n_0}^{k-1} P_j \|_1^2)^{1/2}(\| P_k \|_2^2 - \| P_k \|_1^2)^{1/2}$$

$$\leq v_k.$$

So,

$$\mid (\| \prod_{j=n_0}^{k} P_j \|_1 - \| \prod_{j=n_0}^{k-1} P_j \|_1 \| P_k \|_1) \mid \leq v_k.$$

$$| (\| \prod_{j=n_0}^{k-1} P_j \|_1 \| P_k \|_1 - (\| \prod_{j=n_0}^{k-2} P_j \|_1)(\| P_{k-1} \|_1) \| P_k \|_1) | \le v_{k-1} \| P_k \|_1 \le v_{k-1}$$

$$\vdots \ \vdots \ \vdots$$

$$| (\| \prod_{j=n_0}^{n_0+1} P_j \|_1 \prod_{j=n_0+2}^{k} \| P_j \|_1 - \prod_{j=n_0}^{k} \| P_j \|_1) | \le v_{n_0+1}.$$

On adding the above inequalities:

$$| (\| \prod_{j=n_0}^{k} P_j \|_1 - \prod_{j=n_0}^{k} \| P_j \|_1) | \le \sum_{j=n_0}^{k} v_j$$

Since $\prod_{j=1}^{\infty} \| P_j \|_1 > 0$ and $\sum_{j=1}^{\infty} v_k < \infty$, we see that $\limsup_{k \to \infty} \| \prod_{j=1}^{k} P_j \|_1 > 0$, so by Bourgain's criterion for singularity (i.e., corollary 4.4 with $Q_j = 1$ for all j,) we see that μ is not singular to Lebesgue measure on S^1. □

The only known rank one non-singular T for which U_T has Lebesgue spectrum is the one where the cutting parameter satisfies $\prod_{j=1}^{\infty} p_{l_j,j} > 0$. One can ask if there exists a non-dissipative rank one transformation whose maximal spectral type admits a component equivalent to the Lebesgue measure on S^1. We discuss this question in the light of the above considerations. Fix $0 < \lambda < 1$ and let

$$A_\lambda = \{\sum_{j=0}^{n} a_j z^j : \forall j, 0 \le a_j \le \lambda, \sum_{j=0}^{n} | a_j |^2 = 1, n = 1, 2, \cdots\},$$

$$\alpha_\lambda = \sup_{P \in A_\lambda} \| P \|_1$$

PROPOSITION 6.11. If $\alpha_\lambda = 1$ for some $\lambda, 0 < \lambda < 1$, then there is a non-dissipative non-singular rank one map T such U_T has absolutely continuous part (w.r.t (dz)) in its spectrum.

PROOF. This follows from lemma 6.4 and proposition 6.10.

□

PROPOSITION 6.12. If there is a sequence $P_n, n = 1, 2, \cdots$, of polynomials in A_λ such that $\lim_{n \to \infty} | P_n(z) | = 1$ a.e.(dz), then there is a non-dissipative non-singular rank one T such that the spectrum of U_T has a part equivalent to the Lebesgue measure on S^1.

PROOF. This follows from theorem 6.6.

□

There is a partial converse to proposition 6.11. Let $\mu = \prod_{j=1}^{\infty} | P_j |^2$ be a generalized Riesz product of class (L), with each $P_j \in A_\lambda$, and $P_j's$ dissociated.

PROPOSITION 6.13. If $\frac{d\mu}{dz} > 0$ a.e. (dz), then $\alpha_\lambda = 1$.

PROOF. We know from proposition 5.2.

$$\lim_{k \to \infty} \prod_{j=1}^{k} | P_j(z) | = \frac{d\mu}{dz} \ a.e. \ (dz).$$

Hence there is a sequence $n_1 < n_2 < \cdots$ such that

$$\prod_{j=n_k+1}^{n_{k+1}} | P_j(z) | \to 1 \text{ a.e. } (dz) \text{ as } k \to \infty$$

Since P_j's are dissociated, each finite product is in A_λ, the proposition follows from Fatou's lemma.   □

The following three problems about the class A_λ are thus intimately related to spectral questions about invertible non-singular rank one transformations: (1) is $\sup_{P \in A_\lambda} \| P \|_1 = 1$?, (2) is $\sup_{P \in A_\lambda} M(P) = 1$?   (3) is there a sequence $P_n, n = 1, 2, \cdots$ in A_λ such that $| P_n(z) | \to 1$ a.e. (dz) as $n \to \infty$

Concerning Bourgain's question, it is known that $\sup_{P \in B} \int_{S^1} | P(z) | \, dz \geq \frac{\sqrt{\pi}}{2}$. Indeed, let $P_n, n = 1, 2, \cdots$ be the polynomials as in the generalized Riesz product associated to the measure preserving rank one map. Put $X_{n,j}(z) = z^{jh_n + a_{1,n} + \cdots + a_{n,j}}$. $X_{n,j}$ is a random variables. Since $\|P_n\|_2 = 1$, the random variables P_n are uniformly integrable. In [1], [4] and [5], the authors, proved that there is a subclass of $P_n, n = 1, 2, \cdots$, for which $P_n, n = 1, 2, \cdots$ converges in distribution to the complex Gaussian measure $\mathcal{N}_{\mathbb{C}}(0, 1)$ on $\mathbb{C}$, that is,

$$dz\{P_n \in A\} \xrightarrow[n \to +\infty]{} \int_A \frac{1}{\pi} e^{-|z|^2} dz.$$

Denote by $\mathcal{D}(P_n)$ the distribution of P_n. It follows from the Standard Moment Theorem [12, pp.100] that

$$\|P_n\|_1 = \int |P_n| dz = \int |w| d\mathcal{D}(P_n)(w) \xrightarrow[n \to \infty]{} \int |w| \frac{1}{\pi} e^{-|w|^2} dw = \frac{\sqrt{\pi}}{2},$$

dw is the usual Lebesgue measure on $\mathbb{C}$, that is, $dx \cdot dy = r dr d\theta$.

Let $E = \{z : \frac{d\mu}{dz}(z) > 0\}$, where μ is a generalized Riesz product. We will give an upper estimate of $dz(E)$.

THEOREM 6.14. *Let $\mu = \prod_{j=1}^{\infty} | P_j |^2$ be of class (L). Let $E = \{z : \frac{d\mu}{dz}(z) > 0\}$. If $dz(E) = 1$ then there is a flat sequence of finite subproducts of P_j's. If $dz(E)$ is less than 1, then $dz(E) \leq d$, where d is the liminf of $L^1(S^1, dz)$ norms of all finite subproducts of P_j's.*

PROOF. The first part follows from the discussion above. We consider the second part. Let $a \overset{\text{def}}{=} \sup \|P_{i_1} P_{i_2} \cdots P_{i_l}\|_1$, where the supremum is taken over all finite sequences of increasing natural numbers $i_1 < i_2 < \cdots < i_l$. By Fatou's lemma we know that $\left\| \sqrt{\frac{d\mu}{dz}} \right\|_1 \leq a$. Take an infinite subset $\mathcal{K}_1$ of natural numbers such that its complement $\mathcal{K}_2$ within natural numbers is also infinite. Let μ_1 and μ_2 be the Riesz subproducts of μ over $\mathcal{K}_1$ and $\mathcal{K}_2$ respectively. Then by corollary 5.3

$$\left(\frac{d\mu}{dz}\right)^{\frac{1}{4}} = \left(\frac{d\mu_1}{dz}\right)^{\frac{1}{4}} \left(\frac{d\mu_2}{dz}\right)^{\frac{1}{4}}$$

By Cauchy-Schwarz inequality we get

$$\int_{S^1} \left(\frac{d\mu}{dz}\right)^{\frac{1}{4}} dz \leq \sqrt{a}\sqrt{a} = a$$

In general, by iterating,

(2)
$$\int_{S^1} \left(\frac{d\mu}{dz} \right)^{\frac{1}{2^n}} dz \le a$$

Letting $n \to \infty$, we see that $dz(E) \le a$
Let

$$d \overset{\text{def}}{=} \liminf \left\{ \int_{S^1} \mid P_{i_1} P_{i_2} \cdots P_{i_k} \mid dz : i_1 < i_2 < \cdots i_k, k = 1, 2, \cdots \right\}$$

Now for any $\eta > 0$, considerations leading to equation (2) above can be applied to a suitable subproduct, say μ_1, over a set $\mathcal{K}_1$ of natural numbers, so that

$$dz \left\{ z : \frac{d\mu_1}{dz}(z) > 0 \right\} \le d + \eta.$$

By formula (1) we see that $dz(E) \le d + \eta$. Since η is arbitrary, we have $dz(E) \le d$. $\qquad \square$

In connection with the discussion above, we have the following:

THEOREM 6.15. *Let* $\mu = \prod_{j=1}^{\infty} \mid P_j \mid^2$ *be of class (L) and assume that*

(1) $\|P_j\|_1 \xrightarrow[j \to +\infty]{} c$, $c \in [0, 1[$ *and,*
(2) for any continuous function g *on* $\mathbb{T}$, *we have*

$$\int g|P_j| dz \xrightarrow[j \to +\infty]{} c \int g dz.$$

Then μ *is singular.*

PROOF. The sequence $\|P_j| - 1|$ is bounded in $L^2(dz)$. It follows that there exists a function ϕ in $L^2(dz)$ such that $\|P_j| - 1|$ converge weakly over a subsequence, say $n_j, j = 1, 2, \cdots$, to ϕ (without loss of generality we assume that $n_j = j, j = 1, 2, \cdots$). It is shown in [1] that the measure $\phi(z) dz$ is singular with respect to μ. According to our assumptions, we further have that $\phi(z) dz$ is equivalent to Lebesgue measure. Indeed, for any nonnegative continuous function g on $\mathbb{T}$, we have

$$\int g\|P_j| - 1| dz \ge \int g dz - \int g|P_j| dz.$$

Hence, by taking the limit combined with our assumptions, we get

$$\int g\phi dz \ge (1 - c) \int g dz,$$

which finish the proof of the theorem.

$\qquad \square$

7. Zeros of Polynomials

Consider the polynomial of the type

(R)　　　$P(z) = 1 + z^{h+a_1} + z^{2h+a_1+a_2} + \cdots + z^{(m-1)h+a_1+a_2+\cdots+a_{m-1}}$,

which appears in the generalized Riesz product connected with rank one measure preserving transformation.

It is easy to see that zeros of these polynomials cluster near the unit circle as h tends to ∞. We prove a quantitative result, namely, if w is a zero of this polynomial then

(3)　　　　　　　　$\left(\frac{1}{2}\right)^{\frac{1}{h}} \leq |w| \leq (2)^{\frac{1}{h}}$

To see this we write $|w| = a$. Assume first that $a \leq 1$. Then, since w is a zero of P,

$$a^h + a^{2h} + \cdots + a^{(m-1)h} \geq 1.$$

Equivalently,

$$a^h \frac{(1 - a^{(m-1)h})}{1 - a^h} \geq 1.$$

$$a^h - a^{mh} \geq 1 - a^h$$

$$2a^h \geq 1 + a^{mh} \geq 1$$

which proves the result when $|w| \leq 1$. To prove the second half of the inequality we note that if $|w|$ is greater than 1 then $\frac{1}{|w|} \leq 1$ and $\frac{1}{w}$ is a zero of $P(\frac{1}{z})$ so the second half follows from the first half. A slight improvement of the inequality is possible. If $m = 2$ then all the zeros of P lie on the unit circle. It is easy to show that if $m > 2$ then the equation $x^m - 2x + 1$ has a zero, say b_m, in the open interval $\frac{1}{2} < x < 1$. and one can show that

$$(b_m)^{\frac{1}{h}} \leq |w| \leq \left(\frac{1}{b_m}\right)^{\frac{1}{h}}.$$

However, it is not a very big improvement since one can show that $b_m \to \frac{1}{2}$ as $m \to \infty$.

This simple result tells us that if each P_k has less than ch_{k-1} zeros bigger than 1 in absolute value where c is a positive constant less than one, then $\prod_{k=1}^{\infty} |\alpha_k| = 0$.

We mention that M. Odlyzko and B. Poonen in [23] proved that the zeros of the polynomials with coefficients in $\{0, 1\}$ are contained in the annulus $\frac{1}{\phi} < |z| < \phi$ where ϕ is the golden ratio.

Zeros of polynomials with restricted coefficients has deep and extensive literature. We mention only a result in a recent paper. (P. Brown, T. Erdélyi, F. Littmann [9]). Let

$$K_n = \left\{ \sum_{k=0}^{n} a_k z^k : |a_0| = |a_n| = 1, |a_k| \leq 1 \right\},$$

and let n be so large that $\delta_n = 33\pi \frac{log(n)}{\sqrt{n}} < 1$, then any polynomial in K_n admits at least $8\sqrt{n} \log n$ zeros in δ_n neighborhood of any point of the unit the circle. Thus the derived set, i.e., the set of limit points of the zeros of the polynomials appearing in the generalized Riesz product (R) is the full unit circle.

Acknowledgements. The first author wishes to express his sincere thanks to the National Center for Mathematics and IIT, Mumbai, where a part of this paper was written, for an invitation and warm hospitality.

References

[1] El Houcein El Abdalaoui, *A new class of rank-one transformations with singular spectrum*, Ergodic Theory Dynam. Systems **27** (2007), no. 5, 1541–1555, DOI 10.1017/S0143385706001106. MR2358977 (2008k:37016)

[2] E. H. el Abdalaoui and M. Lemańczyk, *Approximately transitive dynamical systems and simple spectrum*, Arch. Math. (Basel) **97** (2011), no. 2, 187–197, DOI 10.1007/s00013-011-0285-7. MR2820581 (2012g:37006)

[3] e. H. el Abdalaoui, *On the singularity of the spectrum of rank one maps*, preprint Feb. 2013.

[4] Christoph Aistleitner and Markus Hofer, *On the maximal spectral type of a class of rank one transformations*, Dyn. Syst. **27** (2012), no. 4, 515–523. MR2989747

[5] Christoph Aistleitner, *On a problem of Bourgain concerning the L^1-norm of exponential sums*, Math. Z. **275** (2013), no. 3-4, 681–688, DOI 10.1007/s00209-013-1155-8. MR3127032

[6] Peter Borwein and Michael J. Mossinghoff, *Barker sequences and flat polynomials*, Number theory and polynomials, London Math. Soc. Lecture Note Ser., vol. 352, Cambridge Univ. Press, Cambridge, 2008, pp. 71–88, DOI 10.1017/CBO9780511721274.007. MR2428516 (2009k:11040)

[7] P. Borwein, M. J. Mossinghoff, *Wiefrich pairs and Barker sequences II*, Preprint, July 2013.

[8] Gavin Brown and William Moran, *On orthogonality of Riesz products*, Proc. Cambridge Philos. Soc. **76** (1974), 173–181. MR0350319 (50 #2812)

[9] Peter Borwein, Tamás Erdélyi, and Friedrich Littmann, *Polynomials with coefficients from a finite set*, Trans. Amer. Math. Soc. **360** (2008), no. 10, 5145–5154, DOI 10.1090/S0002-9947-08-04605-9. MR2415068 (2009d:30005)

[10] J. Bourgain, *On the spectral type of Ornstein's class one transformations*, Israel J. Math. **84** (1993), no. 1-2, 53–63, DOI 10.1007/BF02761690. MR1244658 (94j:28007)

[11] J. R. Choksi and M. G. Nadkarni, *The maximal spectral type of a rank one transformation*, Canad. Math. Bull. **37** (1994), no. 1, 29–36, DOI 10.4153/CMB-1994-005-4. MR1261554 (95g:28027)

[12] Kai Lai Chung, *A course in probability theory*, 3rd ed., Academic Press, Inc., San Diego, CA, 2001. MR1796326 (2001g:60001)

[13] Mélanie Guenais, *Morse cocycles and simple Lebesgue spectrum*, Ergodic Theory Dynam. Systems **19** (1999), no. 2, 437–446, DOI 10.1017/S0143385799126579. MR1685402 (2000g:28037)

[14] T. Downarowicz and Y. Lacroix, *Merit factors and Morse sequences*, Theoret. Comput. Sci. **209** (1998), no. 1-2, 377–387, DOI 10.1016/S0304-3975(98)00121-2. MR1647483 (99i:11015)

[15] Nathaniel A. Friedman, *Introduction to ergodic theory*, Van Nostrand Reinhold Co., New York-Toronto, Ont.-London, 1970. Van Nostrand Reinhold Mathematical Studies, No. 29. MR0435350 (55 #8310)

[16] Kenneth Hoffman, *Banach spaces of analytic functions*, Dover Publications, Inc., New York, 1988. Reprint of the 1962 original. MR1102893 (92d:46066)

[17] Bernard Host, Jean-François Méla, and François Parreau, *Nonsingular transformations and spectral analysis of measures* (English, with French summary), Bull. Soc. Math. France **119** (1991), no. 1, 33–90. MR1101939 (93d:43002)

[18] Jean-Pierre Kahane, *Sur les polynômes à coefficients unimodulaires* (French), Bull. London Math. Soc. **12** (1980), no. 5, 321–342, DOI 10.1112/blms/12.5.321. MR587702 (82a:30003)

[19] Ivo Klemes and Karin Reinhold, *Rank one transformations with singular spectral type*, Israel J. Math. **98** (1997), 1–14, DOI 10.1007/BF02937326. MR1459845 (99c:28049)

[20] François Ledrappier, *Des produits de Riesz comme mesures spectrales* (French, with English summary), Ann. Inst. H. Poincaré Sect. B (N.S.) **6** (1970), 335–344. MR0296726 (45 #5785)

[21] M. G. Nadkarni, *Spectral theory of dynamical systems*, Birkhäuser Advanced Texts: Basler Lehrbücher. [Birkhäuser Advanced Texts: Basel Textbooks], Birkhäuser Verlag, Basel, 1998. MR1719722 (2001d:37001)

[22] M. G. Nadkarni *some remark on the spectrum of a rank one transformation* preprint Feb. 2012

[23] A. M. Odlyzko and B. Poonen, *Zeros of polynomials with* 0, 1 *coefficients*, Enseign. Math. (2) **39** (1993), no. 3-4, 317–348. MR1252071 (95b:11026)

[24] Donald S. Ornstein, *On the root problem in ergodic theory*, Proceedings of the Sixth Berkeley Symposium on Mathematical Statistics and Probability (Univ. California, Berkeley, Calif., 1970/1971), Univ. California Press, Berkeley, Calif., 1972, pp. 347–356. MR0399415 (53 #3259)

[25] Hervé Queffelec and Bahman Saffari, *On Bernstein's inequality and Kahane's ultraflat polynomials*, J. Fourier Anal. Appl. **2** (1996), no. 6, 519–582, DOI 10.1007/s00041-001-4043-2. MR1423528 (97h:41027)

[26] Friedrich Riesz, *Über die Fourierkoeffizienten einer stetigen Funktion von beschränkter Schwankung* (German), Math. Z. **2** (1918), no. 3-4, 312–315, DOI 10.1007/BF01199414. MR1544321

[27] A. Zygmund, *Trigonometric series: Vols. I, II*, Second edition, reprinted with corrections and some additions, Cambridge University Press, London-New York, 1968. MR0236587 (38 #4882)

NORMANDIE UNIVERSITY, UNIVERSITY OF ROUEN, DEPARTMENT OF MATHEMATICS, LMRS UMR 60 85 CNRS, AVENUE DE L'UNIVERSITÉ, BP.12, 76801 SAINT ETIENNE DU ROUVRAY - FRANCE

E-mail address: elhoucein.elabdalaoui@univ-rouen.fr

URL: http://www.univ-rouen.fr/LMRS/Persopage/Elabdalaoui/

DEPARTMENT OF MATHEMATICS, UNIVERSITY OF MUMBAI, VIDYANAGARI, KALINA, MUMBAI, 400098, INDIA

E-mail address: mgnadkarni@gmail.com

URL: http://insaindia.org/detail.php?id=N91-1080

Contemporary Mathematics
Volume **631**, 2015
http://dx.doi.org/10.1090/conm/631/12603

Diophantine approximation exponents
on homogeneous varieties

Anish Ghosh, Alexander Gorodnik, and Amos Nevo

To S. G. Dani on the occasion of his 65th birthday

ABSTRACT. Recent years have seen very important developments at the inter-
face of Diophantine approximation and homogeneous dynamics. In the first
part of the paper we give a brief exposition of a dictionary developed by Dani
and Kleinbock-Margulis which relates Diophantine properties of vectors to distribution of orbits of flows on the space of unimodular lattices. In the second
part of the paper we briefly describe an extension of this dictionary recently
developed by the authors, which establishes an analogous dynamical correspondence for general lattice orbits on homogeneous spaces. We concentrate
specifically on the problem of estimating exponents of Diophantine approximation by arithmetic lattices acting on algebraic varieties. In the third part
of the paper, we exemplify our results by establishing explicit bounds for the
Diophantine exponent of dense lattice orbits in a number of basic cases. These
include the linear and affine actions on affine spaces, and the action on the
variety of matrices of fixed determinant. In some cases, these exponents are
shown to be best possible.

CONTENTS

1. Introduction

The theory of Diophantine approximation has many deep and fruitful connections with dynamical properties of flows on homogeneous spaces. These connections
have provided many fundamental new insights enriching both fields. For instance,
E. Artin [1] used continued fractions to construct dense geodesics on the modular

---

2010 *Mathematics Subject Classification.* Primary 37A17, 11K60.

Key words and phrases. Diophantine approximation, semisimple algebraic group, homogeneous space, lattice subgroup, automorphic spectrum.

The first author acknowledges support of the Royal Society. The second author acknowledges
support of EPSRC, ERC and RCUK. The third author acknowledges support of ISF.

surface, and it was observed by C. Series [**45**] that classical continued fraction expansions can be constructed as cutting sequences for orbits of the geodesic flow on the modular surface.

This remarkable connection between Diophantine approximation and dynamics also exists in higher dimensions. It was realized by S.G. Dani [**11**] that Diophantine properties of vectors in Euclidean space can be encoded by orbits of a suitable one-parameter flow on the space of unimodular lattices. In particular, he showed that badly approximable vectors correspond to bounded orbits, and singular vectors correspond to divergent orbits. This work has inspired many subsequent investigations exploring properties of flows on homogeneous spaces, and the techniques developed gave rise to the solution of several longstanding open problems in number theory. For instance one can mention such notable advances as the solution of Sprindzhuk's conjecture in the theory of Diophantine approximation with dependent quantities by D. Kleinbock and G. Margulis [**23**] and the computation of the Hausdorff dimension of the set of singular vectors in $\mathbb{R}^2$ by Y. Cheung [**7**].

In this paper, we first discuss the correspondence between Diophantine properties of vectors and recurrence properties of flows developed by Dani, Kleinbock and Margulis. We then put these results in the context of recent works of the authors on Diophantine exponents on homogeneous varieties of semisimple algebraic groups. Finally, we present some new estimates for Diophantine exponents on specific homogeneous varieties, complementing the results of [**19**]. We thank the anonymous referees for very helpful suggestions.

2. Diophantine approximation and the shrinking target property

2.1. Classical Diophantine approximation. It is a well-known theorem of Dirichlet that given a vector $x \in \mathbb{R}^d$, for every $R > 1$ one can find $m \in \mathbb{Z}^d$ and $n \in \mathbb{N}$ such that

$$(2.1) \qquad \left\| x - \frac{m}{n} \right\|_\infty \leq n^{-1} R^{-1/d} \quad \text{and} \quad n \leq R.$$

Here $\| \; \|_\infty$ denotes the maximum norm. Dirichlet introduced his famous pigeonhole or box principle to prove the above theorem. A proof can also be provided using Minkowski's convex body theorem. We refer the reader to [**43**] for details. In particular, it follows from (2.1) that the inequality

$$(2.2) \qquad \left\| x - \frac{m}{n} \right\|_\infty \leq n^{-1-1/d}$$

always has a solution with $m \in \mathbb{Z}^d$ and $n \in \mathbb{N}$. Diophantine properties of vectors can also be studied in the context of linear forms, i.e. given $x \in \mathbb{R}^d$ one can study small values of the linear form

$$|m_1 x_1 + \cdots + m_d x_d + n|$$

for $m = (m_1, \ldots, m_d) \in \mathbb{Z}^d$ and $n \in \mathbb{Z}$. This is referred to as the *linear* setting. These two settings are related by Khinchin's transference principle [**43**].

Dirichlet's theorem provides us with the first examples of *Diophantine exponents*. Let $x, x_0 \in \mathbb{R}^d$. Following Bugeaud and Laurent [**4–6**], we denote by $\omega(x, x_0)$ the supremum of real numbers ω such that there are arbitrarily large real numbers R for which the inequalities

$$(2.3) \qquad \| nx - x_0 - m \|_\infty \leq R^{-\omega} \quad \text{and} \quad n \leq R$$

have a nonzero solution $n \in \mathbb{Z}, m \in \mathbb{Z}^d$. We define $\hat{\omega}(x, x_0)$ to be the supremum of ω such that (2.3) has a nonzero integer solution for *all* sufficiently large positive real numbers R. Taking $x_0 = (0, \ldots, 0)$ gives the uniform homogeneous Diophantine exponents $\omega(x)$ and $\hat{\omega}(x)$. Then Dirichlet's theorem implies that

$$\omega(x) \geq \hat{\omega}(x) \geq 1/d.$$

The exponents ω and $\hat{\omega}$ are related by various transference principles which are still not completely understood in higher dimensions. We refer the reader to the works of Bugeaud and Laurent cited above for details. In this paper, we will study analogues of the exponent $\hat{\omega}$ in the more general context of dense lattice orbits on homogeneous varieties. We begin with an example which is closely related to the above discussion. Let Γ denote a discrete subgroup of the group $\text{Aff}(\mathbb{R}^d)$ of affine transformations of $\mathbb{R}^d$. We equip $\text{Aff}(\mathbb{R}^d)$ with a norm which is a natural way to measure complexity of elements of Γ. Studying effective density of Γ-orbits amounts to estimating Diophantine exponents which are defined as follows.

DEFINITION 2.1. Assume that for $x, x_0 \in \overline{\Gamma x}$, there exist constants $c = c(x, x_0)$ and $\epsilon_0 = \epsilon_0(x, x_0)$ such that for all $\epsilon < \epsilon_0$, the system of inequalities

$$\|\gamma^{-1}x - x_0\|_\infty \leq \epsilon \quad \text{and} \quad \|\gamma\| \leq c\epsilon^{-\kappa}.$$

has a solution $\gamma \in \Gamma$. Define the Diophantine approximation exponent $\kappa_\Gamma(x, x_0)$ as the infimum of $\kappa > 0$ such that the foregoing inequalities have a solution.

Note that κ_Γ is closely related to the exponent $\hat{\omega}$ above and should be viewed as a generalization of $\hat{\omega}$ to the context of lattice actions on homogeneous varieties. Taking $\Gamma = \mathbb{Z}^\times \ltimes \mathbb{Z}^d$ acting on $\mathbb{R}^d$, we see that $\kappa_\Gamma(x, x_0)$ is the infimum of $\kappa > 0$ such that the inequalities

$$\|nx - x_0 - m\|_\infty < cR^{-1/\kappa} \text{ and } n \leq R$$

has a solution for all sufficiently large R.

We will provide new examples of exponents κ_Γ in §3 based on our work [**19**]. For now we return to a discussion of dynamical approaches to the study of ω and $\hat{\omega}$, specifically questions regarding vectors in $\mathbb{R}^d$ for which the general estimates (2.1) and (2.2) can - or cannot - be improved; in particular, one wishes to study the size of sets of such vectors. A vector x is called *badly approximable* if there exists $c > 0$ such that the inequality

(2.4)
$$\left\| x - \frac{m}{n} \right\|_\infty \leq c\, n^{-1-1/d}$$

has no solutions $m \in \mathbb{Z}^d$ and $n \in \mathbb{N}$.

At the other extreme, the vector x is called *singular* if for every $c > 0$ and $R \geq R(c)$, the system of inequalities

(2.5)
$$\left\| x - \frac{m}{n} \right\|_\infty \leq c\, n^{-1}R^{-1/d} \quad \text{and} \quad n \leq R$$

has a solution $m \in \mathbb{Z}^d$ and $n \in \mathbb{N}$. In other words, Dirichlet's theorem can be infinitely improved for such vectors. One can show that a vector is badly approximable (resp. singular) if and only if it also has the analogous property in the sense of the linear setting.

We note that a number $x \in \mathbb{R}$ is badly approximable if and only if its continued fraction expansion has bounded digits, and a number $x \in \mathbb{R}$ is singular if and only if it is rational. These properties are more difficult to characterize in higher dimensions, but it turns out they have a very convenient interpretation based on dynamics of certain flows on the space of unimodular lattices. Denote by $\mathcal{L}_{d+1}$ the space of unimodular lattices in $\mathbb{R}^{d+1}$, which can be identified with the homogeneous space $\mathrm{SL}_{d+1}(\mathbb{Z}) \backslash \mathrm{SL}_{d+1}(\mathbb{R})$. It has an invariant probability measure as well as a metric, inherited from a left invariant metric on $\mathrm{SL}_{d+1}(\mathbb{R})$. The quotient is non-compact and its compact subsets are described by Mahler's compactness criterion. For $x \in \mathbb{R}^d$ we define the lattice

$$\Lambda_x := \{(n, m - nx) : m \in \mathbb{Z}^d, n \in \mathbb{Z}\} \in \mathcal{L}_{d+1}.$$

We consider the action on $\mathcal{L}_{d+1}$ by the one-parameter subgroups

$$g_t := \mathrm{diag}(e^{-t}, e^{t/d}, \ldots, e^{t/d}).$$

The following results follow from the work of Dani:

PROPOSITION 2.2 (Dani [11]). *With notation as above,*

(i) $x \in \mathbb{R}^d$ *is badly approximable if and only if the semiorbit* $\Lambda_x g_t$, $t > 0$, *is bounded in* $\mathcal{L}_{d+1}$.

(ii) $x \in \mathbb{R}^d$ *is singular if and only if the semiorbit* $\Lambda_x g_t$, $t > 0$, *is divergent, i.e. leaves every compact set in* $\mathcal{L}_{d+1}$.

The idea of the proof of Proposition 2.2 is based on the observation that if the lattice $\Lambda_x g_t$ contains a small non-zero vector, this gives a solution of the relevant Diophantine inequalities. Indeed, let

(2.6) $$\Omega(\delta) := \{\Lambda \in \mathcal{L}_{d+1} : \exists z \in \Lambda - \{0\} : \|z\|_\infty < \delta\}.$$

Then if for some $t > 0$, we have $\Lambda_x g_t \in \Omega(\delta)$ with $\delta \in (0, 1)$, the system of inequalities

(2.7) $$\left\| x - \frac{m}{n} \right\|_\infty < \delta e^{-t/d} \quad \text{and} \quad n < \delta e^t$$

has a solution $m \in \mathbb{Z}^d$ and $n \in \mathbb{N}$. By Mahler's compactness criterion, the family of sets $\Omega(\delta)$ form a basis of neighborhoods of infinity of $\mathcal{L}_{d+1}$. Thus, Diophantine properties of the vector x are determined by visits of the semiorbit $\Lambda_x g_t$ to neighborhoods of infinity.

2.2. Schmidt's game and bounded orbits. The points whose g_t semiorbits are bounded (resp. divergent) form a set of measure zero. Nevertheless, the former are quite abundant. In [10, 11] Dani used the above correspondence (i.e. Proposition 2.2 (i)), along with Schmidt's results on his game to show that bounded orbits for certain partially hyperbolic flows on homogeneous spaces have full Hausdorff dimension. Schmidt's game was introduced in [42] and is played in a complete metric space X. Two players, Bob and Alice start with a subset $W \subseteq X$, and two parameters $0 < \alpha, \beta < 1$. The game consists of choosing a sequence of nested closed balls. Bob begins by choosing a ball B_0 and Alice continues by choosing A_0 and so on:

$$B_0 \supset A_0 \supset B_1 \supset A_1 \supset \ldots$$

The radii of the balls are related by the parameters α and β:

$$r(A_i) = \alpha r(B_i) \quad \text{and} \quad r(B_{i+1}) = \beta r(A_i).$$

Alice wins this game if $\bigcap_n B_n$ intersects W. The set W is called (α, β)-winning if Alice can find a winning strategy , α-winning if it is (α, β)-winning for all $0 < \beta < 1$ and winning if it is α-winning for some $\alpha > 0$.

Schmidt games have many nice properties. Most prominently, a winning subset of $\mathbb{R}^d$ is *thick*, i.e. the intersection of a winning set with every open set in $\mathbb{R}^d$ has Hausdorff dimension d. Schmidt showed that badly approximable vectors form an α-winning set for $0 < \alpha \leq 1/2$. Subsequent to Dani's introduction of Schmidt's game in homogeneous dynamics, a general conjecture on abundance of bounded orbits was formulated by Margulis [37] in his Kyoto ICM address, generalizing Dani's results. The conjecture was proved in stages by Kleinbock and Margulis [25] and Kleinbock and Weiss [26]. There has been intense activity in this subject, and Schmidt games and their variations have been used to prove a wide variety of results. Recently in [13], Dani and H. Shah have introduced a new topological variant of Schmidt's game.

2.3. Dani's correspondence and Khinchin's theorem. Proposition 2.2 and Proposition 2.4 below are examples of what Kleinbock and Margulis have termed the *Dani correspondence*. In their paper [24], they further developed this correspondence to handle inequalities of the form

$$(2.8) \qquad \left\| x - \frac{m}{n} \right\|_\infty \leq n^{-1} \psi(n),$$

where ψ is a general nonincreasing function, and to obtain a new proof of Khinchin's theorem[1] using homogeneous dynamics. Recall that Khinchin's theorem states that the set of $x \in \mathbb{R}^d$ for which inequality (2.8) has infinitely many solutions $m \in \mathbb{Z}^d$ and $n \in \mathbb{N}$ has zero (resp. full) measure depending on the convergence (resp. divergence) of the sum

$$\sum_{n=1}^\infty \psi(n)^d.$$

As before, the original question in Diophantine approximation is transfered to a problem about visits of the semiorbit $\Lambda_x g_t$ to a family of shrinking neighbourhoods of infinity. The rate at which these neighbourhoods shrink is determined by the decay rate of the function ψ. Kleinbock and Margulis then use the exponential mixing property of the flow g_t in conjunction with a very general and quantitative form of the Borel–Cantelli lemma due to Sprindzhuk, to establish a zero-one law and deduce the Khinchin-Groshev theorem as a corollary. As another corollary they also obtain logarithm laws for geodesic excursions to shrinking neighborhoods of cusps of locally symmetric spaces, thereby generalizing Sullivan's logarithm law. Further, they established versions of zero-one laws for multi-parameter actions, thereby confirming, in stronger form, a conjecture of Skriganov in the geometry of numbers.

2.4. Diophantine approximation on manifolds. One says that the vector $x \in \mathbb{R}^d$ is *very well approximable* if there exists $\epsilon > 0$ such that the inequality

$$\left\| x - \frac{m}{n} \right\|_\infty \leq m^{-1-1/d-\epsilon}$$

has infinitely many solutions $m \in \mathbb{Z}^d$ and $n \in \mathbb{N}$. The exponent in (2.2) is the best possible, and it follows from the Borel-Cantelli lemma that the set of $x \in \mathbb{R}^d$

---

[1]In fact, their paper deals with systems of linear forms, i.e. the Khinchin-Groshev theorem.

which are very well approximable has zero Lebesgue measure. Note that the above definition and the results discussed in this section pertain to the exponent ω. It is easy to see that $x \in \mathbb{R}^d$ is very well approximable if and only if $\omega(x) = d$. The subject of metric Diophantine approximation on manifolds seeks to study the extent to which generic Diophantine properties are inherited by proper submanifolds of $\mathbb{R}^d$, in particular inheritance of values of Diophantine exponents as above. In 1932, Mahler's investigations into the classification of numbers according to their Diophantine properties, led him to conjecture that almost every point on the curve

$$(x, x^2, \ldots, x^d)$$

is not very well approximable. Mahler's conjecture was resolved by Sprindzhuk who in turn conjectured a more general form of his theorem. Let M be a k-dimensional submanifold of $\mathbb{R}^d$ parametrised by a C^l-map $f : U \to M$. We say that M is nondegenerate if for almost every $x \in U$, the space spanned by the partial derivatives of f up to order l coincides with $\mathbb{R}^d$. The following was a long standing conjecture of Sprindzhuk (in the case of analytic manifolds) and proved by Kleinbock and Margulis[2]:

THEOREM 2.3 (Kleinbock–Margulis [**23**]). *Almost every point on a nondegenerate submanifold $M \subset \mathbb{R}^d$ is not very well approximable.*

The proof once more follows a dynamical route. Namely, it was observed by Kleinbock and Margulis that the property of being very well approximable also has a convenient interpretation in terms of dynamics on the space of unimodular lattices:

PROPOSITION 2.4 (Kleinbock–Margulis [**23**]). *A vector $x \in \mathbb{R}^d$ is very well approximable if and only if there exists $\alpha > 0$ such that $\Lambda_x g_{t_i} \in \Omega(e^{-\alpha t_i})$ for a sequence $t_i \to \infty$.*

Hence, in order to improve the exponent of Diophantine approximation for a vector $x \in \mathbb{R}^d$, one needs to establish that the semiorbit $\Lambda_x g_t$ visits the sequence of exponentially shrinking sets $\Omega(e^{-\alpha t})$ infinitely often. This is a common feature of many chaotic dynamical systems, and is usually called the *shrinking target property*. With the help of Proposition 2.4, the proof of Theorem 2.3 reduces to an analysis of visits of translated submanifolds $\Lambda_{f(U)} g_t$ to the neighborhoods $\Omega(e^{-\alpha t})$. The crucial and the most difficult part of the argument is an explicit estimate on the measure of the set of $x \in U$ such that $\Lambda_{f(x)} g_t \in \Omega(e^{-\alpha t})$. This estimate generalises the non-divergence properties of unipotent flows discovered by Margulis in [**36**] and developed further in quantitative forms by Dani in [**10, 12**].

The relevant result is stated as follows. Let C, α be positive numbers, B an open subset of $\mathbb{R}^k$ and λ denote Lebesgue measure of appropriate dimension. We say that a function $f : B \to \mathbb{R}$ is (C, α)-good on B if for any open ball $J \subset B$ and any $\epsilon > 0$,

$$\lambda(\{x \in J \ : \ |f(x)| < \epsilon\}) \leq \left(\frac{\epsilon}{\sup_{x \in J} |f(x)|} \right)^\alpha \lambda(B).$$

The main property of unipotent flows which allows for nondivergence is precisely the (C, α)-good property. Kleinbock and Margulis showed that more generally, smooth nondegenerate maps also have this property. Let $\mathcal{L}(\mathbb{Z}^d)$ be the poset of

---

[2]In fact, they proved more general *multiplicative* versions of the conjecture.

primitive subgroups of $\mathbb{Z}^d$. For discrete subgroups Λ we define $\|\Lambda\|$ as the norm of the corresponding vector in a suitable wedge product. Now we state the main estimate in [23], which plays crucial role in the proof of Theorem 2.3:

THEOREM 2.5. *Let an open ball $B(x_0, r_0) \subset \mathbb{R}^k, C, \alpha > 0, 0 < \rho < 1/d$ and a continuous map $h : B(x_0, 3^d r_0) \to \mathrm{SL}_d(\mathbb{R})$ be given. We assume that for every $\Delta \in \mathcal{L}(\mathbb{Z}^d)$,*

(1) *the function $x \to \|h(x)\Delta\|$ is (C, α)-good on $B(x_0, 3^d r_0)$,*
(2) *$\sup_{x \in B(x_0, r_0)} \|h(x)\Delta\| \geq \rho$.*

Then for every $\epsilon \in (0, \rho]$,

$$\lambda(\{x \in J \; : \; \mathbb{Z}^d h(x) \notin \Omega(\epsilon)\}) \leq D(d, k) C \left(\frac{\epsilon}{\rho}\right)^\alpha \lambda(B),$$

where $D(d, k)$ is a constant depending on d and k only.

In subsequent work, [28–30], D. Kleinbock has further studied the Diophantine exponents ω in the context of affine subspaces and manifolds. In particular, in [30] it is proved that if an analytic manifold contains a point x_0 such that $\omega(x) \geq c$ then $\omega(x) \geq c$ for almost all x in the manifold. We refer the reader to [31] for an overview of further developments.

We note that in the theory of metric Diophantine approximation on manifolds, one is concerned with approximating points on manifolds by *all rational points in the ambient Euclidean space*. We now turn in the next section to discuss a completely different, but equally natural, question. Namely we will consider Diophantine approximation of a general point on a variety intrinsically, by *rational points lying on the variety itself*.

2.5. Intrinsic Diophantine approximation on algebraic varieties. The question of Diophantine approximation on algebraic varieties by rational points on the variety itself was raised already half a century ago by Serge Lang but the results in this direction are still very scarce. On page 14 of his paper [33], Lang asks if homogeneous and inhomogeneous versions of metric Diophantine approximation can be developed for group varieties and homogeneous varieties over number fields. Specifically, he asks for analogues of Khinchin's theorem in this context. In some cases one can deduce that the set $X(\mathbb{Q})$ of rational points on an algebraic variety X is dense in the set $X(\mathbb{R})$ of real points using a rational parametrization of X (for example, the stereographic projection for the quadratic surfaces), but this approach usually provides poor bounds on Diophantine exponents that depend on the degree of the parametrization map. One can also consider the problem of Diophantine approximation by the set $X(\mathbb{Z}[1/p])$ of $\mathbb{Z}[1/p]$-points in X, or more generally approximation by S-integral points, where S is a set of completions of $\mathbb{Q}$. Here even establishing density is a nontrivial task.

One of the most natural examples of algebraic varieties with rich structure of rational points is given by algebraic groups and their homogeneous spaces. Here several results regarding quantitative density of rational points have been proved. This includes elliptic curves and abelian varieties [46], general homogeneous spaces of semisimple algebraic groups [17, 18], and quadratic surfaces [14, 16, 32, 44]. In the latter two cases, one can also use dynamical correspondences which relate Diophantine properties of points on the varieties to shrinking target properties of orbits for the corresponding dynamical system. Let us now turn to a brief description of these correspondences.

Let X be an algebraic variety in $\mathbb{C}^d$ defined over $\mathbb{Q}$ equipped with a transitive action of a connected almost simple algebraic group $G \subset \mathrm{GL}_d(\mathbb{C})$ defined over $\mathbb{Q}$. For simplicity of exposition, let us consider here the problem of Diophantine approximation in $X(\mathbb{R})$ by rational points in $X(\mathbb{Z}[1/p])$ where p is prime. The dynamical correspondence alluded to above amounts to the observation that since rational points on X can be parametrized using orbits of the group $G(\mathbb{Z}[1/p])$, they can be studied using techniques from the theory of dynamical systems. The relevant dynamical system here is the space

$$Y = G(\mathbb{Z}[1/p]) \backslash (G(\mathbb{R}) \times G(\mathbb{Q}_p))$$

with the action of the group $G(\mathbb{Q}_p)$ by right multiplication. The following proposition can be viewed as an analogue the classical Dani correspondence described in the previous section and is a special case of results in §5 of [**17**]. However, we note that the Dani correspondence is an if and only if statement while the statement below is only in one direction. Moreover, the Dani correspondence is for cuspidal excursions of one parameter groups while the statement below involves the shrinking target problem for larger groups where the target is a point in the space. For a prime p (possibly infinite), we denote by $\mathbb{Q}_p$ the corresponding completion of $\mathbb{Q}$ and define metrics on $X(\mathbb{Q}_p)$ (resp $G(\mathbb{Q}_p)$) as follows:

$$\|x - y\|_p := \max_{1 \le i \le d} |x_i - y_i|_p.$$

PROPOSITION 2.6 (Ghosh–Gorodnik–Nevo [**17**]). *Given* $x \in X(\mathbb{R})$*, there exists* $y_x \in Y$ *and a sequence of neighborhoods* $\mathcal{O}_\epsilon$ *of the identity coset in* Y *such that if*

$$y_x \cdot b \in \mathcal{O}_\epsilon \quad \text{for some } b \in G(\mathbb{Q}_p) \text{ with } \|b\|_p \le R,$$

then the system of inequalities

$$\left\| x - \frac{m}{n} \right\|_\infty \le \epsilon \quad \text{and} \quad n \le c(x)\, R$$

has a solution $\frac{m}{n} \in X(\mathbb{Z}[1/p])$*, with the constant* $c(x)$ *uniform over* x *in compact sets.*

Proposition 2.6 shows that the problem of Diophantine approximation reduces to the shrinking target problem for the orbit $y_x G(\mathbb{Q}_p)$ with respect to the sequence of neighbourhoods $\mathcal{O}_\epsilon$.

Let us assume that the set $X(\mathbb{Z}[1/p])$ is not empty and the group G is isotropic over $\mathbb{Q}_p$. Then the closure $\overline{X(\mathbb{Z}[1/p])}$ in $X(\mathbb{R})$ is open and closed in $X(\mathbb{R})$ (see [**17**]). In particular, when $X(\mathbb{R})$ is connected, $X(\mathbb{Z}[1/p])$ is dense. To measure the quality of Diophantine approximation, let us consider the following exponents for the homogeneous varieties in question.

DEFINITION 2.7. Assume that for $x \in \overline{X(\mathbb{Z}[1/p])}$, there exist constants $c = c(x)$ and $\epsilon_0 = \epsilon_0(x)$, such that for all $\epsilon < \epsilon_0$, the system of inequalities

$$(2.9) \qquad \left\| x - \frac{m}{n} \right\|_\infty \le \epsilon \quad \text{and} \quad n \le c\epsilon^{-\kappa}$$

has a solution $\frac{m}{n} \in X(\mathbb{Z}[1/p])$. Define the *Diophantine approximation exponent* $\kappa_p(x)$ as the infimum of $\kappa > 0$ such that the foregoing inequalities have a solution.

We note that the exponent above is related to the exponent κ_Γ from the introduction. A unified notion of exponent for lattices in locally compact groups encompassing both, can be found in [19].

A lower bound on the exponents $\kappa_p(x)$ can be deduced from the following pigeon-hole argument. Let us introduce the growth exponent of the number of rational points

$$a_p(X) = \sup_{\text{compact } K \subset X(\mathbb{R})} \limsup_{R \to \infty} \frac{\log N_R(K, X(\mathbb{Z}[1/p]))}{\log R}$$

where $N_R(K, X(\mathbb{Z}[1/p]))$ denotes the number of points $\frac{m}{n} \in K \cap X(\mathbb{Z}[1/p])$ such that $n \leq R$. One can show that if the constant $c = c(x)$ is uniform over compact sets in $X(\mathbb{R})$, then

(2.10)
$$\kappa_p(X) \geq \frac{\dim(X)}{a_p(X)}.$$

The problem of establishing upper bounds for $\kappa_p(X)$ is much deeper. Indeed, any such upper bound would explicitly quantify the density of $X(\mathbb{Z}[1/p])$ in $\overline{X(\mathbb{Z}[1/p])}$. In view of Proposition 2.6, this problem can be solved by establishing quantitative equidistribution for orbits of $G(\mathbb{Q}_p)$ in Y. Let us therefore introduce the family of averaging operators $A_R : L^2(Y) \to L^2(Y)$ given by

$$A_R\phi(y) := \frac{1}{|B_R|} \int_{B_R} \phi(yb) \, db$$

where $B_R = \{b \in G(\mathbb{Q}_p) : \|b\|_p \leq R\}$. It was shown in [17] that there exists $\theta > 0$ such that for every $\eta > 0$, there exists $C := C(\eta)$ such that

(2.11)
$$\|A_R(\phi) - P(\phi)\| \leq C |B_R|^{-\theta+\eta} \|\phi\|_2,$$

for every $R \geq R_\eta$ where P is an explicit projection operator on $L^2(Y)$. Let θ_p denote the supremum over θ's for which the estimate (2.11) holds. This parameter provides the crucial input to deduce the following upper bound on the Diophantine exponent:

THEOREM 2.8 (Ghosh, Gorodnik, Nevo [17]). *With notation as above,*

(i) *For almost every* $x \in \overline{X(\mathbb{Z}[1/p])}$,

$$\kappa_p(x) \leq (2\theta_p)^{-1} \frac{\dim(X)}{a_p(G)}.$$

(ii) *For every* $x \in \overline{X(\mathbb{Z}[1/p])}$,

$$\kappa_p(x) \leq \theta_p^{-1} \frac{\dim(X)}{a_p(G)}.$$

Moreover, the constant $c = c(x)$ *in (2.9) is uniform over* x *in compact sets.*

Typically, $a_p(G) \geq a_p(X)$, and in that case, it follows that when $\theta_p = 1/2$, the bound in Theorem 2.8(i) matches with the lower bound (2.10), so that it is best possible. For instance, Theorem 2.8 gives a sharp bound for Diophantine approximation by $\mathbb{Z}[1/p]$-points on the two-dimensional sphere. We note that the exponent θ_p is closely related to the integrability exponents of automorphic representations, which has been extensively studied in relation to the generalised Ramanujan conjectures, and explicit estimates on θ_p are available in a number of cases (see [2, 41] for a detailed account).

We note that a fruitful approach to the problem of Diophantine approximation by the set of all rational points on nonsingular quadratic surfaces has been developed in [14, 16, 32]. In particular, it turns out that the problem of Diophantine approximation on the sphere is related to a shrinking target problem for a suitable one-parameter flow g_t on the space $Y = \Gamma \backslash G$ where $G = \mathrm{SO}(d, 1)$ and Γ is a subgroup of integral matrices in G. We note that Y can be naturally embedded in the space of unimodular lattices $\mathcal{L}_{d+1}$ and one can set $\Omega_Y(\delta) = Y \cap \Omega(\delta)$ where $\Omega(\delta)$ is defined in (2.6). Given a vector x on the d dimensional unit sphere S^d, one can associate a point $y_x \in Y$ such that the following dynamical correspondence holds:

PROPOSITION 2.9 (Kleinbock–Merrill [32]). *Suppose that there exists $t > 0$ such that $y_x g_t \in \Omega_Y(\delta)$. Then the system of inequalities*

$$\left\| x - \frac{m}{n} \right\|_\infty < \frac{2\delta^{1/2} e^{-t/2}}{n^{1/2}} \quad and \quad n < \delta e^t$$

has a solution with $\frac{m}{n} \in S^d$.

For instance, Proposition 2.9 is used to prove the following analogue of the classical Dirichlet theorem (cf. (2.1) in the case of the unit spheres:

THEOREM 2.10 (Kleinbock–Merrill [32]). *There exists $c > 0$ such that for every $x \in S^d$ and $R > 1$, the system of inequalities*

$$\left\| x - \frac{m}{n} \right\|_\infty \leq \frac{c}{n^{1/2} R^{1/2}} \quad and \quad n \leq R$$

has a solution with $\frac{m}{n} \in S^d$.

We also mention that this approach also allows to prove an analogue of the Khinchin–Groshev theorem [32] and to study the set of badly approximable vectors [32] and well approximable vectors [14] in the context of the intrinsic Diophantine approximation.

3. Diophantine exponents for group actions

The problem of establishing Diophantine exponents discussed in the previous sections is an instance of a much more general problem, namely establishing a rate of distribution for dense lattice orbits on homogeneous spaces of the ambient group. In [19] we developed a general approach to establishing quantitative density of orbits which is based on a *duality principle* combined with a *quantitative mean ergodic theorem*. We then establish a dynamical correspondence similar to the correspondences discussed in the previous sections. More explicitly, the duality principle implies that quantitative density of lattice orbits Γx in the homogeneous space Gx can be reduced to the quantitative density of the corresponding orbits of H_x in the space $Y = \Gamma \backslash G$, where H_x denotes the stabilizer of x in G. We note that it would be interesting to develop this theory further and to study analogues of the various results discussed in §2 in the context of dense lattice orbits, for instance notions of badly approximable points, analogues of Schmidt's game etc. We are not aware of any results in this direction, other than those mentioned in §2.

To motivate the discussion by a concrete example, consider a case where the approach mentioned above works especially well, namely when Γ is a lattice in an algebraic subgroup G of the group of affine transformations. In classical inhomogeneous Diophantine approximation, one is interested in minimizing the quantity

$|nx - m + x_0|$ for given $x, x_0 \in \mathbb{R}^d$, as n varies over $\mathbb{Z}$ and m varies over $\mathbb{Z}^d$. This problem can be interpreted as establishing quantitative density for orbits of the semigroup $\Gamma = \mathbb{Z}^{\times} \ltimes \mathbb{Z}^d$ acting on $\mathbb{R}^d$. There has been considerable recent interest in the study of density of orbits for more general groups of affine transformations. One example is $\Gamma = \mathrm{SL}_2(\mathbb{Z})$ acting on the punctured plane, where the Γ-orbit of a point with irrational slope is dense. In [34,35], Laurent and Nogueira have studied effective versions of this density. In [39], Maucourant and Weiss have used effective equidistribution results for horocycle flows to obtain effective results for dense Γ-orbits on the plane, where Γ is an arbitrary lattice in $\mathrm{SL}_2(\mathbb{R})$.

In our discussion below Γ will denote for instance a discrete subgroup of the group $\mathrm{Aff}(\mathbb{R}^d)$ of affine transformations of $\mathbb{R}^d$. We recall the definition of Diophantine exponent for Γ from the introduction.

DEFINITION 3.1. Assume that for $x, x_0 \in \overline{\Gamma x}$, there exist constants $c = c(x, x_0)$ and $\epsilon_0 = \epsilon_0(x, x_0)$ such that for all $\epsilon < \epsilon_0$, the system of inequalities

$$\|\gamma^{-1} x - x_0\|_{\infty} \leq \epsilon \quad \text{and} \quad \|\gamma\| \leq c\epsilon^{-\kappa}.$$

has a solution $\gamma \in \Gamma$. Define the Diophantine approximation exponent $\kappa_{\Gamma}(x, x_0)$ as the infimum of $\kappa > 0$ such that the foregoing inequalities have a solution.

We note that the exponent defined above generalizes the Diophantine exponent for *uniform* approximation by $\mathrm{SL}_2(\mathbb{Z})$-orbits in $\mathbb{R}^2$ as considered by Laurent and Nogueira [34]. In fact, the definition above makes sense for general varieties equipped with a distance dist. In the next section, we present a wide variety of new estimates for Diophantine exponents.

The exponent κ_{Γ} above is naturally related to the exponent κ_p discussed in the previous section (which is associated with dense orbits of the group $\Gamma = G(\mathbb{Z}[\frac{1}{p}])$).

A basic geometric argument leads to a lower bound on the Diophantine exponent for dense lattice orbits on homogeneous varieties. Let us define the growth exponent of Γ-orbits by

$$a_{\Gamma}(x) = \sup_{\text{compact } K \subset X(\mathbb{R})} \limsup_{R \to \infty} \frac{\log N_R(K, x)}{\log R}$$

where $N_R(K, x)$ denotes the number of elements $\gamma \in \Gamma$ such that $\|\gamma\| \leq R$ and $\gamma^{-1} x$ belongs to K. It is not hard to show that for almost every $x_0 \in X$,

$$\kappa_{\Gamma}(x, x_0) \geq \frac{\dim(X)}{a_{\Gamma}(x)}.$$

The quantity $a_{\Gamma}(x)$ could be difficult to estimate in general, but if the variety X is homogeneous, it can be estimated in terms of volume growth of a suitable subgroup. X can then be identified with the homogeneous space G/H where H is a closed subgroup of G. We set

$$a(X) = \limsup_{t \to \infty} \frac{\log m_H(H_t)}{t},$$

where $H_t = \{h \in H; \log \|h\| \leq t\}$ and m_H is a right-invariant Haar measure on H. Then one can show using discreteness of Γ that for every $x \in X$,

$$a_{\Gamma}(x) \leq a(X).$$

In particular, for almost every $x_0 \in X$,

$$(3.1) \qquad \kappa_\Gamma(x, x_0) \geq \frac{\dim(X)}{a(X)}.$$

Thus the fundamental question that arises is to determine when this lower bound is in fact sharp (for almost all x), and in general to give an estimate for the upper bound.

We show that in the above homogeneous setting one can reduce the original problem of quantitative density of Γ-orbits in X to the problem of quantitative density of the corresponding H orbits in the space $Y := \Gamma\backslash G$. More precisely, we have the following dynamical correspondence:

PROPOSITION 3.2 (Ghosh–Gorodnik–Nevo [19]). *Let $x = gH \in X$, $x_0 = g_0H \in X$ and $y = \Gamma g \in Y$, $y_0 = \Gamma g_0 \in Y$. There exists a sequence of neighbourhoods $\mathcal{O}_\epsilon(y_0)$ of y_0 such that if there exists $h \in H$ such that*

$$y \cdot h \in \mathcal{O}_\epsilon(y_0) \quad and \quad \|h\| \leq R,$$

then there exists $\gamma \in \Gamma$ such that

$$\|\gamma^{-1}x - x_0\|_\infty \leq \epsilon \quad and \quad \|\gamma\| \leq c(x, x_0)R,$$

where $c(x, x_0)$ is uniform over x, x_0 in compact sets.

Hence, the problem of establishing upper bounds on $\kappa_\Gamma(x, x_0)$ is closely related to the shrinking target problem for the orbit yH in Y with respect to the family of neighbourhoods $\mathcal{O}_\epsilon(y_0)$. The latter can be approached using a quantitative mean ergodic theorem for the action of H on Y.

Now we assume that the discrete group Γ has finite covolume in G and consider a family of averaging operators $\pi_Y(\beta_t) : L^2(Y) \to L^2(Y)$ defined by

$$\phi(y) := \frac{1}{m_H(H_t)} \int_{H_t} \phi(yh) \, dm_H(h).$$

Let us suppose that there exist $C, \theta > 0$ such that for all sufficiently large t,

$$(3.2) \qquad \left\| \pi_Y(\beta_t)(\phi) - \int_Y \phi \, dm_Y \right\| \leq C_\theta \, m_H(H_t)^{-\theta} \|\phi\|_2,$$

where m_Y denotes the normalised Haar measure on Y. Let $\theta_\Gamma(X)$ denote the supremum over θ's for which the estimate (2.11) holds.

THEOREM 3.3 (Ghosh–Gorodnik–Nevo [19]). *For every $x_0 \in X$ and almost every $x \in X$,*

$$\kappa_\Gamma(x, x_0) \leq (2\theta_\Gamma(X))^{-1}\frac{\dim(X)}{a(X)}.$$

Moreover, the constants $c(x, x_0)$ and $\epsilon_0(x, x_0)$ in Definition 3.1 are uniform over x, x_0 in compact sets.

Looking at the case $\theta_\Gamma(X) = 1/2$ we have the following sample conclusion:

COROLLARY 3.4. *If G and the stability group H are semisimple and non-compact, and the representation of H on $L^2_0(\Gamma\backslash G)$ is a tempered representation of H, then the Diophantine exponent of Γ-action on $X = G/H$ is best possible, and* *is given by*

$$\kappa_\Gamma(x, x_0) = \frac{\dim(X)}{a(X)}$$

for every x_0 and almost every x in X.

We note that the results established in Theorem 3.3 and Corollary 3.4 hold not only for linear and affine action on varieties in $\mathbb{R}^d$, but also for linear and affine actions over other local fields and more generally for actions on general homogeneous spaces of locally compact groups satisfying some natural assumptions. We refer to [19] for general statements of these results. In this general setting one defines the dimension of X as

$$\dim(X) = \limsup_{\epsilon \to 0} \frac{\log m_X(B_\epsilon(x_0))}{\log \epsilon},$$

where $B_\epsilon(x_0)$ is the ϵ-ball around x_0 in X, and m_X is the measure on X induced by the chosen measures on G and H. We note that under the assumption imposed in [19] this limit is independent of x_0.

The bound obtained in Theorem 3.3 for the approximation exponent depends on spectral information pertaining to automorphic representations $L^2(\Gamma \backslash G)$ as well as arithmetic data related to the group and the lattice. As such, it is a highly nontrivial task to compute these parameters in any given example. One of the main advantages in our approach is that this task is feasible in many interesting cases, and in fact often leads to best possible results. This necessarily involves a detailed study of the spectral theory of unitary representations of semisimple groups, and we now turn to demonstrating the method in some cases.

4. Examples of Diophantine exponents

In this section we give some concrete examples of explicit estimates of Diophantine exponents arising in Theorem 3.3, complementing some of the examples presented in [19]. For simplicity of exposition, let now F denote the fields $\mathbb{R}$, $\mathbb{C}$ or $\mathbb{Q}_p$.[3] Let G be a linear algebraic subgroup of the group $\mathrm{SL}_n(F) \ltimes F^n$ considered as a group of affine transformations of F^n. We fix a norm on $\mathbb{R}^n$ and $\mathbb{C}^n$, and a (vector space) norm on $\mathrm{M}_n(\mathbb{R})$ and $\mathrm{M}_n(\mathbb{C})$. In the local field case we take the standard valuation on the field, and the standard maximum norm on the linear space F^n, and on $\mathrm{M}_n(F)$. We view the affine group $\mathrm{SL}_n(F) \ltimes F^n$, $n \geq 2$ as a subgroup of $\mathrm{SL}_{n+1}(F)$, specifically as the stability group of the standard basis vector e_{n+1}, and consider norms on it by restriction from $\mathrm{SL}_{n+1}(F) \subset \mathrm{M}_{n+1}(F)$. Let $X \subset F^n$ be an affine subvariety which is invariant and homogeneous under the G-action, so that $X \simeq G/H$ where H is closed subgroup of G. We define the distance on X by restricting the norm defined on F^n. Let Γ be a lattice subgroup of G such that almost every Γ-orbit is dense in X.

We now proceed to describe several examples of classical Diophantine approximation problems in this context and estimate the exponents $\kappa_\Gamma(x, x_0)$ defined in Definition 3.1.

We remark that any choice of norm on $\mathbb{R}^n$ or $\mathbb{C}^n$ (and hence on X) and on $\mathrm{M}_n(\cdot)$ (and hence on G and H) does not change the estimate of the exponent. We

---

[3] Although we do not treat this case, the methods in [19] are very general and also hold for local fields of positive characteristic. In this setting one also has the advantage of better spectral estimates in certain instances, for example arising from work of Drinfeld and Lafforgue.

will thus choose the norm most convenient for us, namely one whose restriction to H has convenient properties.

For simplicity, we use notation $A \ll B$ if $A \leq c\,B$ for some constant c, and $A \asymp B$ if $c_1\,B \leq A \leq c_2\,B$ for some constants c_1, c_2.

4.1. Homogeneous Diophantine approximation in linear space.

Consider first the classical case of homogeneous Diophantine approximation in the linear action of a lattice subgroup of $\mathrm{SL}_n(F)$ on $F^n \setminus \{0\}$.

PROPOSITION 4.1. *For $n \geq 3$ the exponent of Diophantine approximation for an arbitrary lattice Γ in $G = \mathrm{SL}_n(F)$ acting on $X = F^n \setminus \{0\}$ is estimated by*

$$\frac{n}{(n-1)^2} \leq \kappa_\Gamma(x, x_0) \leq \frac{n}{n-1}$$

for almost every $x, x_0 \in F^n \setminus \{0\}$.

It is also possible to give an upper bound for the Diophantine exponent for lattices in $\mathrm{SL}_2(F)$ acting on the plane $F^2 \setminus \{0\}$. The bound depends on the lattice subgroup in question, and involves different and more elaborate considerations. In the case of the Euclidean plane $\mathbb{R}^2$, the bound obtained for $\mathrm{SL}_2(\mathbb{Z})$ holds for any tempered lattice, namely any lattice Γ for which $L_0^2(\Gamma \setminus G)$ is a tempered representation of G. For further details and results about Diophantine approximation by lattice orbits in the real and complex plane we refer to [19].

REMARK 4.2.   • The only results we are aware of in the literature regarding estimates of the Diophantine exponent for lattice actions on homogeneous varieties are due to Laurent–Nogueira [34, 35] and to Maucourant–Weiss [39]. Laurent and Nogueira established that generically $\kappa_\Gamma(x, x_0) \leq 3$ for $\mathrm{SL}_2(\mathbb{Z})$ acting on the plane by explicitly constructing a sequence of approximants using a suitable continued fractions algorithm. Maucourant and Weiss have also established an (explicit, but not as sharp) upper bound for $\kappa_\Gamma(x, x_0)$ for arbitrary lattice subgroups of $\mathrm{SL}_2(\mathbb{R})$ using effective equidistribution of horocycle flows.
     • We note that for the linear action, determining the exact value of $\kappa_\Gamma(x, x_0)$ generically remains an open problem, for any lattice subgroup, over any field, in any dimension.

PROOF. *Part I : volume estimates.* $F^n \setminus \{0\}$ is a homogeneous space of $\mathrm{SL}_n(F)$, and the stability H group of the standard basis vector e_n is isomorphic to the group $\mathrm{SL}_{n-1}(F) \ltimes F^{n-1}$. We choose a vector space norm on $\mathrm{M}_n(F)$ with the property that its restriction to $h \in H = \mathrm{SL}_{n-1}(F) \ltimes F^{n-1}$, written as $h = (h_1, v_1)$, is given by

$$\|(h_1, v_1)\| = \max\left\{\|h_1\|, \|v_1\|\right\},$$

where $\|\cdot\|$ denotes the Euclidean norm when $F = \mathbb{R}, \mathbb{C}$ and the maximal norm when $F = \mathbb{Q}_p$.

We can now evaluate the volume growth of $H_t = \{h \in H \,;\, \log\|h\| < t\}$. Due to our choice of the norm,

$$m_H(H_t) = \left(\int_{\|h_1\| \leq e^t} dh_1\right) \cdot \left(\int_{\|v_1\| \leq e^t} dv_1\right).$$

The first integral can be estimated using [**15**, Appendix 1] for $F = \mathbb{R}, \mathbb{C}$ or [**20**, Sec. 7] for general local fields. This gives

$$(4.1) \qquad \int_{\|h_1\| \leq e^t} dh_1 \asymp \begin{cases} e^{t((n-1)^2-(n-1))} & \text{for } F = \mathbb{R}, \mathbb{Q}_p, \\ e^{2t((n-1)^2-(n-1))} & \text{for } F = \mathbb{C}. \end{cases}$$

Also clearly,

$$\int_{\|v_1\| \leq e^t} dv_1 \asymp \begin{cases} e^{t(n-1)} & \text{for } F = \mathbb{R}, \mathbb{Q}_p, \\ e^{2t(n-1)} & \text{for } F = \mathbb{C}. \end{cases}$$

Hence,

$$a(X) = \limsup_{t \to \infty} \frac{\log m_H(H_t)}{t} \asymp \begin{cases} (n-1)^2 & \text{for } F = \mathbb{R}, \mathbb{Q}_p, \\ 2(n-1)^2 & \text{for } F = \mathbb{C}. \end{cases}$$

Now it follows from (3.1) that for almost every $x_0 \in X$,

$$\kappa_\Gamma(x, x_0) \geq \frac{\dim(X)}{a(X)} = \frac{n}{(n-1)^2},$$

which proves the lower bound in the proposition.

Part II : spectral estimates. Let π denote the representation of H on $L_0^2(\Gamma \backslash G)$. We proceed to estimate the decay of the operator norm $\pi(\beta_t)$, where β_t are the Haar-uniform averages supported on the subsets H_t. The cases $n \geq 4$ and $n = 3$ require separate arguments.

When $n \geq 4$, we use that for the group $\mathrm{SL}_k(F)$, any unitary representation without invariant vectors is L^p-integrable for $p > 2(k-1)$, provided that $k \geq 3$ (see e.g. [**21**, **40**]). Hence, using the spectral transfer principle [**38**], when $n \geq 4$ any unitary representation σ of $\mathrm{SL}_{n-1}(F)$ without invariant vectors has the property that $\sigma^{\otimes(n-2)}$ is weakly contained in the regular representation of $\mathrm{SL}_{n-1}(F)$. In particular, this statement holds for the representation of $\mathrm{SL}_{n-1}(F)$ of the form $\sigma = \pi|_{\mathrm{SL}_{n-1}(F)}$, for any lattice subgroup $\Gamma \subset \mathrm{SL}_n(F)$. Thus the restriction of the tensor power $\pi^{\otimes(n-2)}$ to the closed subgroup $\mathrm{SL}_{n-1}(F)$ is weakly contained in the regular representation of $\mathrm{SL}_{n-1}(F)$. By [**8**], it follows that the K-finite matrix coefficients of the representation $\pi|_{\mathrm{SL}_{n-1}(F)}$ are dominated along the split Cartan subgroup by a scalar multiple of $\Xi_{n-1}^{1/(n-2)}$, where Ξ_{n-1} denotes the Harish-Chandra function of $\mathrm{SL}_{n-1}(F)$.

We claim that the last estimate also holds when $n = 3$. To prove this, we consider the restriction of the representation π to $H = \mathrm{SL}_2(F) \ltimes F^2$. Since this representation has no F^2-invariant vectors, it follows from the Kazhdan's original argument [**22**] that $\pi|_{\mathrm{SL}_2(F)}$ is weakly contained in the regular representation of $\mathrm{SL}_2(F)$, so that the same estimate as above applies.

Applying the general method of spectral estimates on groups with an Iwasawa decomposition (see [**19**] for details), and using the fact that the sets H_t are bi-invariant by a maximal compact subgroup of $\mathrm{SL}_{n-1}(F)$, we can estimate the operator norm by integrating the bound $\Xi_{n-1}^{1/(n-2)}$ over H_t, where we view the function as independent of the second variable $v_1 \in F^{n-1}$. This gives

$$\|\pi(\beta_t)\| \leq \frac{1}{m_H(H_t)} \int_{H_t} \Xi_{n-1}^{1/(n-2)}(h_1) dh_1 dv_1$$

$$= \frac{1}{m_H(H_t)} m_{F^{n-1}}(B_{e^t}) \int_{\|h_1\| \leq e^t} \Xi_{n-1}^{1/(n-2)}(h_1) \, dh_1$$

Since the function $\Xi_{n-1}^{1/(n-2)}$ is $L^{2(n-2)+\eta}$-integrable for every $\eta > 0$, we can apply the following simple estimate, based on Hölder's inequality for the conjugate exponents $\frac{1}{p} = \frac{1}{2(n-2)+\eta}$ and $\frac{1}{q} = 1 - \frac{1}{2(n-2)+\eta}$:

$$\int_{\|h_1\| \leq e^t} \Xi_{n-1}^{1/(n-2)}(h_1) \, dh_1 \leq \|\Xi_{n-1}^{1/(n-2)}\|_p \cdot \|\chi_{\{\|h_1\| \leq e^t\}}\|_q$$

$$\ll m_{\mathrm{SL}_{n-1}(F)}(\{\|h_1\| \leq e^t\})^{1 - \frac{1}{2(n-2)+\eta}}.$$

Hence, using the above volume estimates, we deduce that

$$\|\pi(\beta_t)\| \ll m_{\mathrm{SL}_{n-1}(F)}(\{\|h_1\| \leq e^t\})^{-\frac{1}{2(n-2)+\eta}} \ll m_H(H_t)^{-\theta + \eta'}$$

for every $\eta' > 0$ with $\theta = \frac{1}{2(n-1)}$. Now it follows from Theorem 3.3 that for almost every $x \in X$,

$$\kappa_\Gamma(x, x_0) \leq \frac{\dim(X)}{2\theta a(X)} = \frac{n}{n-1},$$

as claimed.   □

4.2. Simultaneous approximation on 3-dimensional space.

We consider the standard action of $\mathrm{SL}_3(F)$ on the space $U = F^3 \setminus \{0\}$ and the action of $\mathrm{SL}_3(F)$ on the space $U \times U$ defined by $g(v, w) = (gv, (g^{-1})^t w)$. Given a lattice Γ in $\mathrm{SL}_3(F)$, we are interested in the problem of simultaneous Diophantine approximation in U, namely, we will seek to solve the inequalities

$$\|\gamma v - v'\| \leq \epsilon, \quad \|(\gamma^{-1})^t w - w'\| \leq \epsilon \quad \text{and} \quad \|\gamma\| \leq \epsilon^{-\zeta}.$$

We note that the action on $U \times U$ preserves the standard bilinear form $J(v, w) = \sum_{i=1}^{3} v_i w_i$ and hence each of the subvarieties

$$W_\alpha = \{(v, w) \in U \times U(F) \, ; \, J(v, w) = \alpha\}.$$

We establish the following best possible result for Diophantine exponents, uniformly for all lattices.

PROPOSITION 4.3. *For an arbitrary lattice Γ in $\mathrm{SL}_3(F)$, the exponent of Diophantine approximation for Γ-action on W_α with $\alpha \neq 0$ is given by*

$$\kappa_\Gamma(x, x_0) = 5/2$$

for almost every $x, x_0 \in W_\alpha$.

PROOF. We first observe that the action of $\mathrm{SL}_3(F)$ on W_α with $\alpha \neq 0$ is transitive. Clearly W_α contains the orbit $\mathrm{SL}_3(F) \cdot (e_3, \alpha e_3)$ where $e_3 = (0, 0, 1)^t$. This is in fact an equality, namely if $J(v, w) = \alpha$, then for some $g \in SL_3(F)$ we have $ge_3 = v$ and $(g^{-1})^t \alpha e_3 = w$, or equivalently $g^t w = \alpha e_3$. Indeed the first condition amounts to the third column of g being equal to v, and the second is solved by choosing the first two columns of g to be a basis of $w^\perp$ (under the form J), linearly independent of v, and adjusting one of the basis vectors so that g has determinant 1.

Since the action is transitive, $W_\alpha \simeq \mathrm{SL}_3(F)/H$ where H is the copy of $SL_2(F)$ embedded in $\mathrm{SL}_3(F)$ in the upper left hand corner. It follows from Kazhdan's original argument [**22**] (proving property T) that the unitary representation of H on $L_0^2(\Gamma \setminus \mathrm{SL}_3(F))$ is tempered. Hence, by Corollary 3.4,

$$\kappa_\Gamma(x, x_0) = \frac{\dim(W_\alpha)}{a(W_\alpha)} = 5/2$$

for almost every $x, x_0 \in W_\alpha$. Here we used that $a(W_\alpha) = 2$ when $F = \mathbb{R}, \mathbb{Q}_p$ and $a(W_\alpha) = 4$ when $F = \mathbb{C}$, which follows from the volume estimates (4.1). □

4.3. Inhomogeneous Diophantine approximation. We turn to consider the action of a lattice subgroup Γ of the affine group $\mathrm{SL}_n(F) \ltimes F^n$ acting on F^n.

PROPOSITION 4.4. *For $n \geq 3$ the exponent of Diophantine approximation for an arbitrary lattice Γ in $G = \mathrm{SL}_n(F) \ltimes F^n$ acting on $X = F^n \setminus \{0\}$ is estimated by*

$$(4.2) \qquad \frac{1}{n-1} \leq \kappa_\Gamma(x, x_0) \leq 1$$

for almost every $x, x_0 \in X$.

We also remark that when $n = 2$ and the lattice satisfies a suitable spectral condition, the best possible exponent $\kappa_\Gamma(x, x_0) = \frac{1}{n-1}$ is achieved generically. This applies, for example, to the lattice $\mathrm{SL}_2(\mathbb{Z}) \ltimes \mathbb{Z}^2$ acting on $\mathbb{R}^2$, and the lattice $\mathrm{SL}_2(\mathbb{Z}[i]) \ltimes \mathbb{Z}[i]^2$ acting on $\mathbb{C}^2$. For full details on this matter we refer to [**19**].

PROOF. We note that $X \simeq G/H$ where $H = \mathrm{SL}_n(F)$ is the stabiliser of the origin. We introduce the norm on G as in the proof of Proposition 4.1. Then it follows from (4.1) that

$$(4.3) \qquad a(X) = \limsup_{t \to \infty} \frac{\log m_H(H_t)}{t} \asymp \begin{cases} n^2 - n & \text{for } F = \mathbb{R}, \mathbb{Q}_p, \\ 2(n^2 - n) & \text{for } F = \mathbb{C}. \end{cases}$$

Now the lower bound in (4.5) follows from (3.1). To establish an upper bound, we need to estimate decay of the operator norm of $\pi(\beta_t)$ where π denotes the unitary representation of H on $L_0^2(\Gamma \setminus G)$, and β_t is the Haar uniform average supported on the set H_t. As in the proof of Proposition 4.1, the representation $\pi^{\otimes(n-1)}$ is weakly contained in the regular representation, and using bi-invariance of β_t under the maximal compact subgroup, the estimate of general matrix coefficients by a power of the Harish-Chandra function from [**8**], and Hölder's inequality, we deduce that

$$(4.4) \qquad \|\pi(\beta_t)\| \leq \frac{1}{m_H(H_t)} \int_{H_t} \Xi_n^{1/(n-1)}(h) dm_H(h) \ll m_H(H_t)^{-\frac{1}{2(n-1)+n}}.$$

Hence, the upper bound in (4.5) now follows from Theorem 3.3. □

4.4. The variety of matrices with a fixed determinant. Consider the variety of matrices with a determinant $k \neq 0$:

$$X = \{x \in \mathrm{M}_n(F) \,;\, \det(x) = k\} .$$

The group $G = \mathrm{SL}_n(F) \times \mathrm{SL}_n(F)$ acts transitively on X, via $(g_1, g_2)x = g_1 x g_2^{-1}$. We introduce a norm on G by embedding it diagonally in $\mathrm{SL}_{2n}(F)$. Given a lattice Γ in G, we are interested in estimating the exponents of Diophnatine approximation for Γ-orbits in X, namely, to investigate existence of solutions $(\gamma_1, \gamma_2) \in \Gamma$ of the inequalities

$$\left\|\gamma_1 x \gamma_2^{-1} - x_0\right\| \leq \epsilon \quad \text{and} \quad \|(\gamma_1, \gamma_2)\| \leq \epsilon^{-\kappa}.$$

In this setting we establish the following bounds:

PROPOSITION 4.5. *For $n \geq 3$ the exponent of Diophantine approximation for an arbitrary irreducible lattice Γ in $G = \mathrm{SL}_n(F) \times \mathrm{SL}_n(F)$ acting on X is estimated by*

$$\frac{n+1}{n} \leq \kappa_\Gamma(x, x_0) \leq \frac{n^2 - 1}{2n}$$

for almost every $x, x_0 \in X$.

When $n = 3$ the lower and upper bounds match, so that the optimal exponent is given by $4/3$.

We note that when $n = 2$ it is also possible to obtain estimate of the Diophantine exponent, but these depend on the irreducible lattice chosen. We refer to [19] for these results.

PROOF. We observe that $X \simeq G/H$ where $H = \{(h, h); h \in \mathrm{SL}_n(F)\} \simeq \mathrm{SL}_n(F)$ is the stabiliser of the identity matrix. The growth rate of $m_H(H_t)$ can be estimated as in the previous sections (cf. (4.3)), and $\dim(X) = n^2 - 1$ when $F = \mathbb{R}, \mathbb{Q}_p$ and $\dim(X) = 2(n^2 - 1)$ when $F = \mathbb{C}$. The lower bound follows from (3.1). To prove the upper bound, we need to estimate $\|\pi(\beta_t)\|$ where π denotes the representation of H on $L_0^2(\Gamma \backslash G)$. The bound (4.4) is valid for all unitary representations of $\mathrm{SL}_n(F)$ without invariant vectors. Hence, it applies in our case as well, and it follows from Theorem 3.3 that with $\theta = \frac{1}{2(n-1)}$,

$$(4.5) \qquad\qquad \kappa_\Gamma(x, x_0) \leq \frac{\dim(X)}{2\theta a(X)} = \frac{n^2 - 1}{n}$$

for almost every $x \in X$.

However, in the present situation it is possible to give a better estimate. Indeed, since the lattice is irreducible, the spectral decomposition of the representation of $G = \mathrm{SL}_n(F) \times \mathrm{SL}_n(F)$ in $L_0^2(\Gamma \backslash G)$ involves tensor products $\pi_1 \otimes \pi_2$ of irreducible infinite dimensional representations of $\mathrm{SL}_n(F)$. It follows that the restriction of the associated K-finite matrix coefficients to the diagonally embedded group $H \simeq \mathrm{SL}_n(F)$ is in fact not just in $L^{2(n-1)+\eta}(H)$ but in $L^{n-1+\eta}(H)$ for every $\eta > 0$. For a detailed account of this argument we refer to [19]. It follows that the spectral estimate obtained is

$$\|\pi(\beta_t)\| \ll m_H(H_t)^{-\frac{1}{n-1+\eta}}.$$

Hence, (4.5) holds with $\theta = \frac{1}{n-1}$, and this implies the upper bound in the proposition. $\qquad\square$

References

[1] Emil Artin, *Ein mechanisches system mit quasiergodischen bahnen* (German), Abh. Math. Sem. Univ. Hamburg **3** (1924), no. 1, 170–175, DOI 10.1007/BF02954622. MR3069425

[2] Valentin Blomer and Farrell Brumley, *On the Ramanujan conjecture over number fields*, Ann. of Math. (2) **174** (2011), no. 1, 581–605, DOI 10.4007/annals.2011.174.1.18. MR2811610

[3] Yann Bugeaud, *Approximation by algebraic numbers*, Cambridge Tracts in Mathematics, vol. 160, Cambridge University Press, Cambridge, 2004. MR2136100 (2006d:11085)

[4] Yann Bugeaud and Michel Laurent, *On exponents of homogeneous and inhomogeneous Diophantine approximation*, Mosc. Math. J. **5** (2005), no. 4, 747–766, 972. MR2266457 (2007g:11077)

[5] Yann Bugeaud and Michel Laurent, *Exponents of Diophantine approximation and Sturmian continued fractions* (English, with English and French summaries), Ann. Inst. Fourier (Grenoble) **55** (2005), no. 3, 773–804. MR2149403 (2006f:11078)

[6] Yann Bugeaud and Michel Laurent, *Exponents of Diophantine approximation*, Diophantine geometry, CRM Series, vol. 4, Ed. Norm., Pisa, 2007, pp. 101–121. MR2349650 (2008i:11097)

[7] Yitwah Cheung, *Hausdorff dimension of the set of singular pairs*, Ann. of Math. (2) **173** (2011), no. 1, 127–167, DOI 10.4007/annals.2011.173.1.4. MR2753601 (2011j:22017)

[8] M. Cowling, U. Haagerup, and R. Howe, *Almost L^2 matrix coefficients*, J. Reine Angew. Math. **387** (1988), 97–110. MR946351 (89i:22008)

[9] S. G. Dani, *Bounded orbits of flows on homogeneous spaces*, Comment. Math. Helv. **61** (1986), no. 4, 636–660, DOI 10.1007/BF02621936. MR870710 (88i:22011)

[10] S. G. Dani, *On orbits of unipotent flows on homogeneous spaces*, Ergodic Theory Dynam. Systems **4** (1984), no. 1, 25–34, DOI 10.1017/S0143385700002248. MR758891 (86b:58068)

[11] S. G. Dani, *Divergent trajectories of flows on homogeneous spaces and Diophantine approximation*, J. Reine Angew. Math. **359** (1985), 55–89, DOI 10.1515/crll.1985.359.55. MR794799 (87g:58110a)

[12] S. G. Dani, *On orbits of unipotent flows on homogeneous spaces. II*, Ergodic Theory Dynam. Systems **6** (1986), no. 2, 167–182. MR857195 (88e:58052)

[13] S. G. Dani and Hemangi Shah, *Badly approximable numbers and vectors in Cantor-like sets*, Proc. Amer. Math. Soc. **140** (2012), no. 8, 2575–2587, DOI 10.1090/S0002-9939-2011-11105-5. MR2910746

[14] Cornelia Druțu, *Diophantine approximation on rational quadrics*, Math. Ann. **333** (2005), no. 2, 405–469, DOI 10.1007/s00208-005-0683-x. MR2195121 (2007b:11103)

[15] W. Duke, Z. Rudnick, and P. Sarnak, *Density of integer points on affine homogeneous varieties*, Duke Math. J. **71** (1993), no. 1, 143–179, DOI 10.1215/S0012-7094-93-07107-4. MR1230289 (94k:11072)

[16] L. Fishman, D. Kleinbock, K. Merill, D. Simmons, *Intrinsic Diophantine approximation on manifolds*, preprint arXiv:1405.7650.

[17] Anish Ghosh, Alexander Gorodnik, and Amos Nevo, *Diophantine approximation and automorphic spectrum*, Int. Math. Res. Not. IMRN **21** (2013), 5002–5058. MR3123673

[18] A. Ghosh, A. Gorodnik and A. Nevo, *Metric Diophantine approximation on homogeneous varieties*. Compositio Math., **150** (2014), no.8, 1435–1456.

[19] A. Ghosh, A. Gorodnik and A. Nevo, *Best possible rates of distribution of dense lattice orbits in homogeneous spaces*, preprint, arXiv:1407.2824.

[20] Alex Gorodnik and Barak Weiss, *Distribution of lattice orbits on homogeneous varieties*, Geom. Funct. Anal. **17** (2007), no. 1, 58–115, DOI 10.1007/s00039-006-0583-6. MR2306653 (2008i:37012)

[21] Roger Howe and Eng-Chye Tan, *Nonabelian harmonic analysis*, Universitext, Springer-Verlag, New York, 1992. Applications of SL(2, **R**). MR1151617 (93f:22009)

[22] D. A. Každan, *On the connection of the dual space of a group with the structure of its closed subgroups* (Russian), Funkcional. Anal. i Priložen. **1** (1967), 71–74. MR0209390 (35 #288)

[23] D. Y. Kleinbock and G. A. Margulis, *Flows on homogeneous spaces and Diophantine approximation on manifolds*, Ann. of Math. (2) **148** (1998), no. 1, 339–360, DOI 10.2307/120997. MR1652916 (99j:11083)

[24] D. Y. Kleinbock and G. A. Margulis, *Logarithm laws for flows on homogeneous spaces*, Invent. Math. **138** (1999), no. 3, 451–494, DOI 10.1007/s002220050350. MR1719827 (2001i:37046)

[25] D. Y. Kleinbock and G. A. Margulis, *Bounded orbits of nonquasiunipotent flows on homogeneous spaces*, Sinaĭ's Moscow Seminar on Dynamical Systems, Amer. Math. Soc. Transl. Ser. 2, vol. 171, Amer. Math. Soc., Providence, RI, 1996, pp. 141–172. MR1359098 (96k:22022)

[26] Dmitry Kleinbock and Barak Weiss, *Dirichlet's theorem on Diophantine approximation and homogeneous flows*, J. Mod. Dyn. **2** (2008), no. 1, 43–62. MR2366229 (2008k:11078)

[27] Dmitry Kleinbock, *Some applications of homogeneous dynamics to number theory*, Smooth ergodic theory and its applications (Seattle, WA, 1999), Proc. Sympos. Pure Math., vol. 69, Amer. Math. Soc., Providence, RI, 2001, pp. 639–660, DOI 10.1090/pspum/069/1858548. MR1858548 (2002g:37009)

[28] D. Kleinbock, *Extremal subspaces and their submanifolds*, Geom. Funct. Anal. **13** (2003), no. 2, 437–466, DOI 10.1007/s000390300011. MR1982150 (2004f:11073)

[29] Dmitry Kleinbock, *An extension of quantitative nondivergence and applications to Diophantine exponents*, Trans. Amer. Math. Soc. **360** (2008), no. 12, 6497–6523, DOI 10.1090/S0002-9947-08-04592-3. MR2434296 (2009h:37008)

[30] Dmitry Kleinbock, *An 'almost all versus no' dichotomy in homogeneous dynamics and Diophantine approximation*, Geom. Dedicata **149** (2010), 205–218, DOI 10.1007/s10711-010-9477-8. MR2737689 (2012a:37008)

[31] Dmitry Kleinbock, *Diophantine properties of measures and homogeneous dynamics*, Pure Appl. Math. Q. **4** (2008), no. 1, Special Issue: In honor of Grigory Margulis., 81–97, DOI 10.4310/PAMQ.2008.v4.n1.a3. MR2405996 (2009e:37005)

[32] D. Kleinbock and K. Merrill, *Rational approximation on spheres*. arXiv1301.0989.

[33] Serge Lang, *Report on diophantine approximations*, Bull. Soc. Math. France **93** (1965), 177–192. MR0193064 (33 #1286)

[34] Michel Laurent and Arnaldo Nogueira, *Approximation to points in the plane by SL(2, ℤ)-orbits*, J. Lond. Math. Soc. (2) **85** (2012), no. 2, 409–429, DOI 10.1112/jlms/jdr061. MR2901071

[35] Michel Laurent and Arnaldo Nogueira, *Inhomogeneous approximation with coprime integers and lattice orbits*, Acta Arith. **154** (2012), no. 4, 413–427, DOI 10.4064/aa154-4-5. MR2949877

[36] G. A. Margulis, *The action of unipotent groups in a lattice space* (Russian), Mat. Sb. (N.S.) **86(128)** (1971), 552–556. MR0291352 (45 #445)

[37] Grigorii A. Margulis, *Dynamical and ergodic properties of subgroup actions on homogeneous spaces with applications to number theory*, Proceedings of the International Congress of Mathematicians, Vol. I, II (Kyoto, 1990), Math. Soc. Japan, Tokyo, 1991, pp. 193–215. MR1159213 (93g:22011)

[38] Amos Nevo, *Spectral transfer and pointwise ergodic theorems for semi-simple Kazhdan groups*, Math. Res. Lett. **5** (1998), no. 3, 305–325, DOI 10.4310/MRL.1998.v5.n3.a5. MR1637840 (99e:28030)

[39] François Maucourant and Barak Weiss, *Lattice actions on the plane revisited*, Geom. Dedicata **157** (2012), 1–21, DOI 10.1007/s10711-011-9596-x. MR2893477

[40] Hee Oh, *Uniform pointwise bounds for matrix coefficients of unitary representations and applications to Kazhdan constants*, Duke Math. J. **113** (2002), no. 1, 133–192, DOI 10.1215/S0012-7094-02-11314-3. MR1905394 (2003d:22015)

[41] Peter Sarnak, *Notes on the generalized Ramanujan conjectures*, Harmonic analysis, the trace formula, and Shimura varieties, Clay Math. Proc., vol. 4, Amer. Math. Soc., Providence, RI, 2005, pp. 659–685. MR2192019 (2007a:11067)

[42] Wolfgang M. Schmidt, *Metrical theorems on fractional parts of sequences*, Trans. Amer. Math. Soc. **110** (1964), 493–518. MR0159802 (28 #3018)

[43] Wolfgang M. Schmidt, *Diophantine approximation*, Lecture Notes in Mathematics, vol. 785, Springer, Berlin, 1980. MR568710 (81j:10038)

[44] Eric Schmutz, *Rational points on the unit sphere*, Cent. Eur. J. Math. **6** (2008), no. 3, 482–487, DOI 10.2478/s11533-008-0038-4. MR2425007 (2009c:11112)

[45] Caroline Series, *The modular surface and continued fractions*, J. London Math. Soc. (2) **31** (1985), no. 1, 69–80, DOI 10.1112/jlms/s2-31.1.69. MR810563 (87c:58094)

[46] Michel Waldschmidt, *Density measure of rational points on abelian varieties*, Nagoya Math. J. **155** (1999), 27–53. MR1711387 (2000h:11077)

SCHOOL OF MATHEMATICS, TATA INSTITUTE OF FUNDAMENTAL RESEARCH, MUMBAI, INDIA
E-mail address: ghosh@math.tifr.res.in

SCHOOL OF MATHEMATICS, UNIVERSITY OF BRISTOL, BRISTOL UNITED KINGDOM
E-mail address: a.gorodnik@bristol.ac.uk

DEPARTMENT OF MATHEMATICS, TECHNION IIT, ISRAEL
E-mail address: anevo@tx.technion.ac.il

Contemporary Mathematics
Volume **631**, 2015
http://dx.doi.org/10.1090/conm/631/12604

Ergodicity of principal algebraic group actions

Hanfeng Li, Jesse Peterson, and Klaus Schmidt

Dedicated to Shrikrishna Gopalrao Dani on the occasion of his 65th birthday

ABSTRACT. An *algebraic* action of a discrete group Γ is a homomorphism from Γ to the group of continuous automorphisms of a compact abelian group X. By duality, such an action of Γ is determined by a module $M = \widehat{X}$ over the integer group ring $\mathbb{Z}\Gamma$ of Γ. The simplest examples of such modules are of the form $M = \mathbb{Z}\Gamma/\mathbb{Z}\Gamma f$ with $f \in \mathbb{Z}\Gamma$; the corresponding algebraic action is the *principal algebraic Γ-action* α_f defined by f.

In this note we prove the following extensions of results by Hayes [**2**] on ergodicity of principal algebraic actions: If Γ is a countably infinite discrete group which is not virtually cyclic, and if $f \in \mathbb{Z}\Gamma$ satisfies that right multiplication by f on $\ell^2(\Gamma, \mathbb{R})$ is injective, then the principal Γ-action α_f is ergodic (Theorem 1.3). If Γ contains a finitely generated subgroup with a single end (e.g. a finitely generated amenable subgroup which is not virtually cyclic), or an infinite nonamenable subgroup with vanishing first ℓ^2-Betti number (e.g., an infinite property T subgroup), the injectivity condition on f can be replaced by the weaker hypothesis that f is not a right zero-divisor in $\mathbb{Z}\Gamma$ (Theorem 1.2). Finally, if Γ is torsion-free, not virtually cyclic, and satisfies Linnell's *analytic zero-divisor conjecture*, then α_f is ergodic for every $f \in \mathbb{Z}\Gamma$ (Remark 1.5).

1. Principal Algebraic Group Actions

Let Γ be a countably infinite discrete group with integral group ring $\mathbb{Z}\Gamma$. Every $g \in \mathbb{Z}\Gamma$ is written as a formal sum $g = \sum_\gamma g_\gamma \cdot \gamma$, where $g_\gamma \in \mathbb{Z}$ for every $\gamma \in \Gamma$ and $\sum_{\gamma \in \Gamma} |g_\gamma| < \infty$. The set $\text{supp}(g) = \{\gamma \in \Gamma : g_\gamma \neq 0\}$ is called the *support* of g. For $g = \sum_{\gamma \in \Gamma} g_\gamma \cdot \gamma \in \mathbb{Z}\Gamma$ we denote by $g^* = \sum_{\gamma \in \Gamma} g_\gamma \cdot \gamma^{-1}$ the *adjoint* of g. The map $g \mapsto g^*$ is an *involution* on $\mathbb{Z}\Gamma$, i.e., $(gh)^* = h^*g^*$ for all $g, h \in \mathbb{Z}\Gamma$, where the product fg of two elements $f = \sum_\gamma f_\gamma \cdot \gamma$ and $g = \sum_\gamma g_\gamma \cdot \gamma$ in $\mathbb{Z}\Gamma$ is given by $fg = \sum_{\gamma, \gamma' \in \Gamma} f_\gamma g_{\gamma'} \cdot \gamma\gamma'$.

2010 *Mathematics Subject Classification.* Primary 28D15, 37A25, 20J05.

Key words and phrases. Principal algebraic actions, ergodicity, group cohomology.

The first author was partially supported by the NSF grants DMS-1001625 and DMS-126623, and he would like to thank the Erwin Schrödinger Institute, Vienna, for hospitality and support while some of this work was done,

The second author was partially supported by the NSF grant DMS-1201565 and the Alfred P. Sloan Foundation,

Both the second and third authors would like to thank the University of Buffalo for hospitality and support while some of this work was done.

An *algebraic* Γ-*action* is a homomorphism $\alpha\colon \Gamma \longrightarrow \mathrm{Aut}(X)$ from Γ to the group of (continuous) automorphisms of a compact metrizable abelian group X. If α is an algebraic Γ-action, then $\alpha^\gamma \in \mathrm{Aut}(X)$ denotes the image of $\gamma \in \Gamma$, and $\alpha^{\gamma\gamma'} = \alpha^\gamma \alpha^{\gamma'}$ for every $\gamma, \gamma' \in \Gamma$. The Γ-action α induces an action of $\mathbb{Z}\Gamma$ by group homomorphisms $\alpha^f\colon X \longrightarrow X$, where $\alpha^f = \sum_{\gamma\in\Gamma} f_\gamma \alpha^\gamma$ for every $f = \sum_{\gamma\in\Gamma} f_\gamma \cdot \gamma \in \mathbb{Z}\Gamma$. Clearly, if $f, g \in \mathbb{Z}\Gamma$, then $\alpha^{fg} = \alpha^f \alpha^g$.

Let $\hat{X}$ be the dual group of X. If $\hat{\alpha}^\gamma$ is the automorphism of $\hat{X}$ dual to α^γ, then the map $\hat{\alpha}\colon \Gamma \longrightarrow \mathrm{Aut}(\hat{X})$ satisfies that $\hat{\alpha}^{\gamma\gamma'} = \hat{\alpha}^{\gamma'} \hat{\alpha}^\gamma$ for all $\gamma, \gamma' \in \Gamma$. We write $\hat{\alpha}^f\colon \hat{X} \longrightarrow \hat{X}$ for the group homomorphism dual to α^f and set $f \cdot a = \hat{\alpha}^{f^*} a$ for every $f \in \mathbb{Z}\Gamma$ and $a \in \hat{X}$. The resulting map $(f, a) \mapsto f \cdot a$ from $\mathbb{Z}\Gamma \times \hat{X}$ to $\hat{X}$ satisfies that $(fg) \cdot a = f \cdot (g \cdot a)$ for all $f, g \in \mathbb{Z}\Gamma$ and turns $\hat{X}$ into a module over the group ring $\mathbb{Z}\Gamma$. Conversely, if M is a countable module over $\mathbb{Z}\Gamma$, we set $X = \widehat{M}$ and put $\hat{\alpha}^f a = f^* \cdot a$ for $f \in \mathbb{Z}\Gamma$ and $a \in M$. The maps $\alpha^f\colon \widehat{M} \longrightarrow \widehat{M}$ dual to $\hat{\alpha}^f$, $f \in \mathbb{Z}\Gamma$, define an action of $\mathbb{Z}\Gamma$ by homomorphisms of $\widehat{M}$, which in turn induces an algebraic action α of Γ on $X = \widehat{M}$.

The simplest examples of algebraic Γ-actions arise from $\mathbb{Z}\Gamma$-modules of the form $M = \mathbb{Z}\Gamma/\mathbb{Z}\Gamma f$ with $f \in \mathbb{Z}\Gamma$. Since these actions are determined by principal left ideals of $\mathbb{Z}\Gamma$ they are called *principal algebraic* Γ-*actions*. In order to describe these actions more explicitly we put $\mathbb{T} = \mathbb{R}/\mathbb{Z}$ and define the left and right shift-actions λ and ρ of Γ on $\mathbb{T}^\Gamma$ by setting

$$(1.1) \qquad (\lambda^\gamma x)_{\gamma'} = x_{\gamma^{-1}\gamma'}, \qquad (\rho^\gamma x)_{\gamma'} = x_{\gamma'\gamma},$$

for every $\gamma \in \Gamma$ and $x = (x_{\gamma'})_{\gamma'\in\Gamma} \in \mathbb{T}^\Gamma$. The Γ-actions λ and ρ extend to actions of $\mathbb{Z}\Gamma$ on $\mathbb{T}^\Gamma$ given by

$$(1.2) \qquad \lambda^f = \sum_{\gamma\in\Gamma} f_\gamma \lambda^\gamma, \qquad \rho^f = \sum_{\gamma\in\Gamma} f_\gamma \rho^\gamma$$

for every $f = \sum_{\gamma\in\Gamma} f_\gamma \cdot \gamma \in \mathbb{Z}\Gamma$.

The pairing $\langle f, x \rangle = e^{2\pi i \sum_{\gamma\in\Gamma} f_\gamma x_\gamma}$, $f = \sum_{\gamma\in\Gamma} f_\gamma \cdot \gamma \in \mathbb{Z}\Gamma$, $x = (x_\gamma) \in \mathbb{T}^\Gamma$, identifies $\mathbb{Z}\Gamma$ with the dual group $\widehat{\mathbb{T}^\Gamma}$ of $\mathbb{T}^\Gamma$. We claim that, under this identification,

$$(1.3) \quad X_f := \ker \rho^f = \left\{ x \in \mathbb{T}^\Gamma : \rho^f x = \sum_{\gamma\in\Gamma} f_\gamma \rho^\gamma x = 0 \right\} = (\mathbb{Z}\Gamma f)^\perp \subset \widehat{\mathbb{Z}\Gamma} = \mathbb{T}^\Gamma.$$

Indeed,

$$\langle h, \rho^f x \rangle = \Big\langle h, \sum_{\gamma'\in\Gamma} f_{\gamma'} \rho^{\gamma'} x \Big\rangle = e^{2\pi i \sum_{\gamma\in\Gamma} h_\gamma \sum_{\gamma'\in\Gamma} f_{\gamma'} x_{\gamma\gamma'}}$$
$$= e^{2\pi i \sum_{\gamma\in\Gamma} \sum_{\gamma'\in\Gamma} h_{\gamma\gamma'^{-1}} f_{\gamma'} x_\gamma} = e^{2\pi i \sum_{\gamma\in\Gamma} (hf)_\gamma x_\gamma} = \langle hf, x \rangle$$

for every $h \in \mathbb{Z}\Gamma$ and $x \in \mathbb{T}^\Gamma$, so that $x \in \ker \rho^f$ if and only if $x \in (\mathbb{Z}\Gamma f)^\perp$.

Since the Γ-actions λ and ρ on $\mathbb{T}^\Gamma$ commute, the group $X_f = \ker \rho^f \subset \mathbb{T}^\Gamma$ is invariant under λ, and we denote by α_f the restriction of λ to X_f. In view of this we adopt the following terminology.

DEFINITION 1.1. (X_f, α_f) is the principal algebraic Γ-action defined by $f \in \mathbb{Z}\Gamma$.

In [2] the author calls a countably infinite discrete group Γ *principally ergodic* if every principal algebraic Γ-action α_f, $f \in \mathbb{Z}\Gamma$, is ergodic w.r.t. Haar measure on

X_f and proves that the following classes of groups are principally ergodic: torsion-free nilpotent groups which are not virtually cyclic,[1] free groups on more than one generator, and groups which are not finitely generated.

In order to state our extensions of these results we denote by $\ell^\infty(\Gamma, \mathbb{R}) \subset \mathbb{R}^\Gamma$ the space of bounded real-valued maps $v = (v_\gamma)$ on Γ, where v_γ is the value of v at γ, and we write $\|v\|_\infty = \sup_{\gamma \in \Gamma} |v_\gamma|$ for the supremum norm on $\ell^\infty(\Gamma, \mathbb{R})$. For $1 \le p < \infty$ we set $\ell^p(\Gamma, \mathbb{R}) = \{v = (v_\gamma) \in \ell^\infty(\Gamma, \mathbb{R}) : \|v\|_p = \left(\sum_{\gamma \in \Gamma} |v_\gamma|^p\right)^{1/p} < \infty\}$. By $\ell^p(\Gamma, \mathbb{Z}) = \ell^p(\Gamma, \mathbb{R}) \cap \mathbb{Z}^\Gamma$ we denote the additive subgroup of integer-valued elements of $\ell^p(\Gamma, \mathbb{R})$; for $1 \le p < \infty$, $\ell^p(\Gamma, \mathbb{Z}) = \ell^1(\Gamma, \mathbb{Z})$ is identified with $\mathbb{Z}\Gamma$ by viewing each $g = \sum_\gamma g_\gamma \cdot \gamma \in \mathbb{Z}\Gamma$ as the element $(g_\gamma)_{\gamma \in \Gamma} \in \ell^1(\Gamma, \mathbb{Z})$.

The group Γ acts on $\ell^p(\Gamma, \mathbb{R})$ isometrically by left and right translations: for every $v \in \ell^p(\Gamma, \mathbb{R})$ and $\gamma \in \Gamma$ we denote by $\tilde{\lambda}^\gamma v$ and $\tilde{\rho}^\gamma v$ the elements of $\ell^p(\Gamma, \mathbb{R})$ satisfying $(\tilde{\lambda}^\gamma v)_{\gamma'} = v_{\gamma^{-1}\gamma'}$ and $(\tilde{\rho}^\gamma v)_{\gamma'} = v_{\gamma'\gamma}$, respectively, for every $\gamma' \in \Gamma$. Note that $\tilde{\lambda}^{\gamma\gamma'} = \tilde{\lambda}^\gamma \tilde{\lambda}^{\gamma'}$ and $\tilde{\rho}^{\gamma\gamma'} = \tilde{\rho}^\gamma \tilde{\rho}^{\gamma'}$ for every $\gamma, \gamma' \in \Gamma$.

The Γ-actions $\tilde{\lambda}$ and $\tilde{\rho}$ extend to actions of $\ell^1(\Gamma, \mathbb{R})$ on $\ell^p(\Gamma, \mathbb{R})$ which will again be denoted by $\tilde{\lambda}$ and $\tilde{\rho}$: for $h = (h_\gamma) \in \ell^1(\Gamma, \mathbb{R})$ and $v \in \ell^p(\Gamma, \mathbb{R})$ we set

$$(1.4) \qquad \tilde{\lambda}^h v = \sum\nolimits_{\gamma \in \Gamma} h_\gamma \tilde{\lambda}^\gamma v, \qquad \tilde{\rho}^h v = \sum\nolimits_{\gamma \in \Gamma} h_\gamma \tilde{\rho}^\gamma v.$$

These definitions correspond to the usual convolutions

$$(1.5) \qquad \tilde{\lambda}^h v = h \cdot v, \qquad \tilde{\rho}^h v = v \cdot h^*,$$

where $h \mapsto h^*$ is the involution on $\ell^1(\Gamma, \mathbb{C})$ defined as for $\mathbb{Z}\Gamma$: $h_\gamma^* = \overline{h_{\gamma^{-1}}}$, $\gamma \in \Gamma$, for every $h = (h_\gamma) \in \ell^1(\Gamma, \mathbb{C})$. For $p = 2$, the bounded linear operators $\tilde{\lambda}^h, \tilde{\rho}^h \colon \ell^2(\Gamma, \mathbb{R}) \longrightarrow \ell^2(\Gamma, \mathbb{R})$ in (1.4) can be viewed as elements of the right (resp. left) equivariant group von Neumann algebra of Γ.

THEOREM 1.2. *Let Γ be a countably infinite discrete group which satisfies one of the following conditions:*

(1) *Γ contains a finitely generated amenable subgroup which is not virtually cyclic, or more generally, a finitely generated subgroup with a single end,*

(2) *Γ is not finitely generated,*

(3) *Γ contains an infinite property T subgroup, or more generally, a nonamenable subgroup Γ_0 with vanishing first ℓ^2-Betti number $\beta_1^{(2)}(\Gamma_0) = 0$.*

If $f \in \mathbb{Z}\Gamma$ is not a right zero-divisor, then the principal Γ-action α_f on X_f is ergodic (with respect to the normalized Haar measure of X_f).

THEOREM 1.3. *Let Γ be a countably infinite discrete group which is not virtually cyclic. If $f \in \mathbb{Z}\Gamma$ satisfies that*

$$(1.6) \qquad \ker \tilde{\rho}^{f^*} = \{v \in \ell^2(\Gamma, \mathbb{R}) : \tilde{\rho}^{f^*}(v) = v \cdot f = 0\} = \{0\},$$

then the principal Γ-action α_f on X_f is ergodic.

In view of the hypotheses on f in the Theorems 1.2 and 1.3 it is useful to recall the following result.

---

[1] A discrete group Γ is *virtually cyclic* if it has a cyclic finite-index subgroup. Virtually cyclic groups can obviously not be principally ergodic: if $\Gamma = \mathbb{Z}$, and if $\mathbb{Z}\Gamma$ is identified with the ring of Laurent polynomials $\mathbb{Z}[u^{\pm 1}]$ in the obvious manner, then the principal algebraic $\mathbb{Z}$-action α_f defined by $f = 1 - u$ is trivial — and hence nonergodic — on $X_f = \mathbb{T}$.

PROPOSITION 1.4. *Let Γ be a countably infinite discrete amenable group. For every $f \in \mathbb{Z}\Gamma$ the following conditions are equivalent.*

(1) *f is a right zero-divisor in $\mathbb{Z}\Gamma$,*
(2) *$\{v \in \ell^2(\Gamma, \mathbb{R}) : f^* \cdot v = 0\} \neq \{0\}$,*
(3) *f is a left zero-divisor in $\mathbb{Z}\Gamma$,*
(4) *$\{v \in \ell^2(\Gamma, \mathbb{R}) : f \cdot v = 0\} \neq \{0\}$,*
(5) *$\ker \tilde{\rho}^f = \{v \in \ell^2(\Gamma, \mathbb{R}) : \tilde{\rho}^f(v) = 0\} \neq \{0\}$.*

PROOF. $(4) \Leftrightarrow (5)$: This follows from $(f \cdot v)^* = v^* \cdot f^*$ for all $v \in \ell^2(\Gamma, \mathbb{R})$.

$(2) \Leftrightarrow (3) \Rightarrow (4)$: This is part of [**5**, Proposition 4.16].

$(1) \Leftrightarrow (2)$: Taking $*$ we see that (1) holds if and only if f^* is a left zero-divisor in $\mathbb{Z}\Gamma$. Applying $(3) \Leftrightarrow (4)$ to f^*, we see that the latter condition is equivalent to (2).   □

REMARK 1.5. Linnell's *analytic zero-divisor conjecture* is the conjectural statement that for any torsion-free discrete group Γ and any nonzero $f \in \mathbb{C}\Gamma$, $\ker \tilde{\rho}^{f^*} = \{0\}$ [**6**, Conjecture 1]. Linnell has shown that this conjecture holds for Γ if G_1 is a normal subgroup of Γ, G_2 is a normal subgroup of G_1, Γ is torsion-free, G_2 is free, G_1/G_2 is elementary amenable, and Γ/G_1 is right orderable [**7**, Proposition 1.4].

If a countably infinite, torsion-free, and not virtually cyclic group Γ satisfies Linnell's analytic zero-divisor conjecture, then the principal Γ-action α_f on X_f is ergodic for every $f \in \mathbb{Z}\Gamma$ by Theorem 1.3.

As a corollary to the Theorems 1.2 – 1.3 and Remark 1.5 we obtain the following results by Hayes.

COROLLARY 1.6 ([**2**, Theorem 2.3.6 and Corollary 2.5.5]). *Suppose that Γ satisfies either of the following conditions.*

(1) *Γ is an infinite, torsion-free, nilpotent group not isomorphic to the integers,*
(2) *Γ is the free group with $k \geq 2$ generators.*

Then the principal Γ-action (X_f, α_f) is ergodic for every $f \in \mathbb{Z}\Gamma$.

PROOF. If $f = 0$, then α_f is the left shift-action by Γ on $X_f = \mathbb{T}^\Gamma$, which is obviously ergodic. Suppose therefore that $f \neq 0$. Since Γ is either free or torsion-free nilpotent, $\ker \tilde{\rho}^{f^*} = \{0\}$ by Remark 1.5, so that α_f is ergodic by either Theorem 1.2 or 1.3.   □

Whereas the proofs of these results in [**2**] use structure theory of Γ, the proofs in this paper employ cohomological methods.

2. Cohomological results

Let Γ be a countably infinite discrete group and $\mathcal{M}$ a left $\mathbb{Z}\Gamma$-module. A map $c \colon \Gamma \longrightarrow \mathcal{M}$ is a 1-*cocycle* (or, for our purposes here, simply a *cocycle*) if

$$(2.1) \qquad\qquad c(\gamma\gamma') = c(\gamma) + \gamma c(\gamma')$$

for all $\gamma, \gamma' \in \Gamma$. A cocycle $c \colon \Gamma \longrightarrow \mathcal{M}$ is a *coboundary* (or *trivial*) if there exists a $b \in \mathcal{M}$ such that

$$(2.2) \qquad\qquad c(\gamma) = b - \gamma b$$

for every $\gamma \in \Gamma$.

A finitely generated group G has two ends if and only if it is infinite and virtually cyclic, i.e., if and only if it contains a finite-index subgroup $G' \cong \mathbb{Z}$. Stallings' theorem ([**13**]) implies that a finitely generated group G has a single end whenever it is amenable and not virtually cyclic (see [**8**] for a short proof).

PROPOSITION 2.1. *Let Γ be a countably infinite discrete group and $\Delta \subset \Gamma$ a finitely generated subgroup with a single end. Then every cocycle $c \colon \Delta \longrightarrow \mathbb{Z}\Gamma$ is a coboundary.*

PROOF. By [**3**, Theorem 4.6] if Δ has a single end, then every 1-cocycle $\Delta \longrightarrow \mathbb{Z}\Delta$ is a coboundary.[2] It follows that for each $\gamma \in \Gamma$ there is some $b_\gamma \in \mathbb{Z}[\Delta\gamma]$ such that the restriction of $c(\delta)$ on $\Delta\gamma$ is equal to $b_\gamma - \delta b_\gamma$ for all $\delta \in \Delta$.

For each $\delta \in \Delta$, there is a finite set W_δ of right cosets of Δ in Γ such that the support of $c(\delta)$ is contained in $\bigcup_{\Delta\gamma \in W_\delta} \Delta\gamma$. If F is a finite symmetric set of generators of Δ, then for any $\Delta\gamma \notin \bigcup_{\delta' \in F} W_{\delta'}$, one has $(1 - \delta) \cdot b_\gamma = 0$ for every $\delta \in F$ and hence for every $\delta \in \Delta$. Therefore $c(\delta)$ is equal to 0 on $\Delta\gamma$ for all $\Delta\gamma \notin \bigcup_{\delta' \in F} W_{\delta'}$ and $\delta \in \Delta$. Set $b = \sum_{\Delta\gamma \in \bigcup_{\delta' \in F} W_{\delta'}} b_\gamma \in \mathbb{Z}\Gamma$. Then $c(\delta) = (1 - \delta) \cdot b$ for all $\delta \in \Delta$. $\qquad\square$

Next we prove an analogous result for nonamenable groups with vanishing first ℓ^2-Betti number, e.g., infinite property T groups [**1**, Corollary 6].

PROPOSITION 2.2. *Let Γ be a countably infinite discrete group and $\Delta \subset \Gamma$ a nonamenable subgroup with $\beta_1^{(2)}(\Delta) = 0$. Then every cocycle $c \colon \Delta \longrightarrow \mathbb{Z}\Gamma$ is a coboundary.*

For the proof of Proposition 2.2 we have to discuss cocycles of Γ which take values in a Hilbert space $\mathcal{H}$ carrying a unitary action $U \colon \gamma \mapsto U^\gamma$ of Γ. A map $c \colon \Gamma \longrightarrow \mathcal{H}$ is a 1-*cocycle* for U if

$$(2.3) \qquad c(\gamma\gamma') = c(\gamma) + U^\gamma c(\gamma')$$

for all $\gamma, \gamma' \in \Gamma$, and such a cocycle is a *coboundary* if and only if there exists a $b \in \mathcal{H}$ with

$$(2.4) \qquad c(\gamma) = b - U^\gamma b$$

for every $\gamma \in \Gamma$. The cocycle c is an *approximate coboundary* if there exists a sequence $(c_n)_{n \geq 1}$ of coboundaries $c_n \colon \Gamma \longrightarrow \mathcal{H}$ such that

$$(2.5) \qquad \lim_{n \to \infty} \|c_n(\gamma) - c(\gamma)\| = 0$$

for every $\gamma \in \Gamma$.

The following lemma is well-known (cf. [**11**, Proposition 1.6]). For convenience of the reader, we give a proof here.

LEMMA 2.3. *Let U be a unitary representation of Γ on $\mathcal{H}$ which does not contain the trivial representation weakly. Then every approximate coboundary $c \colon \Gamma \longrightarrow \mathcal{H}$ for U is a coboundary.*

---

[2]The authors are grateful to Andreas Thom for alerting us to this reference.

PROOF. Since U does not weakly contain the trivial representation of Γ, we can find a finite subset $F \subset \Gamma$ and some $\varepsilon > 0$ such that

$$\sum\nolimits_{\delta \in F} \|v - U^\delta v\| \geq \varepsilon \|v\|$$

for all $v \in \mathcal{H}$.

Let c be an approximate coboundary of Γ taking values in $\mathcal{H}$. Let $(b_n)_{n \geq 1}$ be a sequence in $\mathcal{H}$ such that the coboundaries $c_n(\gamma) = b_n - U^\gamma b_n$, $\gamma \in \Gamma$, approximate c in the sense of (2.5). Then

$$\sum\nolimits_{\delta \in F} \|c(\delta)\| = \lim_{n \to \infty} \sum\nolimits_{\delta \in F} \|c_n(\delta)\| \geq \varepsilon \limsup_{n \to \infty} \|b_n\|,$$

and hence

$$\|c(\gamma)\| = \lim_{n \to \infty} \|c_n(\gamma)\| \leq 2 \limsup_{n \to \infty} \|b_n\| \leq 2\varepsilon^{-1} \sum\nolimits_{\delta \in F} \|c(\delta)\|$$

for all $\gamma \in \Gamma$.

For a bounded subset Y of $\mathcal{H}$ and $v \in \mathcal{H}$, set $d(v, Y) = \sup_{y \in Y} \|v - y\|$. Since $\mathcal{H}$ is a Hilbert space, the function $v \mapsto d(v, Y)$ on $\mathcal{H}$ takes a minimal value at exactly one point, namely the Chebyshev center of Y, which we denote by center(Y).

Consider the affine isometric action V of Γ on $\mathcal{H}$ defined by $V^\gamma v = U^\gamma v + c(\gamma)$ for all $\gamma \in \Gamma$ and $v \in \mathcal{H}$. Set $Y = \{c(\gamma') : \gamma' \in \Gamma\}$, and let $\gamma \in \Gamma$. Since $V^\gamma(Y) = Y$, we obtain that $V^\gamma(\text{center}(Y)) = \text{center}(Y)$ and hence that $U^\gamma(\text{center}(Y)) + c(\gamma) = \text{center}(Y)$. Thus $c(\gamma) = \text{center}(Y) - U^\gamma(\text{center}(Y))$ for all $\gamma \in \Gamma$, so that c is a coboundary. $\square$

PROOF OF PROPOSITION 2.2. By [1] in the finitely generated case, and [9, Corollary 2.4] in general, if Δ is nonamenable and $\beta_1^{(2)}(\Delta) = 0$, then every 1-cocycle $\Delta \longrightarrow \ell^2(\Delta, \mathbb{R})$ for the left regular representation is a coboundary. It follows that for each $\gamma \in \Gamma$ there is some $b_\gamma \in \ell^2(\Delta\gamma, \mathbb{R})$ such that the restriction of $c(\delta)$ on $\Delta\gamma$ is equal to $b_\gamma - \delta b_\gamma$ for all $\delta \in \Delta$. Since $c(\delta)$ has finite support for each $\delta \in \Delta$, we conclude that the cocycle $c \colon \Delta \longrightarrow \ell^2(\Gamma, \mathbb{R})$ is an approximate coboundary.

Because Δ is nonamenable, its left regular representation on $\ell^2(\Delta, \mathbb{R})$ does not contain the trivial representation weakly. Since the restriction of the left regular representation of Γ on $\ell^2(\Gamma, \mathbb{R})$ to Δ is a direct sum of copies of the left regular representation of Δ, it does not contain the trivial representation of Δ weakly either. By Lemma 2.3 there exists $v \in \ell^2(\Gamma, \mathbb{R})$ satisfying

$$(2.6) \qquad\qquad c(\delta) = v - \tilde\lambda^\delta v = (1 - \delta)v$$

for every $\delta \in \Delta$.

Since Δ is nonamenable, it is infinite. It follows that that $v \in \ell^2(\Gamma, \mathbb{Z}) = \mathbb{Z}\Gamma$. $\square$

If a subgroup $\Delta \subset \Gamma$ has more than one end then there exist nontrivial cocycles $c \colon \Delta \longrightarrow \mathbb{Z}\Delta$ (cf. [12, 5.2. Satz IV] or [14, Lemma 3.5]), which immediately implies the existence of nontrivial cocycles $c \colon \Delta \longrightarrow \mathbb{Z}\Gamma$. For example, if Δ is the free group on $k \geq 2$ generators, it has nontrivial cocycles. However, Proposition 2.4 below guarantees triviality of cocycles which become trivial under right multiplication by an element $f \in \mathbb{Z}\Gamma$ satisfying (1.6) (cf. Remark 1.5).

PROPOSITION 2.4. *Let Γ be a countably infinite discrete group, $\Delta \subset \Gamma$ a non-amenable subgroup, and let $f \in \mathbb{Z}\Gamma$ satisfy that $\ker \tilde\rho^{f^*} = \{0\}$. If $c \colon \Delta \longrightarrow \mathbb{Z}\Gamma$ is a cocycle such that cf is a coboundary, then c is a coboundary.*

LEMMA 2.5. *Let* Γ *be a countably infinite discrete group,* $\Delta \subset \Gamma$ *a nonamenable subgroup, and let* $f \in \mathbb{Z}\Gamma$. *We write* $\tilde{\lambda}_\Delta$ *for the unitary representation of* Δ *obtained by restricting the left regular representation* $\tilde{\lambda}$ *of* Γ *on* $\ell^2(\Gamma, \mathbb{C})$ *to* Δ.

If $c : \Delta \longrightarrow \ell^2(\Gamma, \mathbb{C})$ *is a cocycle for* $\tilde{\lambda}_\Delta$ *such that* $c \cdot f = \tilde{\rho}^{f^*} c$ *is a coboundary and* $c(\Delta)$ *is contained in the orthogonal complement* V *of* $\ker \tilde{\rho}^{f^*}$ *in* $\ell^2(\Gamma, \mathbb{C})$ *(cf. (1.6)), then* c *is a coboundary.*

PROOF. By assumption there exists a $b \in \ell^2(\Gamma, \mathbb{C})$ such that $(1 - \delta) \cdot b = c(\delta) \cdot f$ for every $\delta \in \Delta$. Let $\tilde{\rho}^{f^*} = UH$ be the polar decomposition [4, Theorem 6.1.2] of $\tilde{\rho}^{f^*}$, where U is a partial isometry on $\ell^2(\Gamma, \mathbb{C})$, $H = \left(\tilde{\rho}^{ff^*}\right)^{1/2} = (\tilde{\rho}^f \tilde{\rho}^{f^*})^{1/2}$, and both U and H lie in the left-equivariant group von Neumann algebra $\mathcal{N}\Gamma$.

Note that $\ker H = \ker \tilde{\rho}^{f^*}$. We write $H = \int_0^{\|\tilde{\rho}^{f^*}\|} \lambda \, dE_\lambda$ for the spectral decomposition of the positive self-adjoint operator H and consider, for each $0 < \varepsilon < \|\tilde{\rho}^{f^*}\|$, the projection operator $P_\varepsilon = P - E_\varepsilon$, where P is the orthogonal projection $\ell^2(\Gamma, \mathbb{C}) \longrightarrow V$. Then one has $P_\varepsilon \to P$ in the strong operator topology as $\varepsilon \searrow 0$.

Put $Q_\varepsilon = U P_\varepsilon U^*$ for every ε with $0 < \varepsilon < \|\rho^{f^*}\|$. Then

$$P_\varepsilon(c(\delta)) \cdot f = \tilde{\rho}^{f^*} P_\varepsilon(c(\delta)) = UH P_\varepsilon(c(\delta)) = U P_\varepsilon H(c(\delta)) = Q_\varepsilon UH(c(\delta))$$

$$= Q_\varepsilon \tilde{\rho}^{f^*}(c(\delta)) = Q_\varepsilon(c(\delta) \cdot f) = Q_\varepsilon((1 - \delta) \cdot b) = (1 - \delta) \cdot Q_\varepsilon(b)$$

for every $\delta \in \Delta$. Since $\|\tilde{\rho}^{f^*} v\| \geq \varepsilon \|v\|$ for every $v \in \text{range}(P_\varepsilon)$, there exists $V_\varepsilon \in \mathcal{N}\Gamma$ vanishing on the orthogonal complement of $\text{range}(\tilde{\rho}^{f^*} P_\varepsilon)$ and satisfying that $V_\varepsilon \tilde{\rho}^{f^*} v = v$ for every $v \in \text{range}(P_\varepsilon)$. Therefore

$$P_\varepsilon(c(\delta)) = V_\varepsilon \tilde{\rho}^{f^*} P_\varepsilon(c(\delta)) = V_\varepsilon Q_\varepsilon((1 - \delta) \cdot b)) = (1 - \delta) \cdot V_\varepsilon Q_\varepsilon(b).$$

The 1-cocycle $\delta \mapsto P_\varepsilon c(\delta) = (1 - \delta) \cdot V_\varepsilon Q_\varepsilon(b)$ for $\tilde{\lambda}_\Delta$ is thus a coboundary. Since $P_\varepsilon(c(\delta)) \to c(\delta)$ in $\ell^2(\Gamma, \mathbb{C})$ as $\varepsilon \searrow 0$ for every $\delta \in \Delta$, we conclude that the 1-cocycle $c : \Delta \longrightarrow \ell^2(\Gamma, \mathbb{C})$ for $\tilde{\lambda}_\Delta$ is an approximate coboundary.

Since Δ is nonamenable, the left regular representation of Δ on $\ell^2(\Delta, \mathbb{C})$ does not weakly contain the trivial representation of Δ. Thus, the representation $\tilde{\lambda}_\Delta$ of Δ on $\ell^2(\Gamma, \mathbb{C})$, as a direct sum of copies of the left regular representation of Δ, does not weakly contain the trivial representation of Δ.

From Lemma 2.3 we conclude that there is some $b \in \ell^2(\Gamma, \mathbb{C})$ satisfying $c(\delta) = (1 - \delta)b$ for every $\delta \in \Delta$. $\qquad \square$

PROOF OF PROPOSITION 2.4. Suppose that $f \in \mathbb{Z}\Gamma$ satisfies (1.6), and that $c : \Delta \longrightarrow \mathbb{Z}\Gamma$ is a 1-cocycle such that cf is a coboundary. Then cf is also a coboundary when c is viewed as an $\ell^2(\Gamma, \mathbb{C})$-valued cocycle for the unitary representation $\tilde{\lambda}_\Delta$ on $\ell^2(\Gamma, \mathbb{C})$. Lemma 2.5 shows that there exists a $b \in \ell^2(\Gamma, \mathbb{C})$ such that $c(\delta) = (1 - \delta) \cdot b$ for every $\delta \in \Delta$. In order to prove that $b \in \mathbb{Z}\Gamma$ we set, for every $\varepsilon > 0$, $F_\varepsilon(b) = \{\gamma \in \Gamma : |b_\gamma| \geq \varepsilon\}$. Then F_ε is finite, and so is the set $\{\delta \in \Delta : |(\delta \cdot b)_\gamma| = |b_{\delta^{-1}\gamma}| \geq \varepsilon\} = \{\delta \in \Delta : \delta^{-1}\gamma \in F_\varepsilon\} = \gamma F_\varepsilon^{-1} \cap \Delta$ for every $\gamma \in \Gamma$. Since Δ is nonamenable, it is infinite, and by varying ε we see that $\lim_{\delta \to \infty}(\delta \cdot b)_\gamma = 0$ for every $\gamma \in \Gamma$. Since $c(\delta)_\gamma = b_\gamma - (\delta \cdot b)_\gamma \in \mathbb{Z}$ we conclude, by letting $\delta \to \infty$, that $b_\gamma \in \mathbb{Z}$ for every $\gamma \in \Gamma$. This completes the proof of the proposition. $\qquad \square$

3. Ergodicity of principal actions

We recall the following result from [**10**, Lemma 1.2 and Theorem 1.6].

THEOREM 3.1. *If α is an algebraic action of a countably infinite discrete group Γ on a compact abelian group X with dual group $\hat{X}$, then α is ergodic if and only if the orbit $\{\hat{\alpha}^\gamma a : \gamma \in \Gamma\}$ is infinite for every nontrivial $a \in \hat{X}$.*

COROLLARY 3.2. *Let Γ be a countably infinite discrete group, $f \in \mathbb{Z}\Gamma$, and let α_f be the principal algebraic Γ-action on the group X_f with Haar measure μ_f (cf. Definition 1.1). For $a \in \mathbb{Z}\Gamma/\mathbb{Z}\Gamma f = \widehat{X_f}$ let $S(a) = \{\gamma \in \Gamma : \gamma \cdot a = a\}$ be its stabilizer.*

Then α_f is ergodic with respect to μ_f if and only if $S(a)$ has infinite index in Γ for every nonzero $a \in \mathbb{Z}\Gamma/\mathbb{Z}\Gamma f$.

PROOF OF THEOREM 1.2. Suppose that $f \in \mathbb{Z}\Gamma$ is not a right zero-divisor, but that α_f is nonergodic. By Corollary 3.2 there exists an $h \in \mathbb{Z}\Gamma$ such that $h \notin \mathbb{Z}\Gamma f$ and the Γ-orbit $D = \{\gamma h + \mathbb{Z}\Gamma f : \gamma \in \Gamma\}$ of $a = h + \mathbb{Z}\Gamma f$ in $\mathbb{Z}\Gamma/\mathbb{Z}\Gamma f$ is finite. We denote by

$$(3.1) \qquad\qquad \Delta = \{\delta \in \Gamma : \delta h - h \in \mathbb{Z}\Gamma f\}$$

the stabilizer of a, which has finite index in Γ by hypothesis, and consider the cocycle $c \colon \Delta \longrightarrow \mathbb{Z}\Gamma$ given by

$$(3.2) \qquad\qquad h - \delta h = c(\delta)f$$

for every $\delta \in \Delta$ (here we are using that f is not a right zero-divisor). If $\Delta_0 \subset \Delta$ is an infinite subgroup on which c is a coboundary then $c(\delta) = b - \delta b$ for some $b \in \mathbb{Z}\Gamma$ and every $\delta \in \Delta_0$. Hence $c(\delta)f = (1 - \delta)bf = (1 - \delta)h$ for every $\delta \in \Delta_0$. Since Δ_0 is infinite, this implies that $h = bf \in \mathbb{Z}\Gamma f$, contrary to our choice of h. In other words, if c is a coboundary when restricted to any infinite subgroup, we run into a contradiction with our assumption that α_f is nonergodic.

PROOF OF (1). If $\Gamma_0 \subset \Gamma$ is a finitely generated subgroup with a single end, then the same is true for its finite-index subgroup $\Delta \cap \Gamma_0$ where Δ is from (3.1). Proposition 2.1 shows that c is a coboundary on $\Delta \cap \Gamma_0$.

As was explained at the beginning of the proof of this theorem this contradicts the non-ergodicity of α_f.

PROOF OF (2). This is [**2**, Theorem 2.4.1]. For convenience of the reader we include the proof. Let $\Gamma_0 \subset \Gamma$ be the subgroup generated by $\mathrm{supp}(h) \cup \mathrm{supp}(f)$. Since Γ is not finitely generated there exists an increasing sequence of subgroups $\Gamma_n \subset \Gamma$, $n \geq 1$, such that Γ_{n+1} is generated over Γ_n by a single element $\gamma_{n+1} \in \Gamma_{n+1} \setminus \Gamma_n$. Put $D = \{\gamma h + \mathbb{Z}\Gamma f : \gamma \in \Gamma\} \subset \mathbb{Z}\Gamma/\mathbb{Z}\Gamma f$ and $D_n = \{\gamma h + \mathbb{Z}\Gamma f : \gamma \in \Gamma_n\} \subset \mathbb{Z}\Gamma/\mathbb{Z}\Gamma f$, $n \geq 0$. Then $|D_0| \leq |D_1| \leq \cdots \leq |D_n| \leq \cdots \leq |D| < \infty$. Hence there exists an $N \geq 0$ with $\gamma_{N+1}h + \mathbb{Z}\Gamma f = \gamma' h + \mathbb{Z}\Gamma f$ for some $\gamma' \in \Gamma_N$. Then $(\gamma_{N+1} - \gamma')h = gf$ for some $g \in \mathbb{Z}\Gamma$. We write $g = g_1 + g_2$ with $\mathrm{supp}(g_1) \subset \Gamma_N$ and $\mathrm{supp}(g_2) \cap \Gamma_N = \varnothing$. Then

$$(3.3) \qquad\qquad \gamma_{N+1}h - g_2 f = g_1 f + \gamma' h.$$

All the terms on the right hand side of (3.3) are supported in Γ_N, whereas the supports of the terms on the left hand side of (3.3) are disjoint from Γ_N. Hence both sides of (3.3) have to vanish, which means that $\gamma_{N+1}h = g_2 f$ and $h \in \gamma_{N+1}^{-1}g_2 f \in$

$\mathbb{Z}\Gamma f$, contrary to our choice of h. As explained above, this contradiction proves the ergodicity of α_f.

PROOF OF (3). If $\Gamma_0 \subset \Gamma$ is a nonamenable subgroup with $\beta_1^{(2)}(\Gamma_0) = 0$, then the same is true for its finite-index subgroup $\Delta \cap \Gamma_0$. By Proposition 2.2, the cocycle $c \colon \Delta \cap \Gamma_0 \longrightarrow \mathbb{Z}\Gamma$ is a coboundary, which leads to a contradiction as in (1).   □

PROOF OF THEOREM 1.3. If Γ is amenable, use Theorem 1.2 (1) or (2). If Γ is nonamenable, combine the argument at the beginning of the proof of Theorem 1.2 with Proposition 2.4.   □

References

[1] Mohammed E. B. Bekka and Alain Valette, *Group cohomology, harmonic functions and the first L^2-Betti number*, Potential Anal. **6** (1997), no. 4, 313–326, DOI 10.1023/A:1017974406074. MR1452785 (98e:20056)

[2] B.R. Hayes, *Ergodicity of nilpotent group actions, Gauss's lemma and mixing in the Heisenberg group*, Senior Thesis, University of Washington, Seattle, 2009.

[3] C. H. Houghton, *Ends of groups and the associated first cohomology groups*, J. London Math. Soc. (2) **6** (1972), 81–92. MR0316595 (47 #5142)

[4] Richard V. Kadison and John R. Ringrose, *Fundamentals of the theory of operator algebras. Vol. II*, Graduate Studies in Mathematics, vol. 16, American Mathematical Society, Providence, RI, 1997. Advanced theory; Corrected reprint of the 1986 original. MR1468230 (98f:46001b)

[5] Hanfeng Li and Andreas Thom, *Entropy, determinants, and L^2-torsion*, J. Amer. Math. Soc. **27** (2014), no. 1, 239–292, DOI 10.1090/S0894-0347-2013-00778-X. MR3110799

[6] Peter A. Linnell, *Zero divisors and $L^2(G)$* (English, with French summary), C. R. Acad. Sci. Paris Sér. I Math. **315** (1992), no. 1, 49–53. MR1172405 (93d:20010)

[7] Peter A. Linnell, *Division rings and group von Neumann algebras*, Forum Math. **5** (1993), no. 6, 561–576, DOI 10.1515/form.1993.5.561. MR1242889 (94h:20009)

[8] S. Moon and A. Valette, *Non-properness of amenable actions on graphs with infinitely many ends*, Ischia group theory 2006, World Sci. Publ., Hackensack, NJ, 2007, pp. 227–233, DOI 10.1142/9789812708670_0020. MR2405942 (2009g:20096)

[9] Jesse Peterson and Andreas Thom, *Group cocycles and the ring of affiliated operators*, Invent. Math. **185** (2011), no. 3, 561–592, DOI 10.1007/s00222-011-0310-2. MR2827095 (2012j:22004)

[10] Klaus Schmidt, *Dynamical systems of algebraic origin*, Progress in Mathematics, vol. 128, Birkhäuser Verlag, Basel, 1995. MR1345152 (97c:28041)

[11] Yehuda Shalom, *Rigidity of commensurators and irreducible lattices*, Invent. Math. **141** (2000), no. 1, 1–54, DOI 10.1007/s002220000064. MR1767270 (2001k:22022)

[12] Ernst Specker, *Die erste Cohomologiegruppe von Überlagerungen und Homotopie-Eigenschaften dreidimensionaler Mannigfaltigkeiten* (German), Comment. Math. Helv. **23** (1949), 303–333. MR0033520 (11,451a)

[13] John R. Stallings, *On torsion-free groups with infinitely many ends*, Ann. of Math. (2) **88** (1968), 312–334. MR0228573 (37 #4153)

[14] Richard G. Swan, *Groups of cohomological dimension one*, J. Algebra **12** (1969), 585–610. MR0240177 (39 #1531)

DEPARTMENT OF MATHEMATICS, CHONGQING UNIVERSITY, CHONGQING 401331, CHINA
and
DEPARTMENT OF MATHEMATICS, SUNY AT BUFFALO, BUFFALO, NEW YORK 14260-2900
E-mail address: hfli@math.buffalo.edu

DEPARTMENT OF MATHEMATICS, VANDERBILT UNIVERSITY, 1326 STEVENSON CENTER,
NASHVILLE, TENNESSEE 37240
E-mail address: jesse.d.peterson@vanderbilt.edu

MATHEMATICS INSTITUTE, UNIVERSITY OF VIENNA, OSKAR-MORGENSTERN-PLATZ 1, A-1090
VIENNA, AUSTRIA
and
ERWIN SCHRÖDINGER INSTITUTE FOR MATHEMATICAL PHYSICS, BOLTZMANNGASSE 9, A-1090
VIENNA, AUSTRIA
E-mail address: klaus.schmidt@univie.ac.at

Contemporary Mathematics
Volume **631**, 2015
http://dx.doi.org/10.1090/conm/631/12605

A note on three problems in metric Diophantine approximation

Victor Beresnevich and Sanju Velani

Dedicated to Shrikrishna Gopalrao Dani
on the occasion of his 65th birthday

ABSTRACT. The use of Hausdorff measures and dimension in the theory of Diophantine approximation dates back to the 1920s with the theorems of Jarník and Besicovitch regarding well-approximable and badly-approximable points. In this paper we consider three inhomogeneous problems that further develop these classical results. Firstly, we obtain a Jarník type theorem for the set $\mathcal{S}_2^\times(\psi; \boldsymbol{\theta})$ of multiplicatively approximable points in the plane $\mathbb{R}^2$. This Hausdorff measure statement does not reduce to Gallagher's Lebesgue measure statement as one might expect and is new even in the homogeneous setting ($\boldsymbol{\theta} = \boldsymbol{0}$). Next, we establish a Jarník type theorem for the set $\mathcal{S}_2^\times(\psi; \boldsymbol{\theta}) \cap \mathcal{C}$ where $\mathcal{C}$ is a non-degenerate planar curve. This completes the Hausdorff theory for planar curves and clarifies a potential oversight in the work of Badziahin and Levesley (2007).Finally, we show that the set $\mathbf{Bad}(i, j; \theta)$ of simultaneously inhomogeneously badly approximable points in $\mathbb{R}^2$ is of full dimension. The underlying philosophy behind the proof has other applications; e.g. towards establishing the inhomogeneous version of Schmidt's Conjecture. The higher dimensional analogues of the planar results are also discussed.

1. Multiplicatively ψ-well approximable points

Throughout $\psi : \mathbb{N} \to [0, +\infty)$ is a non-negative function. We will normally assume that ψ is strictly positive and monotonically decreasing in which case ψ will be referred to as an *approximating function*. Given ψ, a real number x will be called ψ-*well approximable* or simply ψ-*approximable* if there are infinitely many $q \in \mathbb{N}$ such that

$$\|qx\| < \psi(q).$$

Here and throughout $\| \cdot \|$ denotes the distance of a real number to the nearest integer. Let $\mathcal{S}_1(\psi)$ denote the set of all ψ-approximable real numbers. The set

2010 *Mathematics Subject Classification*. Primary 11J83; Secondary 11J13, 11K60.
Key words and phrases. Diophantine approximation, Jarník type theorems, Hausdorff dimension and measure, multiplicative and inhomogeneous simultaneous approximation.
The first author's research was supported by EPSRC grant EP/J018260/1.
The second author's research was supported by EPSRC grants EP/E061613/1, EP/F027028/1 and EP/J018260/1.

$\mathcal{S}_1(\psi)$ is invariant under translations by integers. Hence, we will often restrict x to lie in the unit interval $\mathbb{I} := [0, 1]$.

The well known theorem of Dirichlet states that $\mathcal{S}_1(\psi) = \mathbb{R}$ when $\psi(q) = q^{-1}$. In turn, a rather simple consequence of the Borel-Cantelli lemma from probability theory is that $\mathcal{S}_1(\psi)$ is null (that is of Lebesgue measure zero) whenever $\sum_{q=1}^{\infty} \psi(q) < \infty$. However, Khintchine's theorem [24] tells us that the set $\mathcal{S}_1(\psi)$ is full (that is its complement is of Lebesgue measure zero) whenever $\sum_{q=1}^{\infty} \psi(q) = \infty$ and ψ is monotonic. In order to quantify the size of $\mathcal{S}_1(\psi)$ when it is null, Jarník [22] and Besicovitch [14] pioneered the use of Hausdorff measures and dimension. Throughout, $\dim X$ will denote the Hausdorff dimension of a subset X of $\mathbb{R}^n$ and $\mathcal{H}^s(X)$ the s-dimensional Hausdorff measure (see §1.1.2 for the definition and further details). The modern version of the classical Jarník-Besicovitch theorem (see [7] or [8]) states that for any approximating function ψ

$$(1) \qquad \dim \mathcal{S}_1(\psi) = \min\left\{1, \frac{2}{\tau + 1}\right\} \qquad \text{where} \qquad \tau := \liminf_{q \to \infty} \frac{-\log \psi(q)}{\log q}.$$

In other words, the 'modern theorem' relates the Hausdorff dimension of $\mathcal{S}_1(\psi)$ to the lower order at infinity of $1/\psi$ and up to a certain degree allows us to discriminate between ψ-approximable sets of Lebesgue measure zero. A more delicate measurement of the 'size' of $\mathcal{S}_1(\psi)$ is obtained by expressing the size in terms of Hausdorff measures $\mathcal{H}^s$. With respect to such measures, the modern version of Jarník theorem (see [7] or [8]) states that for any $s \in (0, 1)$ and any approximating function ψ

$$(2) \qquad \mathcal{H}^s\big(\mathcal{S}_1(\psi) \cap \mathbb{I}\big) = \begin{cases} 0 & \text{if } \sum_{q=1}^{\infty} q^{1-s}\psi^s(q) < \infty, \\ \mathcal{H}^s(\mathbb{I}) & \text{if } \sum_{q=1}^{\infty} q^{1-s}\psi^s(q) = \infty. \end{cases}$$

Note that for $0 < s < 1$ we have that $\mathcal{H}^s(\mathbb{I}) = \infty$. However, since $\mathcal{H}^1(\mathbb{I}) = 1$, the statement as written also holds for $s = 1$ due to the aforementioned theorem of Khintchine. Note that it is trivially true for $s > 1$. The upshot is that statement (2) is true for any $s > 0$ and is referred to as the *Khintchine-Jarník theorem*. It is worth pointing out that there is an even more general version of (2) that makes use of more general Hausdorff measures, see [7, 8, 10, 17]. Within this paper we restrict ourselves to the case of s-dimensional Hausdorff measures.

In higher dimensions there are various natural generalizations of $\mathcal{S}_1(\psi)$. Given an approximating function ψ, the point $\mathbf{x} = (x_1, \ldots, x_n) \in \mathbb{R}^n$ will be called ψ-*well approximable* or simply ψ-*approximable* if there are infinitely many $q \in \mathbb{N}$ such that

$$(3) \qquad \max\{\|qx_1\|, \ldots, \|qx_n\|\} < \psi(q)$$

and it will be called *multiplicatively ψ-well approximable* or simply *multiplicatively ψ-approximable* if there are infinitely many $q \in \mathbb{N}$ such that

$$(4) \qquad \|qx_1\| \cdots \|qx_n\| < \psi(q).$$

Denote by $\mathcal{S}_n(\psi)$ the set of ψ-approximable points in $\mathbb{R}^n$ and by $\mathcal{S}_n^{\times}(\psi)$ the set of multiplicatively ψ-approximable points in $\mathbb{R}^n$. On comparing (3) and (4) one easily spots that

$$\mathcal{S}_n(\psi^{1/n}) \subset \mathcal{S}_n^{\times}(\psi).$$

For the sake of clarity, in what follows we will mainly restrict our attention to the case of the plane $\mathbb{R}^2$. The Khintchine-Jarník theorem for $\mathcal{S}_2(\psi)$ (see [7] or [8]) states that for any $s > 0$ and any approximating function ψ

(5)
$$\mathcal{H}^s\left(\mathcal{S}_2(\psi) \cap \mathbb{I}^2\right) = \begin{cases} 0 & \text{if } \sum_{q=1}^{\infty} q^{2-s}\psi^s(q) < \infty, \\ \mathcal{H}^s(\mathbb{I}^2) & \text{if } \sum_{q=1}^{\infty} q^{2-s}\psi^s(q) = \infty. \end{cases}$$

Regarding the Lebesgue case, which corresponds to when $s = 2$, Gallagher [21] showed that the monotonicity of ψ is unnecessary. As a consequence of the Mass Transference Principle [10] we have that (5) holds for any ψ (not necessarily monotonic) and any $s > 0$.

In the multiplicative setup, Gallagher [20] essentially proved that for any approximating function ψ

(6)
$$\mathcal{H}^2\left(\mathcal{S}_2^{\times}(\psi) \cap \mathbb{I}^2\right) = \begin{cases} 0 & \text{if } \sum_{q=1}^{\infty} \psi(q) \log q < \infty, \\ \mathcal{H}^2(\mathbb{I}^2) & \text{if } \sum_{q=1}^{\infty} \psi(q) \log q = \infty. \end{cases}$$

The extra log factor in the above sum accounts for the larger volume of the fundamental domains defined by (4) compared to (3). The recent work [9] has made an attempt to relax the monotonicity assumption on ψ within the multiplicative setting. Our goal in this paper is to investigate the Hausdorff measure theory within the multiplicative setting.

Problem 1: *Determine the Hausdorff measure $\mathcal{H}^s$ of $\mathcal{S}_n^{\times}(\psi)$.*

This problem is somewhat different to the non-multiplicative setting where we have the uniform solution given by (5). First of all, we note that
(7)
if $s \leqslant 1$ then $\mathcal{H}^s(\mathcal{S}_2^{\times}(\psi) \cap \mathbb{I}^2) = \infty$ irrespective of approximating function ψ.

To see this, we observe that for any ψ-approximable number $\alpha \in \mathbb{R}$ the whole line $x_1 = \alpha$ is contained in $\mathcal{S}_2^{\times}(\psi)$. Hence,

(8)
$$\mathcal{S}_1(\psi) \times \mathbb{R} \subset \mathcal{S}_2^{\times}(\psi).$$

It is easy to verify (for example, by using the theory of continued fractions) that $\mathcal{S}_1(\psi)$ is an infinite set for any approximating function ψ and so (8) implies (7). Next, since $\mathcal{S}_2^{\times}(\psi) \subseteq \mathbb{R}^2$, we trivially have that

if $s > 2$ then $\mathcal{H}^s(\mathcal{S}_2^{\times}(\psi) \cap \mathbb{I}^2) = 0$ irrespective of ψ.

The upshot of this and (7) is that when attacking Problem 1, there is no loss of generality in assuming that $s \in (1, 2]$. Furthermore, the Lebesgue case ($s = 2$) is covered by Gallagher's result so we may as well assume that $1 < s < 2$.

Recall that Gallagher's multiplicative statement (6) has the extra 'log factor' in the 'volume' sum compared to the simultaneous statement (5). It is natural to expect the log factor to appear in one form or another when determining the Hausdorff measure $\mathcal{H}^s$ of $\mathcal{S}_2^{\times}(\psi)$ for $s \in (1, 2)$; in other words when s is not an integer and so $\mathcal{H}^s$ is genuinely a fractal measure. This, as we shall soon see, is very far from the truth. The 'log factor' completely disappears! Thus, genuine 'fractal'

Hausdorff measures are insensitive to the multiplicative nature of $\mathcal{S}_2^\times(\psi)$. Indeed, what we essentially have is that

$$\mathcal{H}^s\big(\mathcal{S}_2^\times(\psi)\big) = \mathcal{H}^{s-1}\big(\mathcal{S}_1(\psi)\big).$$

Thus, that for $s < 1$ the s-dimensional Hausdorff measure of both sides of (8) is the same. In short, for any $s \in (0,1)$, the points of $\mathcal{S}_2^\times(\psi)$ that do not lie in $\mathbb{R} \times \mathcal{S}_1(\psi)$ do not contribute any substantial 'mass' in terms of the associated s-dimensional Hausdorff measure.

The ideas and tricks used in our investigation of Problem 1 are equally valid within the more general inhomogeneous setup: given an approximating function ψ and a fixed point $\boldsymbol{\theta} = (\theta_1, \ldots, \theta_n) \in \mathbb{R}^n$, let $\mathcal{S}_n^\times(\psi; \boldsymbol{\theta})$ denote the set of points $(x_1, \ldots, x_n) \in \mathbb{R}^n$ such that there are infinitely many $q \in \mathbb{N}$ satisfying the inequality

$$(9) \qquad \|qx_1 - \theta_1\| \cdots \|qx_n - \theta_n\| < \psi(q).$$

We prove the following inhomogeneous statement.

THEOREM 1. *Let* ψ *be an approximating function,* $\boldsymbol{\theta} = (\theta_1, \theta_2) \in \mathbb{R}^2$ *and* $s \in (1,2)$. *Then*

$$(10) \qquad \mathcal{H}^s\big(\mathcal{S}_2^\times(\psi; \boldsymbol{\theta}) \cap \mathbb{I}^2\big) = \begin{cases} 0 & \text{if } \sum_{q=1}^\infty q^{2-s}\psi^{s-1}(q) < \infty, \\[2mm] \mathcal{H}^s(\mathbb{I}^2) & \text{if } \sum_{q=1}^\infty q^{2-s}\psi^{s-1}(q) = \infty. \end{cases}$$

Remark 1.1. Note that $\mathcal{H}^s(\mathbb{I}^2) = \infty$ when $s < 2$. We reiterate the fact that unlike the Khintchine-Jarník theorem, the statement of Theorem 1 is false when $s = 2$.

Remark 1.2. In higher dimensions, Gallagher's multiplicative statement reads

$$\mathcal{H}^n\big(\mathcal{S}_n^\times(\psi) \cap \mathbb{I}^2\big) = \begin{cases} 0 & \text{if } \sum_{q=1}^\infty \psi(q) \log^{n-1} q < \infty, \\[2mm] \mathcal{H}^2(\mathbb{I}^2) & \text{if } \sum_{q=1}^\infty \psi(q) \log^{n-1} q = \infty. \end{cases}$$

For $n > 2$, the proof of Theorem 1 can be adapted to show that for any $s \in (n-1, n)$

$$\mathcal{H}^s\big(\mathcal{S}_n^\times(\psi; \boldsymbol{\theta}) \cap \mathbb{I}^n\big) = 0 \qquad \text{if} \qquad \sum_{q=1}^\infty q^{n-s}\psi^{s+1-n}(q) \log^{n-2} q < \infty.$$

Thus, for convergence in higher dimensions we lose a log factor from the Lebesgue volume sum appearing in Gallagher's result. This of course is absolutely consistent with the $n = 2$ situation given by Theorem 1. Regarding a divergent statement, the arguments used in proving Theorem 1 can be adapted to show that for any $s \in (n-1, n)$

$$\mathcal{H}^s\big(\mathcal{S}_n^\times(\psi; \boldsymbol{\theta}) \cap \mathbb{I}^n\big) = \mathcal{H}^s(\mathbb{I}^n) \qquad \text{if} \qquad \sum_{q=1}^\infty q^{n-s}\psi^{s+1-n}(q) = \infty.$$

Thus, there is a discrepancy in the above 's-volume' sum conditions for convergence and divergence when $n > 2$. In view of this, it remains an interesting open problem to determine the necessary and sufficient condition for $\mathcal{H}^s\big(\mathcal{S}_n^\times(\psi; \boldsymbol{\theta}) \cap \mathbb{I}^n\big)$ to be zero or infinite in higher dimensions.

1.1. Proof of Theorem 1. To simplify notation the symbols $\ll$ and $\gg$ will be used to indicate an inequality with an unspecified positive multiplicative constant. If $a \ll b$ and $a \gg b$ we write $a \asymp b$, and say that the quantities a and b are comparable. For a real number x, the quantity $\{x\}$ will denote the fractional part of x and $[x]$ the integer part of x.

Without loss of generality, throughout the proof of Theorem 1 we can assume that $\boldsymbol{\theta} = (\theta_1, \theta_2) \in \mathbb{I}^2$.

1.1.1. *A covering of* $\mathcal{S}_2^\times(\psi, \boldsymbol{\theta}) \cap \mathbb{I}^2$. In this section we obtain an effective covering of the set $\mathcal{S}_2^\times(\psi, \boldsymbol{\theta}) \cap \mathbb{I}^2$ that will be used in establishing the convergence case of Theorem 1.

LEMMA 1. *Let* $0 < \varepsilon < 1$, $(x_1, x_2) \in \mathbb{I}^2$, $(\theta_1, \theta_2) \in \mathbb{I}^2$, $q \in \mathbb{N}$ *and*

$$(11) \qquad \prod_{i=1}^{2} \|qx_i - \theta_i\| < \varepsilon .$$

Then there exist $m \in \mathbb{Z}$ *and* $p_1, p_2 \in \{-1, 0, \dots, q\}$ *such that*

$$\|qx_i - \theta_i\| = |qx_i - \theta_i - p_i| \qquad \text{for } i = 1, 2,$$

$$(12) \qquad \|qx_1 - \theta_1\| < 2^m \sqrt{2\varepsilon}, \qquad \|qx_2 - \theta_2\| < 2^{-m} \sqrt{2\varepsilon}$$

and

$$(13) \qquad 2^{|m|} \sqrt{\varepsilon} \leqslant 1 .$$

PROOF. The existence of $p_i \in \{-1, 0, \dots, q\}$ with $|qx_i - p_i - \theta_i| = \|qx_i - \theta_i\|$ is an immediate consequence of the fact that $x_i, \theta_i \in \mathbb{I}$. Thus, the only thing that we need to prove is the existence of m satisfying (12) and (13).

If $\|qx_i - \theta_i\| < \sqrt{2\varepsilon}$ for each $i = 1, 2$ then we can define $m = 0$. In this case (12) is obvious and (13) is a consequence of the fact that $0 < \varepsilon < 1$.

Without loss of generality, assume that $\|qx_1 - \theta_1\| \geqslant \sqrt{2\varepsilon}$ and let $m \in \mathbb{Z}$ be the unique integer such that

$$2^{m-1} \sqrt{2\varepsilon} \leqslant \|qx_1 - \theta_1\| < 2^m \sqrt{2\varepsilon} .$$

Since $\|qx_1 - \theta_1\| \geqslant \sqrt{2\varepsilon}$, we have that $m \geqslant 0$. Furthermore, since $\|qx_1 - \theta_1\| \leqslant 1/2$, we have that $2^m \sqrt{\varepsilon} < 1$ whence (13) follows. The left hand side of (12) holds by the definition of m. To show the right hand side of (13) we use (11). Indeed, we have that

$$2^{m-1} \sqrt{2\varepsilon} \|qx_2 - \theta_2\| \leqslant \prod_{i=1}^{2} \|qx_i - \theta_i\| < \varepsilon$$

whence the right hand side of (13) follows. This completes the proof of the lemma. $\boxtimes$

LEMMA 2. *Let* $\psi : \mathbb{N} \to [0, 1)$ *be decreasing and* $\boldsymbol{\theta} = (\theta_1, \theta_2) \in \mathbb{I}^2$. *Then for any* $\ell \in \mathbb{N}$

$$(14) \quad \mathcal{S}_2^\times(\psi; \boldsymbol{\theta}) \cap \mathbb{I}^2 \subset \bigcup_{t=\ell}^{\infty} \bigcup_{2^t \leqslant q < 2^{t+1}} \bigcup_{\substack{m \in \mathbb{Z} \\ 2^{|m|} \sqrt{\psi(2^t)} \leqslant 1}} \bigcup_{p_1=-1}^{q} \bigcup_{p_2=-1}^{q} S_{\boldsymbol{\theta}}(t, q, m, p_1, p_2) ,$$

where

$$S_{\boldsymbol{\theta}}(t,q,m,p_1,p_2) := \left\{ (x_1,x_2) \in \mathbb{I}^2 : \begin{array}{c} \left| x_1 - \dfrac{p_1 + \theta_1}{q} \right| < \dfrac{2^m \sqrt{2\psi(2^t)}}{2^t} \\[2ex] \left| x_2 - \dfrac{p_2 + \theta_2}{q} \right| < \dfrac{2^{-m} \sqrt{2\psi(2^t)}}{2^t} \end{array} \right\}.$$

PROOF. It is easily verified that

$$\mathcal{S}_2^\times(\psi;\boldsymbol{\theta}) \cap \mathbb{I}^2 = \bigcap_{\ell=1}^{\infty} \bigcup_{t=\ell}^{\infty} \bigcup_{2^t \leqslant q < 2^{t+1}} \left\{ (x_1,x_2) \in \mathbb{I}^2 : \prod_{i=1}^{2} \|qx_i - \theta_i\| < \psi(q) \right\}.$$

Since ψ is decreasing, $\psi(q) \leqslant \psi(2^t)$ for $2^t \leqslant q < 2^{t+1}$. Then, for any $\ell \in \mathbb{N}$

$$(15) \quad \mathcal{S}_2^\times(\psi;\boldsymbol{\theta}) \cap \mathbb{I}^2 \subset \bigcup_{t=\ell}^{\infty} \bigcup_{2^t \leqslant q < 2^{t+1}} \left\{ (x_1,x_2) \in \mathbb{I}^2 : \prod_{i=1}^{2} \|qx_i - \theta_i\| < \psi(2^t) \right\}.$$

For a fixed pair t and q with $2^t \leqslant q < 2^{t+1}$, by Lemma 1 with $\varepsilon = \psi(2^t)$, we get that

$$\left\{ (x_1,x_2) \in \mathbb{I}^2 : \prod_{i=1}^{2} \|qx_i - \theta_i\| < \psi(2^t) \right\} \subset \bigcup_{\substack{m \in \mathbb{Z} \\ 2^{|m|}\sqrt{\psi(2^t)} \leqslant 1}} \bigcup_{p_1 = -1}^{q} \bigcup_{p_2 = -1}^{q} S_{\boldsymbol{\theta}}(t,q,m,p_1,p_2).$$

This together with (15) completes the proof of the lemma. $\boxtimes$

1.1.2. *Hausdorff measure and dimension.* We briefly recall various facts regarding Hausdorff measures that will be used in the course of establishing Theorem 1. Given $\delta > 0$ and a set $X \subset \mathbb{R}^n$, any finite or countable collection $\{B_i\}$ of subsets of $\mathbb{R}^n$ such that

$$X \subset \bigcup_i B_i \qquad (\text{i.e. } \{B_i\} \text{ is a cover for } X)$$

and

$$\operatorname{diam} B_i \leqslant \delta \quad \text{for all } i$$

is called a δ-*cover* of X. Given a real number s, let

$$\mathcal{H}_\delta^s(X) := \inf_{\{B_i\}} \sum_i \operatorname{diam}(B_i)^s,$$

where the infimum is taken over all possible δ-covers $\{B_i\}$ of X. The *s-dimensional Hausdorff measure* $\mathcal{H}^s(X)$ of X is defined to be

$$\mathcal{H}^s(X) := \lim_{\delta \to 0^+} \mathcal{H}_\delta^s(X)$$

and the *Hausdorff dimension* $\dim X$ of X by

$$\dim X := \inf\{s : \mathcal{H}^s(X) = 0\} = \sup\{s : \mathcal{H}^s(X) = \infty\}.$$

The countable collection $\{B_i\}$ is called a *fine cover of* X if for every $\delta > 0$ it contains a subcollection that is a δ-cover of X. The following statement is an immediate and well known consequence of the definition of $\mathcal{H}^s$.

LEMMA 3. *Let $\{B_i\}$ be a fine cover of X and $s > 0$ be such that $\sum_i \operatorname{diam}(B_i)^s < \infty$. Then*

$$\mathcal{H}^s(X) = 0.$$

1.1.3. Proof: the convergence case. We are given that $\sum_{q=1}^{\infty} q^{2-s}\psi(q)^{s-1} < \infty$. As already mentioned, we can assume that $\boldsymbol{\theta} \in \mathbb{I}^2$. The proof will make use of the covering of $\mathcal{S}_2^{\times}(\psi; \boldsymbol{\theta}) \cap \mathbb{I}^2$ given by Lemma 2. The rectangle $S_{\boldsymbol{\theta}}(t, q, m, p_1, p_2)$ arising from this lemma has sides of lengths

$$A := \frac{2^{-|m|+1}\sqrt{2\psi(2^t)}}{2^t} \quad \text{and} \quad B := \frac{2^{|m|+1}\sqrt{2\psi(2^t)}}{2^t}$$

and so can be split into $B/A = 2^{2|m|}$ squares with sidelength A. By Lemma 2, the collection of such squares taken over $t \geqslant \ell$ and over q, p_1, p_2, m as specified in the lemma is a δ-cover of $\mathcal{S}_2^{\times}(\psi; \boldsymbol{\theta}) \cap \mathbb{I}^2$ with $\delta := \sqrt{2}A \to 0$ as $\ell \to \infty$. Therefore, the collection of all such squares, say $\{B_i\}$, is a fine cover of $\mathcal{S}_2^{\times}(\psi; \boldsymbol{\theta}) \cap \mathbb{I}^2$. It follows that

$$\sum_i \operatorname{diam}(B_i)^s \ll \sum_{t=\ell}^{\infty} \sum_{\substack{2^t \leqslant q < 2^{t+1}}} \sum_{\substack{m \in \mathbb{Z} \\ 2^{|m|}\sqrt{\psi(2^t)} \leqslant 1}} \sum_{p_1=-1}^{q} \sum_{p_2=-1}^{q} \left(2^{-|m|} \frac{\sqrt{\psi(2^t)}}{2^t}\right)^s 2^{2|m|}$$

$$\ll \sum_{t=\ell}^{\infty} \sum_{\substack{m \in \mathbb{Z} \\ 2^{|m|}\sqrt{\psi(2^t)} \leqslant 1}} \left(2^{-|m|} \frac{\sqrt{\psi(2^t)}}{2^t}\right)^s 2^{2|m|} \, 2^{3t}$$

$$(16) \qquad \ll \sum_{t=\ell}^{\infty} \left(\frac{\sqrt{\psi(2^t)}}{2^t}\right)^s 2^{3t} \sum_{\substack{m \in \mathbb{Z} \\ 2^{|m|}\sqrt{\psi(2^t)} \leqslant 1}} 2^{(2-s)|m|}.$$

Since $1 < s < 2$, the sum over $m \neq 0$ in the right hand side of (16) is a finite increasing geometric progression, which is easily estimated to give

$$(17) \qquad \sum_{\substack{m \in \mathbb{Z} \\ 2^{|m|}\sqrt{\psi(2^t)} \leqslant 1}} 2^{(2-s)|m|} \ll \left(\frac{1}{\sqrt{\psi(2^t)}}\right)^{2-s} = \left(\sqrt{\psi(2^t)}\right)^{s-2}.$$

Substituting this into (16) gives

$$\sum_i \operatorname{diam}(B_i)^s \ll \sum_{t=\ell}^{\infty} \left(\frac{\sqrt{\psi(2^t)}}{2^t}\right)^s 2^{3t} \left(\sqrt{\psi(2^t)}\right)^{s-2}$$

$$= \sum_{t=\ell}^{\infty} 2^{(3-s)t}\psi(2^t)^{s-1} \ll \sum_{q=1}^{\infty} q^{2-s}\psi(q)^{s-1} < \infty.$$

By Lemma 3, $\mathcal{H}^s(\mathcal{S}_2^{\times}(\psi; \boldsymbol{\theta}) \cap \mathbb{I}^2) = 0$ and thus the proof of the convergence part is complete.

1.1.4. *Proof: the divergence case.* We are given that $\sum_{q=1}^{\infty} q^{2-s}\psi(q)^{s-1} = \infty$. Then, by the inhomogeneous version of Jarník's theorem [15] (see also the remark in [8, §12.1]), it follows that $\mathcal{H}^{s-1}(\mathcal{S}_1(\psi;\theta_1)\cap\mathbb{I}) = \infty$. The observation that led to (8) is equally valid in the inhomogeneous setup; that is to say that

$$\mathcal{S}_1(\psi;\theta_1) \times \mathbb{R} \subset \mathcal{S}_2^{\times}(\psi;\boldsymbol{\theta}).$$

Thus, $\mathcal{H}^s\left(\mathcal{S}_2^{\times}(\psi;\boldsymbol{\theta})\cap\mathbb{I}^2\right) \geqslant \mathcal{H}^s\left((\mathcal{S}_1(\psi;\theta_1)\cap\mathbb{I})\times\mathbb{I}\right)$. Since $\mathcal{H}^{s-1}(\mathcal{S}_1(\psi;\theta_1)\cap\mathbb{I}) = \infty$, the slicing lemma [11, Lemma 4] implies that

$$\mathcal{H}^s\left((\mathcal{S}_1(\psi;\theta_1)\cap\mathbb{I})\times\mathbb{I}\right) = \infty.$$

Hence $\mathcal{H}^s\left(\mathcal{S}_2^{\times}(\psi;\boldsymbol{\theta})\cap\mathbb{I}^2\right) = \infty$ and the proof of Theorem 1 is complete.

2. Diophantine approximation on planar curves

When the coordinates of the approximated point $\mathbf{x} \in \mathbb{R}^n$ are confined by functional relations, we fall into the theory of Diophantine approximation on manifolds [13]. Over the last decade or so, the theory of Diophantine approximation on manifolds has developed at some considerable pace with the catalyst being the pioneering work of Kleinbock & Margulis [25]. For details of this and an overview of the almost complete results regarding $\mathcal{S}_n(\psi)$ restricted to manifolds $\mathcal{M} \subset \mathbb{R}^n$ see [6,8] and references within. However, much less is known regarding multiplicative Diophantine approximation on manifolds. It would be highly desirable to address this imbalance by investigating the following analogue of Problem 1 for manifolds.

Problem 2: *Determine the Hausdorff measure $\mathcal{H}^s$ of $\mathcal{S}_n^{\times}(\psi)\cap\mathcal{M}$.*

Our goal in this paper is to consider the problem in the case $\mathcal{M}$ is a planar curve $\mathcal{C}$ and $\mathcal{H}^s$ is a genuine fractal measure.

THEOREM 2. *Let ψ be any approximating function, $\boldsymbol{\theta} = (\theta_1,\theta_2) \in \mathbb{R}^2$ and $s \in (0,1)$. Let $\mathcal{C}$ be a $C^{(3)}$ curve in $\mathbb{R}^2$ with non-zero curvature everywhere apart from a set of s-dimensional Hausdorff measure zero. Then*

$$(18) \qquad \mathcal{H}^s\left(\mathcal{S}_2^{\times}(\psi;\boldsymbol{\theta})\cap\mathcal{C}\right) = \begin{cases} 0 & \text{if } \sum_{q=1}^{\infty} q^{1-s}\psi^s(q) < \infty, \\ \infty & \text{if } \sum_{q=1}^{\infty} q^{1-s}\psi^s(q) = \infty. \end{cases}$$

Remark 2.1. In [2], the authors prove that $\mathcal{H}^s\left(\mathcal{S}_2^{\times}(\psi)\cap\mathcal{C}\right) = 0$ under the more restrictive assumption that $\sum_{q=1}^{\infty} q^{1-s}\psi^s(q)(\log q)^s < \infty$. Although not an error, in their proof of this homogeneous statement [2, Theorem 1] there is a certain degree of ambiguity in the manner in which a key counting estimate originating from [30, Theorem 1] is applied. More precisely, it is important to stress that the implied constant appearing in inequality (13) associated with [2, Theorem VV] is independent of ψ. This is crucial as it is applied over a countable family of functions $\psi(Q)$ that depend on a parameter $m \in \mathbb{Z}$.

2.1. Proof of Theorem 2.

2.1.1. *Rational points near planar curves.* The proof of Theorem 2 relies on the results obtained in [**12**] regarding the distribution of 'shifted' rational points near planar curves, which we recall here. In view of the metrical nature of Theorem 2, there is no loss of generality in assuming that $\mathcal{C} := \{(x, f(x)) : x \in I\}$ is the graph of a $C^{(3)}$ function $f : I \to \mathbb{R}$ defined on a finite closed interval I and that f'' is continuous and non-vanishing on I. By the compactness of I, there exist positive and finite constants c_1, c_2 such that

$$(19) \qquad c_1 \leqslant |f''(x)| \leqslant c_2 \qquad \text{for all } x \in I.$$

Given $\boldsymbol{\theta} = (\theta_1, \theta_2) \in \mathbb{R}^2$, $\delta > 0$ and $Q \geqslant 1$, consider the set

$$A_{\boldsymbol{\theta}}(Q, \delta) := \left\{ (p_1, q) \in \mathbb{Z} \times \mathbb{N} : \begin{array}{l} Q < q \leqslant 2Q, \ (p_1 + \theta_1)/q \in I \\ \|qf((p_1 + \theta_1)/q) - \theta_2\| < \delta \end{array} \right\}.$$

The function $N_{\boldsymbol{\theta}}(Q, \delta) = \#A_{\boldsymbol{\theta}}(Q, \delta)$ counts the number of rational points $(p_1/q, p_2/q)$ with bounded denominator q such that the shifted points $((p_1 + \theta_1)/q, (p_2 + \theta_2)/q)$ lie within the δ/Q-neighborhood of the curve $\mathcal{C}$. The following result is a direct consequence of Theorem 3 from [**12**] – the inhomogeneous generalization of Theorem 1 from [**30**].

THEOREM 3. *Let $f \in C^{(3)}(I)$ and satisfy (19) and let $\varepsilon > 0$. Then, for any $Q \geqslant 1$ and $0 < \delta \leqslant \frac{1}{2}$ we have that*

$$N_{\boldsymbol{\theta}}(Q, \delta) \ll \delta Q^2 + Q^{1+\varepsilon}$$

where the implied constant is independent of Q and δ.

2.1.2. *Proof: the convergence case.* We are given that $\sum_{q=1}^{\infty} q^{1-s}\psi(q)^s < \infty$. Since ψ is monotonic, we have that

$$(20) \qquad \sum_{t=1}^{\infty} 2^{(2-s)t}\psi(2^t)^s < \infty.$$

Hence

$$(21) \qquad 2^{(2-s)t}\psi(2^t)^s < 1 \qquad \text{for all sufficiently large } t.$$

As in the proof of Theorem 1, without loss of generality we assume that $\boldsymbol{\theta} = (\theta_1, \theta_2) \in \mathbb{I}^2$ and moreover that $\mathcal{C} \subset \mathbb{I}^2$. By Lemma 2, for any $\ell \in \mathbb{N}$ we have that

$$(22) \quad \mathcal{S}_2^{\times}(\psi; \boldsymbol{\theta}) \cap \mathcal{C} \subset \bigcup_{t=\ell}^{\infty} \bigcup_{2^t \leqslant q < 2^{t+1}} \bigcup_{\substack{m \in \mathbb{Z} \\ 2^{|m|}\sqrt{\psi(2^t)} \leqslant 1}} \bigcup_{p_1=-1}^{q} \bigcup_{p_2=-1}^{q} \mathcal{C} \cap S_{\boldsymbol{\theta}}(t, q, m, p_1, p_2).$$

Using condition (19), it is easily verified that

$$(23) \qquad \operatorname{diam}(\mathcal{C} \cap S_{\boldsymbol{\theta}}(t, q, p_1, p_2, m)) \ll 2^{-|m|}\frac{\sqrt{\psi(2^t)}}{2^t}$$

and that whenever $\mathcal{C} \cap S_{\boldsymbol{\theta}}(t, q, p_1, p_2, m) \neq \emptyset$ we have that $(p_1, q) \in A_{\boldsymbol{\theta}}(Q, \delta)$ with

$$(24) \qquad \delta \asymp 2^{|m|}\sqrt{\psi(2^t)} \qquad \text{and} \qquad Q = 2^t,$$

where t, q, p_1, p_2, m are constrained as in (22) and the implied constants in (23) and (24) depend on c_1 and c_2 that appear in (19). By (22), the collection of all such sets

$\mathcal{C} \cap S_\theta(t, q, m, p_1, p_2)$ is a fine cover of $\mathcal{S}_2^\times(\psi; \boldsymbol{\theta}) \cap \mathcal{C}$. By Theorem 3 with $\varepsilon := s/4$, we have that

$$N_{\boldsymbol{\theta}}(Q, \delta) \ll 2^{|m|} \sqrt{\psi(2^t)} \, 2^{2t} + 2^{(1+s/4)t}$$

and so the s-dimensional volume of the above fine cover is

$$\ll \sum_{t=1}^{\infty} \sum_{\substack{m \in \mathbb{Z} \\ 2^{|m|} \sqrt{\psi(2^t)} \leqslant 1}} \left(2^{-|m|} \frac{\sqrt{\psi(2^t)}}{2^t} \right)^s \left(2^{|m|} \frac{\sqrt{\psi(2^t)}}{2^t} 2^{3t} + 2^{(1+s/4)t} \right)$$

$$\ll \sum_{t=1}^{\infty} \left(\frac{\sqrt{\psi(2^t)}}{2^t} \right)^{s+1} 2^{3t} \sum_{\substack{m \in \mathbb{Z} \\ 2^{|m|} \sqrt{\psi(2^t)} \leqslant 1}} 2^{(1-s)|m|} \quad +$$

$$+ \sum_{t=1}^{\infty} \sum_{\substack{m \in \mathbb{Z} \\ 2^{|m|} \sqrt{\psi(2^t)} \leqslant 1}} \left(2^{-|m|} \frac{\sqrt{\psi(2^t)}}{2^t} \right)^s 2^{(1+s/4)t}$$

$$\overset{(17)}{\ll} \sum_{t=1}^{\infty} \left(\frac{\sqrt{\psi(2^t)}}{2^t} \right)^{s+1} 2^{3t} \left(\sqrt{\psi(2^t)} \right)^{s-1} \quad +$$

$$+ \sum_{t=1}^{\infty} \left(\frac{\sqrt{\psi(2^t)}}{2^t} \right)^s 2^{(1+s/4)t}$$

$$\overset{(21)}{\ll} \sum_{t=1}^{\infty} 2^{(2-s)t} \psi(2^t)^s + \sum_{t=1}^{\infty} 2^{-ts/4} \overset{(20)}{<} \infty.$$

By Lemma 3, $\mathcal{H}^s(\mathcal{C} \cap \mathcal{S}_2^\times(\psi; \boldsymbol{\theta})) = 0$ and the proof of the convergence part of Theorem 2 is complete.

2.1.3. *Proof: the divergence case.* We are given that $\sum_{q=1}^{\infty} q^{1-s} \psi(q)^s = \infty$. Then, by the inhomogeneous version of Jarník's theorem [**15**] (see also the remark in [**8**, §12.1]), we have that $\mathcal{H}^s(\mathcal{S}_1(\psi; \theta_1) \cap I) = \infty$. The same observation that led to (8), gives rise to the following obvious inclusion

$$X := \left\{ (x, f(x)) : x \in \mathcal{S}_1(\psi; \theta_1) \cap I \right\} \subset \mathcal{C} \cap \mathcal{S}_2^\times(\psi; \boldsymbol{\theta}).$$

Since $f \in C^{(1)}$, we have that f is locally bi-Lipshitz and thus the map $x \mapsto (x, f(x))$ preserves s-dimensional Hausdorff measure. Therefore,

$$\mathcal{H}^s \left(\mathcal{S}_2^\times(\psi; \boldsymbol{\theta}) \cap \mathcal{C} \right) \geqslant \mathcal{H}^s(X) = \mathcal{H}^s(\mathcal{S}_1(\psi; \theta_1) \cap I) = \infty$$

and so completes the proof of Theorem 2.

3. Inhomogeneous badly approximable points

A real number x is said to be *badly approximable* if there exists a positive constant $c(x)$ such that

$$\|qx\| > c(x) \, q^{-1} \quad \forall \, q \in \mathbb{N}.$$

Here and throughout $\| \cdot \|$ denotes the distance of a real number to the nearest integer. It is well know that the set **Bad** of badly approximable numbers is of Lebesgue measure zero. However, a result of Jarník [23] states that

$$\dim \mathbf{Bad} = 1 \ . \tag{25}$$

Thus, in terms of dimension the set of badly approximable numbers is maximal; it has the same dimension as the real line.

In higher dimensions there are various natural generalizations of **Bad**. Restricting our attention to the plane $\mathbb{R}^2$, given a pair of real numbers i and j such that

$$0 \leqslant i, j \leqslant 1 \quad \text{and} \quad i + j = 1 \ , \tag{26}$$

a point $(x_1, x_2) \in \mathbb{R}^2$ is said to be (i, j)-*badly approximable* if there exists a positive constant $c(x_1, x_2)$ such that

$$\max\{ \ \|qx_1\|^{1/i} \ , \ \|qx_2\|^{1/j} \ \} \ > \ c(x_1, x_2) \ q^{-1} \quad \forall \ q \in \mathbb{N} \ .$$

Denote by $\mathbf{Bad}(i, j)$ the set of (i, j)-badly approximable points in $\mathbb{R}^2$. If $i = 0$, then we use the convention that $x^{1/i} := 0$ and so $\mathbf{Bad}(0, 1)$ is identified with $\mathbb{R} \times \mathbf{Bad}$. That is, $\mathbf{Bad}(0, 1)$ consists of points (x_1, x_2) with $x_1 \in \mathbb{R}$ and $x_2 \in \mathbf{Bad}$. The roles of x_1 and x_2 are reversed if $j = 0$. It easily follows from classical results in the theory of metric Diophantine approximation that $\mathbf{Bad}(i, j)$ is of (two-dimensional) Lebesgue measure zero. Building upon the work of Davenport [16], it is shown in [29] that

$$\dim \mathbf{Bad}(i, j) = 2 \ . \tag{27}$$

For alternative proofs and various strengthenings see [3, 19, 26–28]. In particular, a consequence of the main result in [3] is that the intersection of any finite number of (i, j)-badly approximable sets is of full dimension. Obviously this implies that the intersection of any two such sets is non-empty and thus establishes a conjecture of Wolfgang Schmidt dating back to the eighties. Most recently, Jinpeng An [1] has shown that the set $\mathbf{Bad}(i, j)$ is winning in the sense of Schmidt games and thus the intersection of any countable number of (i, j)-badly approximable sets is of full dimension.

The goal in this paper is to obtain the analogue of (27) within the inhomogeneous setup.

Problem 3: *Find an analogue of* (27) *for inhomogeneous approximation.*

For $\boldsymbol{\theta} = (\theta_1, \theta_2) \in \mathbb{R}^2$, let $\mathbf{Bad}(i, j; \boldsymbol{\theta})$ denote the set of points $(x_1, x_2) \in \mathbb{R}^2$ such that

$$\max\{ \ \|qx_1 - \theta_1\|^{1/i} \ , \ \|qx_2 - \theta_2\|^{1/j} \ \} \ > \ c(x_1, x_2) \ q^{-1} \quad \forall \ q \in \mathbb{N} \ .$$

Naturally, given $\theta \in \mathbb{R}$ the inhomogeneous generalisation of the one-dimensional set **Bad** is the set

$$\mathbf{Bad}(\theta) := \{x \in \mathbb{R} : \exists \ c(x) > 0 \ \text{ so that } \|qx - \theta\| \ > \ c(x) \ q^{-1} \quad \forall \ q \in \mathbb{N}\}$$

and so, for example, $\mathbf{Bad}(0, 1; \theta)$ is identified with $\mathbb{R} \times \mathbf{Bad}(\theta)$. It is straightforward to deduce that $\mathbf{Bad}(i, j; \boldsymbol{\theta})$ is of measure zero from the inhomogeneous version of Khintchine's theorems with varying approximating functions in each co-ordinate.

We will prove the following full dimension statement which represents the inhomogeneous analogue of (27).

THEOREM 4. *Let i, j satisfy (26) and $\boldsymbol{\theta} = (\theta_1, \theta_2) \in \mathbb{R}^2$. Then*

$$\dim \mathbf{Bad}(i, j; \boldsymbol{\theta}) = 2 \,.$$

The basic philosophy behind the proof is simple and is likely to be applicable to other situations where the goal is to generalize a known homogenous badly approximable statement to the inhomogeneous setting – see Remark 3.5 below. The key is to exploit the known homogeneous 'intervals construction' proof and use the power of subtraction; namely

(homogeneous construction) + $(\boldsymbol{\theta} - \boldsymbol{\theta} = \mathbf{0})$ $\implies$ (inhomogeneous statement).

Before moving onto the proof of Theorem 4, several remarks are in order.

Remark 3.1. For i and j fixed, the proof can be easily modified to deduce that the intersection of any finite number of $\mathbf{Bad}(i, j; \boldsymbol{\theta})$ sets is of full dimension. In fact, by making use of standard trickery (such as the argument that proves that the countable intersection of winning sets is winning) one can actually deduce that for any countable sequence $\boldsymbol{\theta}_t \in \mathbb{R}^2$

$$\dim \left(\cap_{t=1}^{\infty} \mathbf{Bad}(i, j; \boldsymbol{\theta}_t) \right) = 2 \,.$$

Remark 3.2. In another direction, the proof can be adapted to obtain the following more general form of Theorem 4 in which the inhomogeneous factor $\boldsymbol{\theta}$ depends on (x_1, x_2). More precisely, let $\boldsymbol{\theta} = (\theta_1, \theta_2) : \mathbb{R}^2 \to \mathbb{R}^2$ and let $\mathbf{Bad}(i, j; \boldsymbol{\theta})$ denote the set of points (x_1, x_2) such that

$$\max\{ \, \|qx_1 - \theta_1(x_1, x_2)\|^{1/i} \, , \, \|qx_2 - \theta_2(x_1, x_2)\|^{1/j} \, \} \, > \, c(x_1, x_2) \, q^{-1} \quad \forall \, q \in \mathbb{N} \,.$$

Then, if $\theta_1 = \theta_1(x_1)$ and $\theta_2 = \theta_2(x_2)$ are Lipshitz functions of one variable, we have that

$$\dim \mathbf{Bad}(i, j; \boldsymbol{\theta}) = 2 \,.$$

As an example, this statement implies that there is a set of $(x_1, x_2) \in \mathbb{R}^2$ of Hausdorff dimension 2 such that

$$\max\{ \, \|qx_1 - x_1^2\|^{1/i} \, , \, \|qx_2 - x_2^3\|^{1/j} \, \} \, > \, c(x_1, x_2) \, q^{-1} \quad \forall \, q \in \mathbb{N} \,.$$

It is worth pointing out that in the case $i = j$, the statement is also true if $\theta_1 = \theta_1(x_1, x_2)$ and $\theta_2 = \theta_2(x_1, x_2)$ are Lipshitz functions of two variables.

Remark 3.3. There is no difficulty in establishing the higher dimension analogue of Theorem 4. For any $\boldsymbol{\theta} = (\theta_1, \dots, \theta_n) \in \mathbb{R}^n$ and n–tuple of real numbers $i_1, \dots, i_n \geq 0$ such that $\sum i_r = 1$, denote by $\mathbf{Bad}(i_1, \dots, i_n; \boldsymbol{\theta})$ the set of points $(x_1, \dots, x_n) \in \mathbb{R}^n$ for which there exists a positive constant $c(x_1, \dots, x_n)$ such that

$$\max\{ \, \|qx_1 - \theta_1\|^{1/i_1} \, , \dots, \, \|qx_n - \theta_n\|^{1/i_n} \, \} \, > \, c(x_1, \dots, x_n) \, q^{-1} \quad \forall \quad q \in \mathbb{N}.$$

By modifying the proof of Theorem 4, in the obvious way, it is easy to show that

$$\dim \mathbf{Bad}(i_1 \dots, i_n; \boldsymbol{\theta}) = n \,.$$

Moreover, the various proofs of the homogeneous results obtained in [**28**] regarding $\mathbf{Bad}(i_1 \dots, i_n) \cap \Omega$, where Ω is some 'nice' fractal set (essentially, the support set of an absolutely friendly, Ahlfors regular measure) can be adapted to give the corresponding inhomogeneous statements without any serious difficulty. We have decided to restrict ourselves to proving Theorem 4 since it already contains the necessary ingredients to obtain the inhomogeneous statement from the homogeneous proof.

Remark 3.4. In the symmetric case $i_1 = \ldots = i_n = 1/n$, our Theorem 4 and indeed its generalizations mentioned in the previous remark are covered by the work of Einsiedler & Tseng [**18**, Theorem 1.1]. They actually deal with badly approximable systems of linear forms and show that the intersection of such sets with the support set Ω of an absolutely friendly measure is winning in the sense of Schmidt games. We mention in passing, that Einsiedler & Tseng proved their results roughly at the same time as us, but for some mystical reason, it has taken us over four years to present our work. Indeed the second author had a useful discussion with Einsiedler regarding their preprint at the conference 'The Diverse Faces of Arithmetic' in honour of the late Graham Everest in 2009.

Remark 3.5. The basic philosophy behind the proof of Theorem 4 can be exploited to yield the inhomogeneous strengthening of Schmidt's Conjecture. More precisely, we are able show that any inhomogeneous $\mathbf{Bad}(i, j; \boldsymbol{\theta})$ set is winning and thus

$$\dim \left(\cap_{t=1}^{\infty} \mathbf{Bad}(i_t, j_t; \boldsymbol{\theta}_t) \right) = 2 \, .$$

Furthermore, it is possible to show that the intersection of $\mathbf{Bad}(i, j; \boldsymbol{\theta})$ with any non-degenerate planar curve $\mathcal{C}$ is winning as is the intersection with any straight line satisfying certain natural Diophantine conditions. The former implies that

$$\dim \left(\cap_{t=1}^{\infty} \mathbf{Bad}(i_t, j_t; \boldsymbol{\theta}_t) \cap \mathcal{C} \right) = 1 \, ,$$

which strengthens even the homogeneous results obtained in [**4**, **5**] that solve an old problem of Davenport. These winning results will be the subject of a forthcoming joint paper with Jinpeng An.

3.1. Proof of Theorem 4. Throughout, we fix $i, j > 0$ satisfying (26) and $\boldsymbol{\theta} = (\theta_1, \theta_2) \in \mathbb{R}^2$. The situation when either $i = 0$ or $j = 0$ is easier and will be omitted.

Since $\mathbf{Bad}(i, j; \boldsymbol{\theta}) \subseteq \mathbb{R}^2$, we obtain for free the upperbound result:

$$\dim \mathbf{Bad}(i, j; \boldsymbol{\theta}) \leqslant 2 \, .$$

Thus, the proof reduces to establishing the complementary lowerbound. With this in mind, for a fixed constant $c > 0$ let

$$\mathbf{Bad}_c(i, j; \boldsymbol{\theta}) := \{(x_1, x_2) \in \mathbb{R}^2 : \max\{\|qx_1 - \theta_1\|^{1/i}, \|qx_2 - \theta_2\|^{1/j}\} > c/q \quad \forall q \in \mathbb{N}\} \, .$$

Clearly $\mathbf{Bad}_c(i, j; \boldsymbol{\theta}) \subset \mathbf{Bad}(i, j; \boldsymbol{\theta})$ and

$$\mathbf{Bad}(i, j; \boldsymbol{\theta}) = \bigcup_{c>0} \mathbf{Bad}_c(i, j; \boldsymbol{\theta}) \, .$$

Geometrically, the set $\mathbf{Bad}_c(i, j; \boldsymbol{\theta})$ consists of points $(x_1, x_2) \in \mathbb{R}^2$ which avoid all rectangles centred at shifted rational points $((\theta_1 - p_1)/q, (\theta_2 - p_2)/q)$ with sidelengths $2c^i q^{-(1+i)}$ and $2c^j q^{-(1+j)}$. The sides are taken to be parallel to the coordinate axes. The overall strategy is to construct a 'Cantor–type' subset $\mathbf{K}_c^{\boldsymbol{\theta}} (= \mathbf{K}_c^{\boldsymbol{\theta}}(i, j))$

of $\mathbf{Bad}_c(i,j;\boldsymbol{\theta})$ with the property that $\dim \mathbf{K}_c^{\boldsymbol{\theta}} \to 2$ as $c \to 0$. This together with the fact that

$$\dim \mathbf{Bad}(i,j;\boldsymbol{\theta}) \geqslant \dim \mathbf{Bad}_c(i,j;\boldsymbol{\theta}) \geqslant \dim \mathbf{K}_c^{\boldsymbol{\theta}}$$

implies the required lower bound result.

To obtain the desired Cantor type set $\mathbf{K}_c^{\boldsymbol{\theta}}$, we adapt the homogeneous construction of $\mathbf{K}_c = \mathbf{K}_c^0$ given in [29, §3.1] that is at the heart of establishing (27); that is to say Theorem 4 with $\boldsymbol{\theta} = \mathbf{0}$.

3.1.1. *The homogeneous construction.* Let $R \geqslant 11$ be an integer and $c > 0$ be given by

$$(28) \qquad c := 8^{-1/i} R^{-2(1+i)/i}.$$

It is established in [29, §3.1], by induction on $n \geqslant 0$, the existence of a nested collection $\mathcal{F}_n$ of closed rectangles $F_n := I_n \times J_n$ with the property that for all points $(x_1, x_2) \in F_n$ the following (homogeneous) condition is satisfied:

$$(\mathrm{H}) \qquad \max\{ \|qx_1\|^{1/i}, \|qx_2\|^{1/j} \} > cq^{-1} \qquad \forall\ 0 < q < R^n.$$

The lengths of the sides I_n and J_n of F_n are given by

$$(29) \quad |I_n| := \tfrac{1}{4} R^{-(1+i)(n+1)} \quad \text{and} \quad |J_n| := \tfrac{1}{4} R^{-(1+j)(n+1)} \quad (n \geqslant 0).$$

Without loss of generality assume that $0 < i \leqslant j < 1$ so that the rectangles F_n are long and thin unless $i = j$ in which case the rectangles are obviously squares.

The crux of the induction is as follows. We work within the closed unit square and start by subdividing the square into closed rectangles F_0 of size $I_0 \times J_0$ – starting from the bottom left hand corner of the unit square (i.e. the origin). Denote by $\mathcal{F}_0$ the collection of rectangles F_0. For $n = 0$, condition (H) is trivially satisfied for any rectangle $F_0 \subset \mathcal{F}_0$, since there are no integers q satisfying $0 < q < 1$. Given $\mathcal{F}_n$ satisfying condition (H), we wish to construct a nested collection $\mathcal{F}_{n+1}$ for which the condition is satisfied for $n + 1$. Suppose F_n is a good rectangle; that is, all points $(x_1, x_2) \in F_n$ satisfy condition (H). In short, $F_n \in \mathcal{F}_n$. Now partition F_n into rectangles $F_{n+1} := I_{n+1} \times J_{n+1}$ – starting from the bottom left hand corner of F_n. From (29), it follows that there are $[R^{1+i}] \times [R^{1+j}]$ rectangles in the partition. Since they are nested, anyone of these rectangles will satisfy condition (H) for $n+1$ if for any point (x_1, x_2) in F_{n+1} the inequality

$$(30) \qquad \max\{ \|qx_1\|^{1/i}, \|qx_2\|^{1/j} \} > c\,q^{-1}$$

is satisfied for

$$(31) \qquad R^n \leqslant q < R^{n+1}.$$

With q in this 'denominator' range, suppose there exists a bad rational pair $(p_1/q, p_2/q)$ so that (30) is violated, in other words

$$|x_1 - p_1/q| \leqslant c^i\, q^{-(1+i)} \qquad \text{and} \qquad |x_2 - p_2/q| \leqslant c^j\, q^{-(1+j)}$$

for some point in F_n and therefore in some F_{n+1}. Such F_{n+1} rectangles are bad in the sense that they do not satisfy condition (H) for $n+1$ and those that remain are good. The upshot of the 'Stage 1' argument in [29, §3.1] is that there are at most

$$(32) \qquad 3 \left[\frac{|J_n|}{|J_{n+1}|} \right] \leqslant 3R^{1+j}$$

bad F_{n+1} rectangles in F_n. Hence, out of the potential $[R^{1+i}] \times [R^{1+j}]$ rectangles, at least

$$(R^{1+i} - 1)(R^{1+j} - 1) - 3R^{1+j} > R^3(1 - 5R^{-(1+i)})$$

are good F_{n+1} rectangles in F_n. Now choose exactly $[R^3 (1 - 5 R^{-(1+i)})]$ of these good rectangles and denote this collection by $\mathcal{F}(F_n)$. Finally, define

$$\mathcal{F}_{n+1} := \bigcup_{F_n \subset \mathcal{F}_n} \mathcal{F}(F_n) .$$

Thus, given the collection $\mathcal{F}_n$ for which condition (H) is satisfied for n, we have constructed a nested collection $\mathcal{F}_{n+1}$ for which condition (H) is satisfied for $n + 1$. This completes the proof of the induction step and so the construction of the Cantor-type set

$$\mathbf{K_c} := \bigcap_{n=0}^{\infty} \mathcal{F}_n .$$

3.1.2. *Bringing the inhomogeneous approximation into play.* The idea is to merge the inhomogeneous approximation constraints into the above homogeneous construction. In short, this involves creating a subcollection $\mathcal{F}_n^{\theta}$ of $\mathcal{F}_n$ so that for all points $(x_1, x_2) \in F_n$ with $F_n \subseteq \mathcal{F}_n^{\theta}$, both the (homogeneous) condition (H) and the following (inhomogeneous) condition are satisfied:

(I) $\max\{ \|qx_1 - \theta_1\|^{1/i} , \|qx_2 - \theta_2\|^{1/j} \} > c_* q^{-1}$ $\forall \ 0 < q < R^{n-d}$,

where

(33) $$c_*^i := \tfrac{1}{8} R^{-(1+i)(d+2)} \qquad \text{and} \qquad d := \left[\frac{3}{i} \right] .$$

For $n = 0$, condition (I) is trivially satisfied for any rectangle $F_0 \subset \mathcal{F}_0$, since there are no integers q satisfying $0 < q < 1$. Put $\mathcal{F}_0^{\theta} := \mathcal{F}_0$. Now suppose $\mathcal{F}_n^{\theta} \subseteq \mathcal{F}_n$ has been constructed and for each F_n in $\mathcal{F}_n^{\theta}$ construct the collection $\mathcal{F}(F_n)$ as before. Then by definition, each $F_{n+1} \in \mathcal{F}(F_n)$ satisfies condition (H) for $n+1$. The aim is to construct a subcollection $\mathcal{F}^{\theta}(F_n)$ such that for each F_{n+1} in $\mathcal{F}^{\theta}(F_n)$ condition (I) for $n + 1$ is also satisfied; in other words, for any point (x_1, x_2) in F_{n+1} the inequality

(34) $$\max\{ \|qx_1 - \theta_1\|^{1/i} , \|qx_2 - \theta_2\|^{1/j} \} > c_* q^{-1}$$

is satisfied for

(35) $$R^{n-d} \leqslant q < R^{n+1-d} .$$

With q satisfying (35), suppose there exists a bad shifted rational pair $((\theta_1 + p_1)/q, (\theta_2 + p_2)/q)$ so that (34) is violated, in other words

$$|x_1 - (\theta_1 + p_1)/q| \leqslant c_*^i q^{-(1+i)} \qquad \text{and} \qquad |x_2 - (\theta_2 + p_2)/q| \leqslant c_*^j q^{-(1+j)}$$

for some point (x_1, x_2) in F_n. Then, in view of (35), it follows that

(36) $$|x_1 - (\theta_1 + p_1)/q| \leqslant c_*^i R^{-(1+i)(n-d)} \leqslant \tfrac{1}{2} |I_{n+1}|$$

if

(37) $$c_*^i \leqslant \tfrac{1}{8} R^{-(1+i)(d+2)} .$$

Similarly,

(38) $|x_2 - (\theta_2 + p_2)/q| \leqslant \frac{1}{2}|J_{n+1}|$

if

(39) $c_*^j \leqslant \frac{1}{8}R^{-(1+j)(d+2)}$.

Observe that (37) implies (39). In view of (33), we have equality in (37), thus (36) and (38) are satisfied and it follows that any bad shifted rational pair gives rise to at most four bad rectangles F_{n+1} in $\mathcal{F}(F_n)$; i.e. rectangles for which (I) is not satisfied for $n+1$. Now suppose there exist two bad shifted rational pairs, say $((\theta_1 + p_1)/q, (\theta_2 + p_2)/q)$ and $((\theta_1 + \tilde{p}_1)/\tilde{q}, (\theta_2 - \tilde{p}_2)/\tilde{q})$. Then, for any $(x_1, x_2) \in F_n$ we have that

$$|x_1 - (\theta_1 + p_1)/q| \overset{(36)}{\leqslant} \frac{1}{2}|I_{n+1}| + |I_n| < 2|I_n|$$

$$\implies \quad |qx_1 - \theta_1 - p_1| < q\,2|I_n| \overset{(35)}{\leqslant} 2R^{n+1-d}|I_n|$$

and

$$|x_2 - (\theta_2 + p_2)/q| \overset{(38)}{\leqslant} \frac{1}{2}|J_{n+1}| + |J_n| < 2|J_n|$$

$$\implies \quad |qx_2 - \theta_2 - p_2| < q\,2|J_n| \overset{(35)}{\leqslant} 2R^{n+1-d}|J_n| .$$

Similarly, we obtain that

$$|\tilde{q}x_1 - \theta_1 - \tilde{p}_1| < 2R^{n+1-d}|I_n| \quad \text{and} \quad |\tilde{q}x_2 - \theta_2 - \tilde{p}_2)| < 2R^{n+1-d}|J_n| .$$

Let $q_* := |q - \tilde{q}|$ and observe that

(40) $0 < q_* < R^{n+1-d} < R^n$

It now follows that

$$|q_*x_1 - (p_1 + \tilde{p}_1)| = |(qx_1 - \theta_1 - p_1) - (\tilde{q}x_1 - \theta_1 - \tilde{p}_1)|$$

$$\leqslant |qx_1 - \theta_1 - p_1| + |\tilde{q}x_1 - \theta_1 - \tilde{p}_1|$$

$$\leqslant 4R^{n+1-d}|I_n| \leqslant R^{-d(1+i)}q_*^{-i} \overset{(33)}{\leqslant} R^{-3(1+i)}q_*^{-i}$$

(41) $\overset{(28)}{\leqslant} c^i q_*^{-i}$.

Similarly,

$$|q_*x_2 - (p_2 + \tilde{p}_2)| = |(qx_2 - \theta_2 - p_2) - (\tilde{q}x_2 - \theta_2 - \tilde{p}_2)|$$

$$\leqslant |qx_2 - \theta_2 - p_2| + |\tilde{q}x_2 - \theta_1 - \tilde{p}_2|$$

$$\leqslant 4R^{n+1-d}|J_n| \leqslant R^{-d(1+j)}q_*^{-j} \overset{(33)}{\leqslant} R^{-\frac{(3-i)}{i}(1+j)}q_*^{-j}$$

(42) $\overset{(28)}{\leqslant} c^j q_*^{-j}$.

The upshot of inequalities (40), (41) and (42) is that the homogeneous condition (H) is not satisfied for points in F_n. This contradicts the fact that $F_n \in \mathcal{F}_n^{\theta} \subseteq \mathcal{F}_n$. In turn, this implies that there exists at most one bad shifted rational pair that gives rise to at most 4 bad F_{n+1} rectangles amongst those in $\mathcal{F}(F_n)$. In other words, at least

$$\#\mathcal{F}(F_n) - 4 \;=\; [R^3\,(1\,-\,5\,R^{-(1+i)})]\,-\,4 \;>\; R^3(1-6R^{-(1+i)})$$

of the F_{n+1} rectangles satisfy both conditions (H) and (I) for $n+1$. Now choose exactly $[R^3\,(1\,-\,6\,R^{-(1+i)})]$ of these good rectangles and denote this collection by $\mathcal{F}^{\theta}(F_n)$. Finally, define

$$\mathcal{F}_{n+1}^{\theta} \;:=\; \bigcup_{F_n \subset \mathcal{F}_n^{\theta}} \mathcal{F}^{\theta}(F_n) \qquad \text{and} \qquad \mathbf{K}_c^{\theta} \;:=\; \bigcap_{n=0}^{\infty} \mathcal{F}_n^{\theta}.$$

3.1.3. *The finale.* It remains to show that

$$\dim \mathbf{K}_c^{\theta} \;\to\; 2 \qquad \text{as} \qquad c \to 0.$$

This involves essentially following line by line the arguments set out in [**29**, §3.2 and §3.3]. The details are left to the reader.

$\boxtimes$

Acknowledgements. SV would like to thank Dani for his support during the early years of his mathematical life – basically when it mattered most! Naturally, an enormous thanks to the delightful and magnificent Bridget, Ayesha and Iona. The younger two are curious and challenging for all the right reasons – long may you remain that way and good luck with high school! This brings me onto their wonderful teachers during the last two years at primary school who have instilled them with great belief and have been superb role models for both children and adults. For this and much much more, I would like to take this opportunity to thank Mr Youdan, Mr Middleton and Mrs Davison. I hope you continue to inspire!

References

[1] J. An. Two dimensional badly approximable vectors and Schmidt's game. Pre-print: arXiv:1204.3610.

[2] D. Badziahin and J. Levesley, *A note on simultaneous and multiplicative Diophantine approximation on planar curves*, Glasg. Math. J. **49** (2007), no. 2, 367–375, DOI 10.1017/S0017089507003722. MR2347267 (2009a:11151)

[3] D. Badziahin, A. Pollington, and S. Velani, *On a problem in simultaneous Diophantine approximation: Schmidt's conjecture*, Ann. of Math. (2) **174** (2011), no. 3, 1837–1883, DOI 10.4007/annals.2011.174.3.9. MR2846492 (2012k:11105)

[4] D. Badziahin, S. Velani. Badly approximable points on planar curves and a problem of Davenport. Pre-print: arXiv:1301.4243.

[5] V. Beresnevich. Badly approximable points on manifolds. Pre-print: arXiv:1304.0571.

[6] V. Beresnevich, *Rational points near manifolds and metric Diophantine approximation*, Ann. of Math. (2) **175** (2012), no. 1, 187–235, DOI 10.4007/annals.2012.175.1.5. MR2874641

[7] V. Beresnevich, V. Bernik, M. Dodson, and S. Velani, *Classical metric Diophantine approximation revisited*, Analytic number theory, Cambridge Univ. Press, Cambridge, 2009, pp. 38–61. MR2508636 (2010c:11082)

[8] V. Beresnevich, D. Dickinson, and S. Velani, *Measure theoretic laws for lim sup sets*, Mem. Amer. Math. Soc. **179** (2006), no. 846, x+91, DOI 10.1090/memo/0846. MR2184760 (2007d:11086)

[9] V. Beresnevich, A. Haynes, and S. Velani, *Multiplicative zero-one laws and metric number theory*, Acta Arith. **160** (2013), no. 2, 101–114, DOI 10.4064/aa160-2-1. MR3105329

[10] V. Beresnevich and S. Velani, *A mass transference principle and the Duffin-Schaeffer conjecture for Hausdorff measures*, Ann. of Math. (2) **164** (2006), no. 3, 971–992, DOI 10.4007/annals.2006.164.971. MR2259250 (2008a:11090)

[11] V. Beresnevich and S. L. Velani. Schmidt's theorem, Hausdorff measures and slicing. *Int. Math. Res. Not.*, 2006:1–24, article ID 48794, 2006.

[12] V. V. Beresnevich, R. C. Vaughan, and S. L. Velani, *Inhomogeneous Diophantine approximation on planar curves*, Math. Ann. **349** (2011), no. 4, 929–942, DOI 10.1007/s00208-010-0548-9. MR2777039 (2012c:11146)

[13] V. I. Bernik and M. M. Dodson, *Metric Diophantine approximation on manifolds*, Cambridge Tracts in Mathematics, vol. 137, Cambridge University Press, Cambridge, 1999. MR1727177 (2001h:11091)

[14] A. S. Besicovitch, *Sets of Fractional Dimensions (IV): On Rational Approximation to Real Numbers*, J. London Math. Soc. **S1-9**, no. 2, 126, DOI 10.1112/jlms/s1-9.2.126. MR1574327

[15] Y. Bugeaud, *An inhomogeneous Jarník theorem*, J. Anal. Math. **92** (2004), 327–349, DOI 10.1007/BF02787766. MR2072751 (2005f:11166)

[16] H. Davenport, *A note on Diophantine approximation. II*, Mathematika **11** (1964), 50–58. MR0166154 (29 #3432)

[17] D. Dickinson and S. L. Velani, *Hausdorff measure and linear forms*, J. Reine Angew. Math. **490** (1997), 1–36, DOI 10.1515/crll.1997.490.1. MR1468922 (98e:11094)

[18] M. Einsiedler and J. Tseng, *Badly approximable systems of affine forms, fractals, and Schmidt games*, J. Reine Angew. Math. **660** (2011), 83–97. MR2855820 (2012m:11095)

[19] L. Fishman, *Schmidt's game, badly approximable matrices and fractals*, J. Number Theory **129** (2009), no. 9, 2133–2153, DOI 10.1016/j.jnt.2009.02.005. MR2528057 (2010j:11123)

[20] P. Gallagher, *Metric simultaneous diophantine approximation*, J. London Math. Soc. **37** (1962), 387–390. MR0157939 (28 #1167)

[21] P. X. Gallagher, *Metric simultaneous diophantine approximation. II*, Mathematika **12** (1965), 123–127. MR0188154 (32 #5593)

[22] I. Jarník : Sur les approximations diophantiennes des nombres p–adiques. *revista Ci Lima.* **47**, 489–505.

[23] V. Jarník. A contribution to the metric theory of Diophantine approximations. *Prace math.-fiz.*, 36:91–106, 1929. (In Polish).

[24] A. Khintchine, *Einige Sätze über Kettenbrüche, mit Anwendungen auf die Theorie der Diophantischen Approximationen* (German), Math. Ann. **92** (1924), no. 1-2, 115–125, DOI 10.1007/BF01448437. MR1512207

[25] D. Y. Kleinbock and G. A. Margulis, *Flows on homogeneous spaces and Diophantine approximation on manifolds*, Ann. of Math. (2) **148** (1998), no. 1, 339–360, DOI 10.2307/120997. MR1652916 (99j:11083)

[26] D. Kleinbock and B. Weiss, *Badly approximable vectors on fractals*, Israel J. Math. **149** (2005), 137–170, DOI 10.1007/BF02772538. Probability in mathematics. MR2191212 (2008d:11079)

[27] D. Kleinbock and B. Weiss, *Modified Schmidt games and Diophantine approximation with weights*, Adv. Math. **223** (2010), no. 4, 1276–1298, DOI 10.1016/j.aim.2009.09.018. MR2581371 (2011b:11103)

[28] S. Kristensen, R. Thorn, and S. Velani, *Diophantine approximation and badly approximable sets*, Adv. Math. **203** (2006), no. 1, 132–169, DOI 10.1016/j.aim.2005.04.005. MR2231044 (2007i:11106)

[29] A. Pollington and S. Velani, *On simultaneously badly approximable numbers*, J. London Math. Soc. (2) **66** (2002), no. 1, 29–40, DOI 10.1112/S0024610702003265. MR1911218 (2003i:11095)

[30] R. C. Vaughan and S. Velani, *Diophantine approximation on planar curves: the convergence theory*, Invent. Math. **166** (2006), no. 1, 103–124, DOI 10.1007/s00222-006-0509-9. MR2242634 (2007i:11107)

DEPARTMENT OF MATHEMATICS, UNIVERSITY OF YORK, HESLINGTON, YORK, YO10 5DD, ENGLAND

E-mail address: victor.beresnevich@york.ac.uk

DEPARTMENT OF MATHEMATICS, UNIVERSITY OF YORK, HESLINGTON, YORK, YO10 5DD, ENGLAND

E-mail address: sanju.velani@york.ac.uk

Contemporary Mathematics
Volume **631**, 2015
http://dx.doi.org/10.1090/conm/631/12606

Dynamical invariants for group automorphisms

Richard Miles, Matthew Staines, and Thomas Ward

ABSTRACT. We discuss some of the issues that arise in attempts to classify automorphisms of compact abelian groups from a dynamical point of view. In the particular case of automorphisms of one-dimensional solenoids, a complete description is given and the problem of determining the range of certain invariants of topological conjugacy is discussed. Several new results and old and new open problems are described.

1. Introduction

Automorphisms and rotations of compact abelian metric groups provide the simplest examples of dynamical systems, and their special structure makes them amenable to detailed analysis. In particular, their strong homogeneity properties allow global invariants (notably the topological entropy) to be calculated locally.

Here we wish to discuss the problem of classifying compact group automorphisms viewed as dynamical systems in the most naive sense. Denote by $\mathcal{G}$ the space of all pairs (G, T), where T is an automorphism of a compact metric abelian group G, and let $\sim$ denote a dynamically meaningful notion of equivalence between two such systems. Then the problem we wish to discuss has three facets:

- *Classification*: can the quotient space $\mathcal{G}/\sim$ be described?
- *Range*: for some invariant π of such systems, what is

$$\{\pi(G, T) \mid (G, T) \in \mathcal{G}\}?$$

- *Fibre*: for a given value f of some invariant or collection of invariants π, what can be said about

$$\{(G, T) \in \mathcal{G} \mid \pi(G, T) = f\}?$$

We will describe some of the issues that arise in formulating these questions, recalling in particular some of the well-known difficulties in the classification problem both from an ergodic theory and a set theory point of view. For the range problem we will describe some recent work and present some new arguments concerned with orbit growth in particular. Many but not all of the problems identified here are well-known.

2010 *Mathematics Subject Classification*. Primary 37C35; Secondary 37P35, 11J72.
Key words and phrases. Topological entropy, zeta function, group automorphism, Mertens' theorem, one-solenoid, Lehmer's problem, Mertens' theorem, orbit Dirichlet series, Bernoulli shift, Ornstein theory.

The same questions make sense for group actions. Given a countable group Γ, there is an associated space $\mathcal{G}_\Gamma$ comprising all pairs (G, T), where $T : \Gamma \to \operatorname{Aut}(G)$ is a representation of Γ by automorphisms of the compact metric abelian group G. While our emphasis is not on any setting beyond $G = \mathbb{Z}$, one or two examples will be given to indicate where the theory has new features for other group actions.

An element $(G, T) \in \mathcal{G}$ (by assumption) carries a rotation-invariant metric giving the topology, a σ-algebra $\mathcal{B}_G$ of Borel sets, and a probability measure (Haar measure) m_G preserved by T. The most natural notion of equivalence between (G_1, T_1) and (G_2, T_2) in $\mathcal{G}$ is a commutative diagram

$$
\begin{array}{ccc}
G_1 & \xrightarrow{\ T_1\ } & G_1 \\
\phi \downarrow & & \downarrow \phi \\
G_2 & \xrightarrow[\ T_2\ ]{} & G_2
\end{array}
$$

where the equivariant map ϕ is required to be:

- a continuous isomorphism of groups, giving the equivalence of *algebraic isomorphism*;
- a homeomorphism, giving the equivalence of *topological conjugacy*; or
- an almost-everywhere defined isomorphism between the measure spaces

$$(G_1, \mathcal{B}_{G_1}, m_{G_1})$$

and

$$(G_2, \mathcal{B}_{G_2}, m_{G_2}),$$

giving the equivalence of *measurable isomorphism*.

Clearly algebraic isomorphism implies topological conjugacy. Entropy arguments show that topological conjugacy implies measurable isomorphism. An easy consequence of the type of constructions discussed later — and an instance of the type of question one might ask about the structure of $\mathcal{G}/\sim$ — is that each measurable isomorphism equivalence class splits into uncountably many topological conjugacy classes. We have not included the equivalence relation that is in some ways the most natural in dynamics (*finitary isomorphism*, essentially measurable isomorphism by a map that is continuous off an invariant null set) because there is little that can be said about it beyond those examples where it is almost self-evident how to construct such a map.

We will also be interested in various invariants for $(G, T) \in \mathcal{G}$. The most significant of these is the entropy,

$$
(1.1) \qquad h(T) = \lim_{\varepsilon \searrow 0} \lim_{n \to \infty} -\frac{1}{n} \log m_G \left(\bigcap_{j=0}^{n-1} T^{-j} B_\varepsilon(0) \right)
$$

where $B_\varepsilon(0)$ denotes an open metric ball around the identity of G. A manifestation of the homogeneity of the dynamics of group automorphisms is that this quantity coincides with the topological entropy and with the measure-theoretic entropy with respect to Haar measure (see Bowen [12] or [17]), and is therefore an invariant of any of the notions of equivalence above.

Topological conjugacy also preserves closed orbits and periodic points. This combinatorial data is all contained in the dynamical zeta function

$$\zeta_T(z) = \exp \sum_{n \geqslant 1} \frac{\mathsf{F}_T(n)}{n} z^n, \qquad (1.2)$$

where $\mathsf{F}_T(n) = |\{g \in G \mid T^n g = g\}|$, with radius of convergence

$$1/\limsup_{n \to \infty} \mathsf{F}_T(n)^{1/n}.$$

DEFINITION 1. For elements $(G, T), (G', T') \in \mathcal{G}$ write

$$(G, T) \sim_h (G', T')$$

if $h(T) = h(T')$, and

$$(G, T) \sim_\zeta (G', T')$$

if $\zeta_T(z) = \zeta_{T'}(z)$ as formal power series.

The primary growth rate measures are

$$p(T) = \lim_{n \to \infty} \frac{1}{n} \log \mathsf{F}_T(n)$$

if this exists, and

$$\pi_T(X) = \sum_{n \leqslant X} \mathsf{O}_T(n),$$

where $\mathsf{O}_T(n)$ denotes the number of closed orbits of length n. As we will see, for many (in several precise senses, most) examples the limit defining p will not exist, and a more averaged measure of orbit growth comes from the dynamical Mertens' theorem, giving asymptotics for the quantity

$$\mathsf{M}_T(N) = \sum_{n \leqslant N} \frac{\mathsf{O}_T(n)}{e^{h(T)n}} = \sum_{n \leqslant N} \sum_{d \mid n} \frac{\mu(n/d) \, \mathsf{F}_T(d)}{n e^{h(T)n}},$$

where the expression on the right is simply obtained by Möbius inversion, as

$$\mathsf{F}_T(n) = \sum_{d \mid n} d \, \mathsf{O}_T(d).$$

In general, for an action T of a countable group Γ, the periodic point data for the action is a map from the space of finite-index subgroups of Γ to $\mathbb{N} \cup \{\infty\}$, associating to each subgroup $L < \Gamma$ the cardinality of the set of L-periodic points

$$\mathsf{F}_T(L) = \{x \in X \mid T_\ell x = x \text{ for all } \ell \in L\}. \qquad (1.3)$$

There is an extension of the dynamical zeta function to the setting of $\mathbb{Z}^d$-actions due to Lind [41], but it does not carry all the periodic point data unless $d = 1$. Some of the questions considered here have been studied for full-shift actions of nilpotent groups in [47], and for specific examples of $\mathbb{Z}^2$-actions by automorphisms of compact connected groups in [48].

Several of the questions we will discuss involve adopting a point of view on what is a natural family of dynamical systems, and what is a dynamically meaningful notion of equivalence. In this context it is reasonable to restrict attention to ergodic group automorphisms throughout, and in some cases the more interesting questions arise if we also insist that the group be connected. From this point of view the natural starting point is to study automorphisms of one-dimensional solenoids, and much of what we will discuss will be concerned with these. A remarkable

234   RICHARD MILES, MATTHEW STAINES, AND THOMAS WARD

consequence of the work of Markus and Meyer is that all of these seemingly exotic one-dimensional solenoids appear generically in Hamiltonian flows on compact
symplectic manifolds of sufficiently high dimension [44] (we thank Alex Clark for
pointing this out). Thus, despite the seemingly strange phenomena and arithmetic
questions that arise here, the dynamical systems discussed here arise naturally in
smooth dynamical situations. Another reason to restrict much of our attention to
one-dimensional solenoids is cowardice. In one natural direction of extension, the
S-integer construction builds families of compact group automorphisms starting
with algebraic number fields or function fields of positive characteristic (see [15] for
the details), and the arithmetic questions that arise become more intricate but are
in principle amenable to the same methods in those settings. Another natural direction is to simply replace 'one' by $d \geqslant 2$, and study automorphisms of d-dimensional
solenoids. In this complete generality the difficulties are more formidable, and this
is discussed further at the end of Section 5.1.

The assumption of ergodicity is a mild requirement and avoids degeneracies –
automorphisms of compact abelian groups cannot be non-ergodic in dynamically interesting ways. Ergodicity guarantees (completely) positive entropy, and the structure of compact group automorphisms with zero entropy has been described by
Seethoff [57].

2. Algebraic isomorphism

Deciding if (G_1, T_1) and (G_2, T_2) are algebraically isomorphic is not a dynamical question at all. The equivalence implies that $\phi : G_1 \to G_2$ is an isomorphism of groups, so we are asking if the groups are isomorphic to a single group G,
and then if the two elements T_1 and $\phi^{-1}T_2\phi$ are conjugate as elements of the
group $\mathrm{Aut}(G)$ of continuous automorphisms of G. If the systems are expansive
(or satisfy the descending chain condition) and connected, then the problem of determining conjugacy in the automorphism group is algorithmically decidable by a
result of Grunewald and Segal [22] (see Kitchens and Schmidt [28, Sec. 6]). In
general this cannot be expected: if the group is infinite-dimensional then elements
of the automorphism group in general correspond to matrices of infinite rank.

EXAMPLE 2. Toral automorphisms are conjugate if the corresponding matrices
have the same characteristic polynomial and share an additional number-theoretic
invariant related to ideal classes in the splitting field of the characteristic polynomial
(see Latimer and Macduffee [33] for the details). This additional invariant is already
visible even on the 2-torus. For example, if

$$A = \begin{pmatrix} 3 & 10 \\ 1 & 3 \end{pmatrix}, B = \begin{pmatrix} 3 & 5 \\ 2 & 3 \end{pmatrix}$$

then it is easy to check that if $Q \in \mathrm{M}_2(\mathbb{Z})$ has $QA = BQ$ we have $5 \mid \det Q$.

A flavour of the sort of examples that will arise in this setting is given by the
next problem.

PROBLEM 1. Given a monic polynomial $\chi \in \mathbb{Z}[x]$ of degree d, describe the
combinatorial properties of the following poset. At the bottom level $n = 0$ there is
a vertex for each of the finitely many algebraic conjugacy classes of the set

$$M_\chi = \{A \in \mathrm{GL}_d(\mathbb{Z}) \mid \det(tI_d - A) = \chi(t)\},$$

which may be enumerated using data from the ideal class groups of the splitting fields of the various irreducible factors of χ using [33]. At level $n \geqslant 1$ there is a vertex for each of the conjugacy classes of the elements of M_χ over the ring $\mathbb{Z}[\frac{1}{p_j} : j \leqslant n]$ where $p_1, p_2, \ldots$ are the rational primes in their natural order. The edges reflect conjugacy classes merging in the larger group at each level, and the top of the poset has a vertex for each conjugacy class of the set M_χ in $\mathrm{GL}_d(\mathbb{Q})$, which are described by the rational canonical forms of matrices with characteristic polynomial χ.

For example, Problem 1 is readily solved for Example 2 as follows. The polynomial $\chi(x) = x^2 - 6x - 1$ is irreducible over $\mathbb{Q}$ with splitting field $\mathbb{Q}(\sqrt{10})$, with ideal class group isomorphic to $\mathbb{Z}/2\mathbb{Z}$. The representative matrices in Example 2 are not conjugate over $\mathbb{Z}[\frac{1}{2}]$ and $\mathbb{Z}[\frac{1}{6}]$, but become conjugate over $\mathbb{Z}[\frac{1}{30}]$, resulting in the diagram in Figure 1.

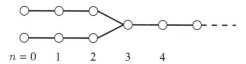

$$n = 0 \quad 1 \quad 2 \quad 3 \quad 4$$

FIGURE 1. Conjugacy for the characteristic polynomial $x^2 - 6x - 1$.

3. Topological conjugacy

As we will see, for connected groups this notion of equivalence collapses to algebraic isomorphism. In fact there is a rigidity phenomenon at work, which means that in many situations not only does the existence of a topological conjugacy force there to be an algebraic isomorphism, it is required to be algebraic itself.

THEOREM 3 (Adler and Palais [2]). *If $T : \mathbb{T}^d \to \mathbb{T}^d$ and $S : G \to G$ are topologically conjugate ergodic automorphisms via a homeomorphism ϕ from $\mathbb{T}^d$ to G, then $G \cong \mathbb{T}^d$ and the map ϕ is itself an automorphism of $\mathbb{T}^d$ composed with rotation by a fixed point of S.*

This extends in several ways to other connected finite-dimensional groups, and to other group and semi-group actions (see Clark and Fokkink [16] and Bhattacharya and the third author [10]). Notice that the phenomena is not universal, and in particular requires the topological entropy to be finite. The next example shows that in infinite entropy situations there is no topological rigidity.

EXAMPLE 4. If $\theta : \mathbb{T} \to \mathbb{T}$ is any homeomorphism, then the map $\Theta : \mathbb{T}^\mathbb{Z} \to \mathbb{T}^\mathbb{Z}$ defined by $(\Theta((x_n)_{n\in\mathbb{Z}}))_k = \theta(x_k)$ for all $k \in \mathbb{Z}$ is a topological conjugacy between the left shift on $\mathbb{T}^\mathbb{Z}$ and itself.

EXAMPLE 5. In contrast to the topological rigidity in Theorem 3 (as pointed out by Kitchens and Schmidt [28]) topological conjugacy on zero-dimensional groups preserves very little of the algebraic structure.

(1) A finite abelian group G has an associated shift automorphism $(G^\mathbb{Z}, \sigma_G)$ where $(\sigma((x_n)))_k = x_{k+1}$ for all $k \in \mathbb{Z}$. Clearly $(G^\mathbb{Z}, \sigma_G)$ will be topologically conjugate to $(H^\mathbb{Z}, \sigma_H)$ if and only if $|G| = |H|$. Thus no part of the internal structure of the group alphabet is preserved by topological conjugacy.

(2) Kitchens [**29**] proves that if G is zero-dimensional and T is expansive, meaning that there is an open neighbourhood U of the identity with

$$\bigcap_{n \in \mathbb{Z}} T^{-n} U = \{e\},$$

and has a dense orbit, then (G, T) is topologically conjugate to a full shift on $e^{h(T)}$ symbols. In the same paper the problem of classifying such automorphisms up to simultaneous algebraic isomorphism and topological conjugacy is also addressed but not completely solved. It is shown that for any given entropy there are only finitely many equivalence classes for simultaneous algebraic isomorphism and algebraic conjugacy. If the entropy is $\log p$ for some prime p then it is shown that there is only one equivalence class, namely that containing $(\mathbb{F}_p^{\mathbb{Z}}, \text{shift})$.

(3) For other group actions entirely new rigidity phenomena appear. Given a finite abelian group G, we may define an action σ_G of $\mathbb{Z}^2$ on the compact group

$$X_G = \{x = (x_{n,m}) \in G^{\mathbb{Z}^2} \mid x_{n,m} + x_{n+1,m} = x_{n,m+1} \text{ for all } n, m \in \mathbb{Z}\}$$

by shifting. Röttger [**54**], extending a partial result in [**68**], shows that the periodic point data in the sense of (1.3) for the system determines the structure of G, and so in particular if (X_G, σ_G) is topologically conjugate to (X_H, σ_H) then $G \cong H$.

(4) For the case $G = \mathbb{Z}/2\mathbb{Z}$ in the setting of X_G as in (3), the only homeomorphisms $X_G \to X_G$ commuting with the action of σ_G are elements of the action itself by [**65**].

PROBLEM 2 (Kitchens [**29**]). Classify expansive ergodic automorphisms of compact abelian zero-dimensional groups up to simultaneous algebraic isomorphism and topological conjugacy.

Example 5(3) is associated in a natural way to the polynomial

$$1 + u_1 - u_2 \in \mathbb{Z}[u_1^{\pm 1}, u_2^{\pm 1}]$$

since the condition defining the elements of X_G may be thought of as requiring that the element of $G^{\mathbb{Z}^2}$ is annihilated by $1 + u_1 - u_2$, with u_1 corresponding to a shift in the first coordinate and u_2 to a shift in the second coordinate.

PROBLEM 3. Associate to a polynomial $f \in \mathbb{Z}[u_1^{\pm 1}, u_2^{\pm 1}]$ and a finite abelian group G the shift action $\sigma_{G,f}$ on the space $X_{G,f}$ of elements of $G^{\mathbb{Z}^2}$ annihilated by f. Does the periodic point data for $\sigma_{G,f}$ determine the structure of G?

4. Measurable isomorphism

This is an opaque equivalence relation, because as measure spaces any two infinite compact metric abelian groups are isomorphic. A consequence of the homogeneity mentioned above is that the entropy $h(T)$ is an invariant of measurable isomorphism. A much deeper fact is that there is an abstract model as an independent identically distributed stochastic process for any ergodic group automorphism. These independent identically distributed processes – Bernoulli shifts – in turn are classified up to measurable isomorphism by their entropy (see Theorem 8 below).

EXAMPLE 6. Example 5 gave some instances of topological rigidity for actions of larger groups. There is a remarkable theory of rigidity for measurable isomorphism of algebraic $\mathbb{Z}^d$ actions, and we refer to the survey of Schmidt [56] and the references therein for this.

THEOREM 7. *Let $(G,T) \in \mathcal{G}$ be an ergodic compact group automorphism. Then there is a probability vector $(p_1, p_2, \dots)$ defining a measure μ_p on $\mathbb{N}$ with the property that (G,T) is measurably isomorphic to the shift map σ on the space $\mathbb{N}^{\mathbb{Z}}$ preserving the product measure $\prod_{n \in \mathbb{Z}} \mu_p$. Moreover, $h(T)$ is equal to the measure-theoretic entropy $\sum_{n \in \mathbb{N}} p_n \log p_n$ of σ with respect to this measure. If $h(T) < \infty$ then we may assume that $p_k = 0$ for all but finitely many k, or equivalently we may replace the alphabet $\mathbb{N}$ with a finite set.*

This was proved by Katznelson [27] in the case $G = \mathbb{T}^d$, and shown in general by Lind [39] and independently by Miles and Thomas [45] and Aoki [4], using abstract characterizations of the property of being measurably isomorphic to a Bernoulli shift developed by Ornstein [51]. It means that an ergodic compact group automorphism is measurably indistinguishable from an abstract independent identically distributed stochastic process. Notice that the space $A^{\mathbb{Z}}$ for a finite alphabet A is a compact metric space in a natural way, so the question of whether or not the measurable isomorphism in Theorem 7 may be chosen to be finitary – continuous off a null set – in the case of finite entropy arises. Little is known in general, but Smorodinsky (unpublished) pointed out that a necessary condition is exponential recurrence – for a given non-empty open set O, the Haar measure of the set of points first returning to O after n iterations of the automorphism decays exponentially in n. Lind [40] showed that any ergodic group automorphism is exponentially recurrent, so all that is known is that this particular property does not preclude the possibility that ergodic group automorphisms are finitarily isomorphic to Bernoulli shifts. Rudolph has given a characterization of the property of being finitarily isomorphic to a Bernoulli shift [55].

PROBLEM 4. Determine when an ergodic group automorphism is finitarily isomorphic to a Bernoulli shift.

THEOREM 8 (Ornstein [50]). *Bernoulli shifts with the same entropy are measurably isomorphic.*

Theorems 7 and 8 together mean that the space $\mathcal{G}/\sim$ for the equivalence of measurable isomorphism embeds into $\mathbb{R}_{>0}$. There is a representative Bernoulli shift for each measurable isomorphism class of compact group automorphisms – but is there a compact group automorphism for each Bernoulli shift? Equivalently, is there a compact group automorphism for each possible entropy?

In order to describe what is known about this problem, let $f \in \mathbb{Z}[x]$ be a polynomial with integer coefficients, with factorization $f(z) = \prod_{i=1}^{d}(z - \lambda_i)$ over $\mathbb{C}$. Then the *logarithmic Mahler measure of f* is the quantity

$$m(f) = \int_0^1 \log |f(e^{2\pi i s})| \mathrm{d}s = \sum_{|\lambda_i| > 1} \log |\lambda_i| = \sum_{i=1}^d \log^+ |\lambda_i|.$$

Thus Kronecker's lemma is the statement that $m(f) = 0$ if and only if f is a product of cyclotomic polynomials. Lehmer [36] raised the question of whether there could be small positive values of the Mahler measure. He found the smallest value known to date, corresponding to the polynomial

$$f(x) = x^{10} + x^9 - x^7 - x^6 - x^5 - x^4 - x^3 + x + 1$$

with Mahler measure approximately $\log 1.176$. Write

$$L = \inf\{m(f) \mid m(f) > 0\}.$$

Yuzvinskiĭ [73] computed the entropy of endomorphisms of solenoids (see Lind and the third author [42] for the history of earlier results and a simpler approach), and this led to a complete description of the range of possible values. These relate to the dynamics of group automorphisms in several ways.

THEOREM 9 (Lind [38]). *The infinite torus* $\mathbb{T}^\infty$ *has ergodic automorphisms with finite entropy if and only if* $L = 0$.

An expansive automorphism is guaranteed to have finite entropy, but unfortunately Hastings [24] shows that no automorphisms of the infinite torus can be expansive.

THEOREM 10 (Lind [39]). *The set of possible entropies of ergodic automorphisms of compact groups is* $(0, \infty]$ *if* $L = 0$, *and is the countable set*

$$\{m(f) \mid f \in \mathbb{Z}[x], m(f) > 0\}$$

if $L > 0$.

This means that the cardinality of the quotient space $\mathcal{G}/\sim$ is uncountable or countable depending on the solution to Lehmer's problem. This is now a problem of some antiquity, and means that we do not know if group automorphisms are special or not when viewed as measure-preserving transformations among the systems measurably isomorphic to Bernoulli shifts.

PROBLEM 5 (Lehmer [36]). Determine if $L = \inf\{m(f) \mid m(f) > 0\} > 0$.

We refer to Boyd [13] for an overview of Lehmer's problem. Expressed in dynamical systems, Theorem 10 may be used to express the same question as follows.

PROBLEM 6 (Lind [39]). Determine if it is possible to construct a compact group automorphism whose entropy is not of the form $m(f)$ for some $f \in \mathbb{Z}[x]$.

Attempting to describe the fibres (that is, the equivalence classes) for measurable isomorphism requires answering an already difficult problem in number theory, the *inverse Mahler measure problem*.

PROBLEM 7. Given $f \in \mathbb{Z}[x]$, characterize the set $\{g \in \mathbb{Z}[x] \mid m(g) = m(f)\}$.

We refer to Boyd [14] for an overview of this problem, and to Staines [59] for recent work and references.

To conclude this section, it is worth noting that for other group actions the same entropy range problem, that is, the problem of determining the set

(4.1) $\{h(T) \mid (G, T) \in \mathcal{G}_\Gamma\} \subset \mathbb{R}$,

can turn out to be considerably simpler than providing an answer to Lehmer's problem corresponding to the case $\Gamma = \mathbb{Z}$. For example, Björklund and Miles [11, Th. 4.1] show that if Γ is any countably infinite amenable group with arbitrarily large finite normal subgroups, then the entropy range (4.1) is $[0, \infty]$. In contrast, for $\Gamma = \mathbb{Z}^d$, the entropy range problem again reduces to Lehmer's problem by [**37**]. Further discussion of the entropy range problem for group actions may be found in [**11**].

We are grateful to an anonymous referee for pointing out the remarkable countable collection of *periods*, introduced by Kontsevich and Zagier [**31**] and raising the following question.

PROBLEM 8. What is the relation between the various natural versions of the entropy range (fixing neither, one, or both of the acting group Γ and the compact group G) and the set of periods?

5. One-solenoids and subgroups of the rationals

Some of the classification and range problems we wish to discuss are already interesting in the special case of one-dimensional solenoids. A compact metric abelian group is called a one-solenoid if it is connected and has topological dimension one. Equivalently, a one-solenoid is a group whose dual or character group is a subgroup of $\mathbb{Q}$. In this section we briefly recall the well-known description of these subgroups.

5.1. Subgroups of the rationals. Subgroups of $\mathbb{Q}$ are readily classified (see Baer [**6**], or Beaumont and Zuckerman [**8**] for a modern account). Let $H \leqslant \mathbb{Q}$ be an additive subgroup, and $x \in H \setminus \{0\}$. Write $\mathbb{P} = \{2, 3, 5, 7, \ldots\}$ for the set of rational primes. For each prime $p \in \mathbb{P}$, the *p-height* of x is

$$k_p(x) = \sup\{n \in \mathbb{N} \mid p^n y = x \text{ has a solution } y \in H\} \in \mathbb{N} \cup \{\infty\}$$

and the *characteristic* of x is the sequence

$$k(x) = (k_p(x))_{p \in \mathbb{P}} \in (\mathbb{N} \cup \{\infty\})^{\mathbb{N}}.$$

Two sequences $(a_p), (b_p) \in (\mathbb{N} \cup \{\infty\})^{\mathbb{N}}$ *belong to the same type* if $a(p) = b(p)$ for all but finitely many p, and for any p with $a(p) \neq b(p)$, both $a(p)$ and $b(p)$ are finite. Notice that if $x, y \in H \setminus \{0\}$ then $k(x)$ and $k(y)$ belong to the same type, allowing us to define the *type of H*, $k(H)$, to be the type of any non-zero element of H.

On the other hand, given any sequence $(k_p)_{p \in \mathbb{P}}$ with $k_p \in \mathbb{N} \cup \{\infty\}$, we may define

$$H\left((k_p)_{p \in \mathbb{P}}\right) = \{\tfrac{a}{b} \in \mathbb{Q} \mid a, b \in \mathbb{Z}, \gcd(a, b) = 1, \operatorname{ord}_p(b) \leqslant k_p \text{ for all } p \in \mathbb{P}\},$$

an additive subgroup of $\mathbb{Q}$.

THEOREM 11 (Baer [**6**]). *Any subgroup of $\mathbb{Q}$ is of the form $H\left((k_p)_{p \in \mathbb{P}}\right)$ for some sequence $(k_p) \in (\mathbb{N} \cup \{\infty\})^{\mathbb{N}}$. Two subgroups of $\mathbb{Q}$ are isomorphic if and only if they are of the same type.*

Clearly the only self-homomorphisms of $H = H\left((k_p)\right)$ are the maps $x \mapsto \frac{a}{b} x$ with $\gcd(a, b) = 1$ and $k_p(H) = \infty$ for any prime p dividing b. This gives a description of all continuous automorphisms of one-solenoids: $(G, T) \in \mathcal{G}$ has G connected with topological dimension 1 if and only if G is dual to a group $H\left((k_p)_{p \in \mathbb{P}}\right)$ and T is dual to a map $x \mapsto \frac{a}{b} x$ with $\gcd(a, b) = 1$ and with $k_p = \infty$ for any prime p dividing a or b. Writing $\mathcal{S}(H)$ for the set of primes p with $k_p(H) = \infty$, Theorem 11 shows in particular that $G \cong G'$ implies that $\mathcal{S}(\widehat{G}) = \mathcal{S}(\widehat{G'})$.

EXAMPLE 12. Some of the diversity of subgroups may be seen in the following examples.

(1) $H\left((\infty)_{p\in\mathbb{P}}\right) = \mathbb{Q}$.
(2) $H\left((0)_{p\in\mathbb{P}}\right) = \mathbb{Z}$.
(3) $H\left((\infty,\infty,0,0,\dots)\right) = \mathbb{Z}[\frac{1}{6}]$.
(4) $H\left((0,1,1,1,1,\dots)\right)$ is the subgroup of rationals with odd square-free denominator.
(5) For a set $\mathcal{S}\subset\mathbb{P}$ of primes, there is an associated subgroup of $\mathcal{S}$-integers

$$R_{\mathcal{S}} = \{x\in\mathbb{Q} \mid |x|_p \leqslant 1 \text{ for all } p\notin\mathcal{S}\} = \mathbb{Z}[\tfrac{1}{p} : p\in\mathcal{S}] = H\left((k_p)_{p\in\mathbb{P}}\right)$$

where $k_p = \infty$ if $p\in S$ and 0 otherwise.
(6) Let $\omega = (\omega_j)\in\{H,T\}^{\mathbb{N}}$ be the outcome of an infinitely repeated fair coin toss, and let

$$k_{p_j} = \begin{cases} \infty & \text{if } j = H; \\ 0 & \text{if } j = T. \end{cases}$$

Then $H\left((k_p)_{p\in\mathbb{P}}\right)$ is a 'random' $\mathcal{S}$-integer subgroup.

Solenoids of higher topological dimension, dual to subgroups of $\mathbb{Q}^d$ with d greater than 1, present peculiar difficulties unless there are additional assumptions. A solenoid that carries an expansive automorphism has a prescribed structure (Lawton [34], [35]; see also Kitchens and Schmidt [28, Sec. 5] for this result in the context of a more general treatment of finiteness conditions on group automorphisms) rendering it amenable to analysis; solenoids dual to rings of $\mathcal{S}$-integers in number fields also have a prescribed structure, and this is used by Chothi *et al.* to study orbit growth in these systems [15], [19] and [20]. Around the same time as Baer's classification of subgroups of $\mathbb{Q}$, Kurosh [32] and Malcev [43] described systems of complete invariants for subgroups of $\mathbb{Q}^d$, but determining when the invariants correspond to isomorphic groups is intractable for $d > 1$. The difficulties encountered here now have a precise formulation in terms of descriptive set theory, and Thomas [62] proves the remarkable result that the classification problem for subgroups of $\mathbb{Q}^{n+1}$ is strictly more difficult than the same problem for $\mathbb{Q}^n$ for $n \geqslant 1$ (we refer to the survey by Thomas [63] for an overview of these results).

5.2. Fixed points and entropy on one-solenoids. There are several routes to a formula for $\mathsf{F}_T(n)$ if T is an automorphism of a one-solenoid: a group-theoretic argument due to England and Smith [18], a general valuation-theoretic approach due to Miles [46] which is independent of the characteristic and the topological dimension, and in the $\mathcal{S}$-integer case an adelic argument due to Chothi, Everest *et al.* [15]. Here we adapt the adelic approach of Weil [71] to cover all subgroups, because this geometrical approach also evaluates the entropy (the harmonic analysis used here is not strictly adelic, but closer to the constructions in Tate's thesis [61], reproduced as [60]). For completeness we explain in Lemma 14 the basic idea from [15, Lem. 5.1], and refer to Hewitt and Ross [26] for background on the duality theory between discrete abelian groups and compact abelian groups.

The dual of a one-solenoid may be identified with a subgroup of $\mathbb{Q}$, and we now describe such a group using a direct limit of fractional ideals over a ring of $\mathcal{S}$-integers (see Example 12(5)) based on the concept of types introduced in the previous section. Let $R_{\mathcal{S}}$ be a ring of $\mathcal{S}$-integers. A set of rational primes $\mathcal{P}$ disjoint from $\mathcal{S}$ corresponds to a set of non-zero prime ideals of $R_{\mathcal{S}}$. Let $k = (k_p)_{p\in\mathcal{P}}$ be a sequence of positive integers and define a submodule of the R_S-module $\mathbb{Q}$ by $H = \sum_Q c_Q R_S$, where $c_Q = \prod_{p \in Q} p^{-k_p}$ and Q runs over all finite subsets of $\mathcal{P}$. The group H is isomorphic to R_S if and only if $\mathcal{P}$ is finite, otherwise H may be expressed using a direct limit $\mathrm{inj}\lim_{i \to \infty} H_i$ of R_S-modules, where

$$H_i = p(1)^{-k_{p(1)}} \cdot \dots \cdot p(i)^{-k_{p(i)}} R_S,$$

and $p(1), p(2), \dots$ is an enumeration of $\mathcal{P}$. By Theorem 11, for any given type we may choose a representative subgroup $H = H((k_p)_{p \in \mathbb{P}})$ and note that the set

$$\mathcal{S}(H) = \{p \in \mathbb{P} \mid k_p = \infty\}$$

is independent of the choice of representative. We set

$$\mathcal{P}(H) = \{p \in \mathbb{P} \mid 0 < k_p < \infty\}$$

and

$$k(H) = (k_p)_{p \in \mathcal{P}(H)},$$

and use this data as above to obtain an $R_{\mathcal{S}(H)}$-module having the same type as H. Any other choice of representative subgroup produces an isomorphic $R_{\mathcal{S}(H)}$-module. This description now gives us a canonical way to view a one-solenoid.

Since there is a short exact sequence of discrete groups

$$\{0\} \longrightarrow \mathbb{Z} \longrightarrow R_{\mathcal{S}(H)} \longrightarrow \sum_{p \in \mathcal{S}(H)} \mathbb{Z}[\tfrac{1}{p}]/\mathbb{Z} \longrightarrow \{0\},$$

and the dual group of $\mathbb{Z}[\tfrac{1}{p}]/\mathbb{Z}$ is isomorphic to the ring of p-adic integers $\mathbb{Z}_p$, via duality, there is a short exact sequence of compact groups

$$(5.1) \qquad \{0\} \longrightarrow \prod_{p \in \mathcal{S}(H)} \mathbb{Z}_p \longrightarrow \widehat{R_{\mathcal{S}(H)}} \longrightarrow \mathbb{T} \longrightarrow \{0\}.$$

Therefore, $\widehat{R_{\mathcal{S}(H)}}$ is a central extension of $\mathbb{T}$ by a cocycle

$$(5.2) \qquad w : \mathbb{T} \times \mathbb{T} \longrightarrow \prod_{p \in \mathcal{S}(H)} \mathbb{Z}_p,$$

and a simple explicit calculation along the lines of [64] shows that we may take

$$w(s,t) = -\lfloor s + t \rfloor ((1, 1, \dots)).$$

Furthermore, there is an explicit geometrical description of the solenoid $\widehat{R_{\mathcal{S}(H)}}$ as follows.

LEMMA 13. *The diagonal map* $\delta : x \mapsto (x, x, \dots)$ *embeds* $R_{\mathcal{S}(H)}$ *as a discrete and co-compact subgroup of the restricted directed product*

$$\mathbb{A}_{\mathcal{S}(H)} = \left\{ (x_\infty, (x_p)) \in \mathbb{R} \times \prod_{p \in \mathcal{S}(H)} \mathbb{Q}_p \mid x_p \in \mathbb{Z}_p \text{ for all but finitely many } p \right\},$$

which is a locally compact topological group. The set

$$F = [0, 1) \times \prod_{p \in \mathcal{S}(H)} \mathbb{Z}_p$$

is a fundamental domain for $\delta(R_{\mathcal{S}(H)})$ *in* $\mathbb{A}_{\mathcal{S}(H)}$, *and*

$$\widehat{R_{\mathcal{S}(H)}} \cong \mathbb{A}_{\mathcal{S}(H)} / \delta(R_{\mathcal{S}(H)}).$$

PROOF. This is a special case of the construction introduced in [**61**]. See also [**53**] for an accessible account. □

Recall that a subgroup $\Gamma < X$ in a locally compact topological group is called a *uniform lattice* if Γ is discrete and the quotient space X/Γ is compact in the quotient topology, and a measurable set $F \subset X$ is a *fundamental domain* for Γ if it contains exactly one representative of each coset of Γ. The *module* of a homomorphism

$$A : X \to X$$

is the quantity $m(AU)/m(U)$ for an open set U of finite measure, where m is a choice of Haar measure on X.

LEMMA 14. *Let Γ be a uniform lattice in a locally compact abelian group X, let F be a fundamental domain for Γ, and let m denote the Haar measure on X normalized to have $m(F) = 1$. Let $\widetilde{A} : X \to X$ be a continuous surjective homomorphism with $\widetilde{A}(\Gamma) \subset \Gamma$, and let $A : X/\Gamma \to X/\Gamma$ be the induced map. If $\ker A$ is discrete, then*

$$\mathrm{mod}_X(\widetilde{A}) = |\ker A| = m(\widetilde{A}F).$$

PROOF. Choose F so that it contains an open neighborhood U of the identity in X (this is possible since Γ is discrete in X). Since X/Γ is compact, $\ker A$ is finite and so there is a disjoint union $A^{-1}V = V_1 \sqcup \cdots \sqcup V_{|\ker A|}$ if V is a sufficiently small neighbourhood of the identity in X/Γ, with each V_i an open neighbourhood of a point in the fibre $A^{-1}(0_{X/\Gamma})$. Since A is measure–preserving,

$$m\left(A^{-1}V\right) = m\left(V\right).$$

If U and V are sufficiently small, then the quotient map $x \mapsto x + \Gamma$ is a homeomorphism between U and V, and so

$$m\left(\widetilde{A}U\right) = m\left(AV\right) = |\ker A| m\left(V\right) = |\ker A| m\left(U\right);$$

furthermore, since $U(0_X) \subset F$, $m(\widetilde{A}F) = |\ker A|$. □

EXAMPLE 15. A simple situation to which Lemma 14 may be applied is endomorphisms of the torus.

(1) Taking $\Gamma = \mathbb{Z} < X = \mathbb{R}$, $\widetilde{A}(x) = bx$ for some $b \in \mathbb{Z} \setminus \{0\}$, and m to be Lebesgue measure with $m\left(F = [0, 1)\right) = 1$, we see that

$$|\ker(x \mapsto bx \pmod 1)| = |b|.$$

In particular, it follows that if $T : \mathbb{R}/\mathbb{Z} \to \mathbb{R}/\mathbb{Z}$ is the map $x \mapsto ax$ for some a in $\mathbb{Z} \setminus \{0, \pm 1\}$ then $\mathsf{F}_T(n) = |a^n - 1|$ and so

$$\zeta_T(z) = \frac{1 - z}{1 - az}$$

if $a > 0$, and

$$\zeta_T(z) = \exp\left(\sum_{n \geqslant 1} \frac{|a|^n}{n} z^n + \sum_{n \geqslant 1} \frac{1}{n} z^n - 2 \sum_{n \geqslant 1} \frac{1}{2n} z^{2n}\right) = \frac{1 + z}{1 - |a|z}$$

if $a < 0$.

(2) As seen above, some care is needed in dealing with the distinction between the expressions $|a^n - 1|$ and $|a|^n - 1$: for an automorphism T of the d-torus $G = \mathbb{T}^d$ defined by an integer matrix $A_T \in \mathrm{GL}_d(\mathbb{Z})$ the same argument gives the formula

$$\mathsf{F}_T(n) = |\det(A_T^n - I)|,$$

and *a priori* an argument is needed to show that $\zeta_T(z)$ is a rational function of z because of the absolute value. This is discussed in Smale [58, Prop. 4.15], and an elementary algorithmic way to compute the zeta function in integer arithmetic is given by Baake, Lau and Paskunas [5].

We can also use Lemma 14 to find a formula for the periodic points of an automorphism of a one-solenoid, recovering in different notation the result of [18] and the one-dimensional case of [46].

PROPOSITION 16. *If* $T : \widehat{H} \to \widehat{H}$ *is an automorphism of a one-solenoid dual to the map* $x \mapsto rx$ *on* $H \leqslant \mathbb{Q}$, *then*

$$\mathsf{F}_T(n) = |r^n - 1| \times \prod_{p \in \mathcal{S}(H)} |r^n - 1|_p.$$

PROOF. Since $x \mapsto rx$ is an automorphism of H, $p \in \mathcal{S}(H)$ whenever $|r|_p \neq 1$. Lemmas 13 and 14 show that if T' is the automorphism dual to $x \mapsto rx$ on $\mathbb{A}_{\mathcal{S}(H)}$, then

$$\mathsf{F}_{T'}(n) = \mathrm{mod}_{\mathbb{A}_{\mathcal{S}(H)}}(x \mapsto rx) = |r^n - 1| \times \prod_{p \in \mathcal{S}(H)} |r^n - 1|_p.$$

The set of points of period n is dual to the group $H/(r^n - 1)H$, and multiplication by p^{k_p} for any $p \in \mathcal{P}(H)$ remains non-invertible in the direct limit H, so

$$H/(r^n - 1)H \cong R_{\mathcal{S}(H)}/(r^n - 1)R_{\mathcal{S}(H)}$$

for each $n \geqslant 1$, showing the proposition. □

The geometric viewpoint using adeles also allows the entropy to be computed easily using (1.1); this calculation is originally due to Abramov [1]. Write

$$\log^+(t) = \max\{0, \log t\}$$

for $t > 0$.

PROPOSITION 17 (Abramov's formula). *If* $T : \widehat{H} \to \widehat{H}$ *is an automorphism of a one-solenoid dual to the map* $x \mapsto rx$ *on* $H \leqslant \mathbb{Q}$, *then*

$$h(T) = \sum_{p \in \mathbb{P} \cup \{\infty\}} \log^+ |r|_p = \log \max\{|a|, |b|\}$$

where $r = \frac{a}{b}$ *with* $\gcd(a, b) = 1$.

PROOF. This formula is explained under the assumption that $H = R_{\mathcal{S}(H)}$ by the following argument. First, the calculation (1.1) may be performed in the covering space $\mathbb{A}_{\mathcal{S}(H)}$ because the quotient map

$$\mathbb{A}_{\mathcal{S}(H)} \to \widehat{R_{\mathcal{S}(H)}}$$

is a local isometry (we refer to [**17**, Prop. 4.7] for the details of this general lifting principle). Secondly, a small metric ball around 0 in the covering space $\mathbb{A}_{\mathcal{S}(H)}$ takes the form

$$(-\varepsilon, \varepsilon) \times \prod_{p \in \mathcal{S}(H)} p^{n_p} \mathbb{Z}_p,$$

and the action $\widetilde{T_r}$ of multiplication by r^{-1} on each of the coordinates $\mathbb{R}$ or $\mathbb{Q}_p$ gives

$$\bigcap_{j=0}^{n-1} T_r^{-j}(-\varepsilon, \varepsilon) = \begin{cases} (-\varepsilon, \varepsilon) & \text{if } |r| \leqslant 1; \\ (-\varepsilon|r|^{-(n-1)}, \varepsilon|r|^{-(n-1)}) & \text{if } |r| > 1, \end{cases}$$

on the quasi-factor $\mathbb{R}$, and

$$\bigcap_{j=0}^{n-1} T_r^{-j} p^{n_p} \mathbb{Z}_p = \begin{cases} p^{n_p} \mathbb{Z}_p & \text{if } |r|_p \leqslant 1; \\ |r|_p^{-(n-1)} p^{n_p} \mathbb{Z}_p & \text{if } |r| > 1, \end{cases}$$

showing the claimed formula.

In the general case an argument is needed to ensure that the entropy is not changed by passing from $R_{\mathcal{S}(H)}$ to the direct limit H, and we refer to [**42**, Prop. 3.1] for the details. $\qquad\square$

5.3. Equivalence relations for one-solenoids. We now consider how the equivalence relations from Definition 1 behave for the subspace $\mathcal{G}_1 \subset \mathcal{G}$ consisting of pairs (G, T) where $T : G \to G$ is an automorphism of a one-solenoid. Thus we may assume that G is dual to a group $H = H((k_p))$, and $T = T_{a/b}$ is dual to the map $x \mapsto \frac{a}{b}x$ with $\gcd(a, b) = 1$ on H, with the property that any prime p dividing ab has $k_p = \infty$.

EXAMPLE 18. It is clear that the entropy and the zeta function do not together determine an element of $\mathcal{G}_1$.

(1) We have $(\widehat{\mathbb{Z}[\frac{1}{2}]}, T_2) \sim_\zeta (\widehat{\mathbb{Z}[\frac{1}{2}]}, T_{1/2})$ and $(\widehat{\mathbb{Z}[\frac{1}{2}]}, T_2) \sim_h (\widehat{\mathbb{Z}[\frac{1}{2}]}, T_{1/2})$.

(2) By varying the group instead of the map, much larger joint equivalence classes may be found. As pointed out by Miles [**46**], a consequence of Proposition 16 and Proposition 17 is that the set

$$\left\{ (G, T) \in \mathcal{G}_1 \mid \zeta_T(z) = \tfrac{1-z}{1-2z}, h(T) = \log 2 \right\}$$

is uncountable. To see this, let $\{S_\lambda\}_{\lambda \in \Lambda}$ be an uncountable set of infinite subsets of $\mathbb{P}$, all containing 2, with the property that $|S_\lambda \bigtriangleup S_\nu| = |\mathbb{N}|$ for all $\lambda \neq \nu$. Associate to each S_λ the subgroup $H_\lambda = H((k_p^{(\lambda)}))$ where

$$k_p^{(\lambda)} = \begin{cases} 0 & \text{if } p \notin S_\lambda, \\ 1 & \text{if } p \in S_\lambda \setminus \{2\}, \\ \infty & \text{if } p = 2. \end{cases}$$

Then the automorphism dual to $x \mapsto 2x$ on H_λ has zeta function $\frac{1-z}{1-2z}$, and by Theorem 11 these are all algebraically, and hence topologically, distinct systems.

LEMMA 19. *If* $(G, T) \sim_\zeta (G', T')$ *then* $\mathcal{S}(G) = \mathcal{S}(G')$.

PROOF. Assume that $(G, T) \in \mathcal{G}_1$ with $T = T_{a/b}$ and G dual to $H((k_p))$. We claim that

$$(5.3) \qquad \{p \in \mathbb{P} \mid p \big| \, \mathsf{F}_T(n) \text{ for some } n \geqslant 1\} = \{p \in \mathbb{P} \mid p \notin \mathcal{S}(G) \text{ and } p \nmid ab\}.$$

Since by hypothesis T is an automorphism, we have

$$p \big| ab \implies p \in \mathcal{S}(G)$$

so (5.3) gives the lemma. To see (5.3), notice that by Proposition 16 no prime in $\mathcal{S}(G)$ can divide any $\mathsf{F}_T(n)$, and any prime $p \notin \mathcal{S}(G)$ will divide $a^{p-1} - b^{p-1}$ and so will divide some $\mathsf{F}_T(n)$. $\qquad \square$

PROBLEM 9. The case of endomorphisms is slightly different, because for example $(\widehat{\mathbb{Z}}, T_2) \sim_\zeta (\widehat{\mathbb{Z}[\frac{1}{2}]}, T_2) \sim_\zeta (\widehat{\mathbb{Z}[\frac{1}{2}]}, T_{1/2})$. Formulate a version of Lemma 19 for endomorphisms of one-solenoids.

The case of subrings, or equivalently of $\mathcal{S}$-integer subgroups of $\mathbb{Q}$, has distinctive features. Let $\overline{\mathcal{G}_1}$ denote the collection of pairs $(G, T) \in \mathcal{G}_1$ with the property that $\widehat{G}$ is a subring of $\mathbb{Q}$. This means that $k_p(G)$ is 0 or ∞ for any prime p.

PROPOSITION 20. *On the space* $\overline{\mathcal{G}_1}$,

(1) $\sim_h$ *has uncountable equivalence classes;*
(2) $\sim_\zeta$ *has countable equivalence classes;*
(3) *the joint relation* $\sim_h$ *and* $\sim_\zeta$ *has finite equivalence classes.*

PROOF. The pair $(\widehat{\mathbb{Z}[\frac{1}{ab}]}, T_{a/b})$ has entropy $\log \max\{|a|, |b|\}$, and by Abramov's formula for any subset $\mathcal{S} \subset \mathbb{P} \setminus \{p \mid p|ab\}$ of primes the pair $(\widehat{R_\mathcal{S}[\frac{1}{ab}]}, T_{a/b})$ has the same entropy, showing (1).

By Lemma 19, the $\sim_\zeta$ equivalence class determines the set of primes of infinite height, and so the only parameter that can change is the rational $\frac{a}{b}$ defining the map $T_{a/b}$, and there are only countably many of these.

For (3), notice that if $(G_2, T_{a/b})$ has entropy h then

$$\max\{|a|, |b|\} \leqslant \exp h,$$

so there are only finitely many choices for the rational r defining the map T_r with a given entropy. Lemma 19 shows that there is no further choice for a fixed zeta function. $\qquad \square$

EXAMPLE 21. Some of the diversity possible in Proposition 20 is illustrated via simple examples.

(1) For any $r \in \mathbb{Q} \setminus \{0, \pm 1\}$ the system $(\widehat{\mathbb{Q}}, T_r)$ has zeta function $\frac{1}{1-z}$, showing that the $\sim_\zeta$ equivalence class may be infinite.
(2) The $\sim_\zeta$ equivalence class may be infinite in a less degenerate way. Let G be the group dual to $\mathbb{Z}_{(3)}$, the subgroup obtained by inverting all the primes except 3. Then Proposition 16 shows that

$$\mathsf{F}_{T_r}(n) = |r^n - 1|_3^{-1}$$

by the product formula

$$(5.4) \qquad \prod_{p \in \mathbb{P} \cup \{\infty\}} |t|_p = |t| \times \prod_{p \in \mathbb{P} : |t|_p \neq 1} |t|_p = 1$$

for all $t \in \mathbb{Q} \setminus \{0\}$. It follows that $(G, T_2) \sim_\zeta (G, T_r)$ if $r = \frac{a}{b}$ where (a, b) is of the form

$$\left(\tfrac{9k+3m}{2} + 2, \tfrac{9k-3m}{2} + 1\right)$$

or

$$\left(\tfrac{9k+3m}{2} + 4, \tfrac{9k-3m}{2} + 2\right)$$

for integers m, k of the same parity chosen with a, b coprime and positive, or of the form

$$\left(\tfrac{9k+3m+5}{2}, \tfrac{9k-3m+1}{2}\right)$$

or

$$\left(\tfrac{9k+3m+7}{2}, \tfrac{9k-3m+5}{2}\right)$$

for integers m, k of opposite parity chosen with a, b coprime and positive.

(3) The $\sim_\zeta$ equivalence class may be finite. For example, if $(G, T) \in \widehat{\mathcal{G}_1}$ has zeta function $\frac{1-z}{1-2z}$, then we claim that (G, T) can only be $(\widehat{\mathbb{Z}[\frac{1}{2}]}, T_2)$ or $(\widehat{\mathbb{Z}[\frac{1}{2}]}, T_{1/2})$. To see this, notice first that these both have the claimed zeta function, and apply Lemma 19 to deduce that any element of $\widehat{\mathcal{G}_1}$ with the same zeta function has the form $(\widehat{\mathbb{Z}[\frac{1}{2}]}, T_{a/b})$. Since the map $T_{a/b}$ is an automorphism, the only prime dividing ab is 2, so $\frac{a}{b}$ is $\pm 2^k$ for some $k \in \mathbb{Z}$ and we may use Proposition 16 to calculate

$$\zeta_T(z) = \begin{cases} \frac{1-z}{1-2^k z} & \text{if } T = T_{2^k}; \\ \frac{1+z}{1-2^k z} & \text{if } T = T_{-2^k}. \end{cases}$$

The constraints seen in Example 21(3) hold more generally for systems with the property that $|\mathcal{S}(\widehat{G})|$ is finite.

PROPOSITION 22. *Let* $(G_1, T_r), (G_2, T_s) \in \mathcal{G}_1$ *have* $|\mathcal{S}(\widehat{G_1})| < \infty$ *and*

$$(G_1, T_r) \sim_\zeta (G_2, T_s).$$

Then $r = s$ *or* $r = s^{-1}$.

PROOF. First notice that $\mathcal{S}(\widehat{G_2}) = \mathcal{S}(\widehat{G_1})$ by Lemma 19. Denote this common set of primes by $\mathcal{S}$. Write $r = \frac{a_1}{b_1}$, $s = \frac{a_2}{b_2}$ with $\gcd(a_i, b_i) = 1$ and $a_i > 0$ for $i = 1, 2$. Since $\zeta_{T_r} = \zeta_{T_r^{-1}}$ and $\zeta_{T_s} = \zeta_{T_s^{-1}}$, without loss of generality we may assume that $a_i > |b_i|$ for $i = 1, 2$. Then, using Proposition 16 (see [**15**, Th 6.1] for the details), it follows that

$$\log a_1 = \lim_{n \to \infty} \frac{1}{n} \log \mathsf{F}_{T_r}(n) \text{ and } \log a_2 = \lim_{n \to \infty} \frac{1}{n} \log \mathsf{F}_{T_s}(n),$$

so that $a_1 = a_2$. Let $a = a_1 = a_2$,

$$\mathcal{T}_i = \{p \in \mathcal{S} \mid |a|_p < |b_i|_p\} \text{ and } \mathcal{T}'_i = \{p \in \mathcal{S} \mid |a|_p > |b_i|_p\},$$

for $i = 1, 2$. Notice that since $\gcd(a, b_i) = 1$, $|b_i|_p = 1$ for all $p \in \mathcal{T}_i$, and $|a|_p = 1$ for all $p \in \mathcal{T}'_i$, $i = 1, 2$. Hence

$$(5.5) \qquad \prod_{p \in \mathcal{T}_i \cup \mathcal{T}'_i} \left| \left(\tfrac{a}{b_i}\right)^n - 1 \right|_p = \prod_{p \in \mathcal{T}'_i} \left| \left(\tfrac{a}{b_i}\right)^n - 1 \right|_p = \prod_{p \in \mathcal{T}'_i} |b_i|_p^{-n} = |b_i|^n,$$

where the last equality follows by the Artin product formula, as $|b_i|_p = 1$ for all $p \in \mathbb{P} \setminus \mathcal{T}'_i$.

For each $p \in \mathcal{S}\backslash(\mathcal{T}_i \cup \mathcal{T}_i')$, $|\frac{a}{b_i}|_p = 1$, that is, $\frac{a}{b_i}$ is a unit in the valuation ring $\mathbb{Z}_{(p)}$. Let $m_i(p)$ denote the multiplicative order of $\frac{a}{b_i}$ in the residue field $\mathbb{Z}_{(p)}/(p)$, and note that if $m_i(p) \nmid n$, then $|(\frac{a}{b_i})^n - 1|_p = 1$. Therefore, whenever n is coprime to

$$m = \prod_{i=1,2} \prod_{p \in \mathcal{S}\backslash(\mathcal{T}_i \cup \mathcal{T}_i')} m_i(p),$$

for both $i = 1$ and $i = 2$ we have

(5.6)
$$\prod_{p \in \mathcal{S}\backslash(\mathcal{T}_i \cup \mathcal{T}_i')} \left| \left(\frac{a}{b_i}\right)^n - 1 \right|_p = 1.$$

Substituting (5.5) and (5.6) into the formula given by Proposition 16 yields

$$\mathsf{F}_{T_r}(n) = |a^n - b_1^n| \text{ and } \mathsf{F}_{T_s}(n) = |a^n - b_2^n|,$$

whenever $\gcd(n,m) = 1$. Since $a > 0$ and $a > |b_1|$, it follows that $b_1 = \pm b_2$.   □

COROLLARY 23. *The $\sim_\zeta$ equivalence class of $(G, T_{a/b})$ in the subset of $\mathcal{G}_1$ with $\mathcal{S}(\widehat{G})$ finite has cardinality $2^{\omega(a)} + 2^{\omega(b)}$, where as usual $\omega(k)$ denotes the number of distinct prime divisors of k.*

6. Counting closed orbits

The most transparent combinatorial invariant of topological conjugacy is the count of periodic orbits, which may be captured using generating functions like the dynamical zeta function. For a class of ergodic group automorphisms with polynomially bounded growth in the number of closed orbits, a more natural generating function is given by the orbit Dirichlet series [20]. A consequence of the structure theorem for expansive automorphisms of connected groups due to Lawton [35] is the following.

THEOREM 24 (Lawton [35]). *If $T : G \to G$ is an ergodic expansive automorphism of a compact connected group, then $\zeta_T(z)$ is a rational function of z.*

It is clear that not every function can be a zeta function of a map – in particular the coefficients in (1.2) must be non-negative. In fact more is true, and it is shown in [52] that a function $\zeta(z) = \exp \sum_{n \geqslant 1} \frac{a_n}{n} z^n$ is the dynamical zeta function of some map if and only if

(6.1)
$$0 \leqslant \sum_{d|n} \mu\left(\frac{n}{d}\right) a_d \equiv 0 \pmod{n}$$

for all $n \geqslant 1$. A beautiful remark of Windsor is that the same condition is equivalent to being the dynamical zeta function of a C^∞ diffeomorphism of the 2-torus [72]. Thus, for example,

$$\frac{e^{-z^2}}{(1-2z)} = \exp\left(2z + z^2 + \frac{2^3}{3}z^3 + \frac{2^4}{4}z^4 + \frac{2^5}{5}z^5 + \cdots\right)$$

has non-negative coefficients but cannot be a dynamical zeta function of any map. England and Smith [18] characterized in combinatorial terms the property of being the dynamical zeta function of an automorphism of a one-dimensional solenoid, and Moss [49] considered the more general question of when a function could be the dynamical zeta function of a group automorphism. Clearly the non-negativity and congruence condition in (6.1) is not sufficient, since the sequence $\mathsf{F}_T = (\mathsf{F}_T(n))_{n \geqslant 1}$

for a group automorphism must be a divisibility sequence. Moss showed that adding this further condition is also not sufficient, even for linear recurrence sequences.

LEMMA 25 (Moss [**49**]). *The function* $f(z) = \frac{1}{(1-z)(1-z^5)}$ *is the dynamical zeta function of the permutation* $\tau = (1)(23456)$ *on the set* $\{1, 2, 3, 4, 5, 6\}$, *which has the property that* F_τ *is a divisibility sequence, but* f *is not the zeta function of any group automorphism.*

PROOF. The given permutation τ has

$$\mathsf{F}_\tau = (1, 1, 1, 1, 6, 1, 1, 1, 1, 6, \ldots),$$

which is a divisibility sequence satisfying the linear recurrence relation

$$u_{n+5} = u_n$$

for $n \geqslant 1$ with the initial conditions

$$u_1 = u_2 = u_3 = u_4 = 1, u_5 = 6.$$

To see that f cannot be the dynamical zeta function of an abelian group automorphism, notice that if there is a group automorphism $T : G \to G$ with $\zeta_T(z) = f(z)$ then there is such a realization with $|G| = 6$. In this case T is an automorphism of G with $T^5 = I$, the identity. If some $g \in G \setminus \{0\}$ has orbit under T of cardinality less than 5, then there are integers $0 \leqslant m < n \leqslant 4$ with $T^m x = T^n x$, which implies that $T^{n-m} x = x$ and so $x = 0$ since $|n - m| \leqslant 4$. It follows that any non-identity element of G has an orbit of length 5 under the automorphism T, and so in particular every non-identity element of G has the same order, which is impossible. $\square$

England and Smith [**18**] showed that

$$f(z) = \exp \sum_{n \geqslant 1} \frac{a_n}{n} z^n$$

is the dynamical zeta function of a group automorphism dual to $x \mapsto \frac{m}{n} x$ on a one-solenoid if and only if

(6.2) $a_k \big| n^k - m^k$ for all $k \geqslant 1$

and

(6.3) $\gcd\left(a_k, \frac{n^\ell - m^\ell}{a_\ell}\right) = 1$ for all $k \neq \ell$.

They also gave an example (which does not seem to be correct) of a group automorphism with irrational zeta function for an automorphism of a one-solenoid; there are now several ways to see that these must exist. In [**66**] it is shown that there are uncountably many distinct zeta functions of automorphisms dual to $x \mapsto 2x$ on a one-solenoid (and hence most in cardinality are irrational functions); Everest, Stangoe and the third author [**21**, Lem. 4.1] showed that the map dual to $x \mapsto 2x$ on $\mathbb{Z}[\frac{1}{6}]$ has a natural boundary at $|z| = \frac{1}{2}$. In order to relate the characterization from [**18**] to the S-integer constructions of Chothi, Everest *et al.* [**15**] outlined in Proposition 16, we show that they are equivalent.

LEMMA 26. *Given coprime integers* m, n, *a sequence* (a_k) *of positive integers satisfies* (6.2) *and* (6.3) *if and only if there is a subset of the primes* $\mathcal{S} \subset \mathbb{P}$ *for which*

(6.4) $$a_k = |n^k - m^k| \prod_{p \in \mathcal{S}} |n^k - m^k|_p$$

for all $k \geqslant 1$.

PROOF. Clearly the existence of a set $\mathcal{S} \subset \mathbb{P}$ with (6.4) implies (6.2) and (6.3). Assume (6.2) and (6.3), let

$$\mathcal{S} = \{p \in \mathbb{P} \mid p | \tfrac{n^\ell - m^\ell}{a_\ell} \text{ for some } \ell \geqslant 1\},$$

and write $b_k = |n^k - m^k| \prod_{p \in \mathcal{S}} |n^k - m^k|_p$. If some prime power p^e divides some a_k, then $p^e | n^k - m^k$ so by (6.3) $p^e | b_k$, and so $a_k | b_k$ for all $k \geqslant 1$. If some prime power p^e divides some b_k, then $p^e | n^k - m^k$ and $p \notin \mathcal{S}$ and hence $p^e | a_k$. Thus $b_k | a_k$ for all $k \geqslant 1$. $\qquad\square$

Miles gave a general characterization for finite entropy automorphisms on finite-dimensional groups. In order to state this, recall from Weil [71] that a global field (or A-field) is an algebraic number field or a function field of transcendence degree one over a finite field. For a global field K we write $\mathbb{P}(K)$ for the set of places of K, and $\mathbb{P}_\infty(K)$ for the set of infinite places of K

THEOREM 27 (Miles [46]). *If $T : G \to G$ is an ergodic automorphism of a finite-dimensional compact abelian group with finite entropy, then there exist global fields $K_1, \ldots, K_n$, sets of finite places $P_i \subset \mathbb{P}(K_i)$ and elements $\xi_i \in K_i$, no one of which is a root of unity for $i = 1, \ldots, n$, such that*

$$\mathsf{F}_T(k) = \prod_{i=1}^{n} \prod_{v \in P_i} |\xi_i^k - 1|_v^{-1}.$$

We now turn to another range problem, concerned with the growth in orbits. This may be thought of as a combinatorial analogue of Lehmer's problem.

THEOREM 28 (Ward [70]). *For any $C \in [0, \infty]$ there is a compact group automorphism $T : G \to G$ with*

$$\frac{1}{n} \log \mathsf{F}_T(n) \longrightarrow C$$

as $n \to \infty$.

Theorem 28 exploits known bounds for the appearance of primes in arithmetic progressions, and the proof works because the growth rate targeted is exponential (the case $C = 0$ is trivial). The faster the growth rate sought, the easier the proof becomes (as may be seen in the proof of Theorem 29). Constructing compact group automorphisms with prescribed growth rates of periodic points slower than exponential is a far more delicate problem, and we refer to recent work of Haynes and White [25] for the current state of the problem.

6.1. Sets of primes and constructions in one-solenoids. The examples constructed in Theorem 28 are not really satisfactory, since the groups used are zero-dimensional and (worse) the automorphisms are not ergodic. As a result it is natural to ask about the range of various invariants for ergodic automorphisms of connected groups. For rapid growth and a crude approximation, it is possible to exhibit growth in periodic orbits in a controlled way on connected groups, as illustrated in the next result (the proof of which we postpone briefly).

THEOREM 29. *Let $\theta = (\theta_n)$ be a sequence of positive integers with*

$$\frac{n}{\log \theta_n} \longrightarrow 0$$

as $n \to \infty$. Then there is an automorphism $T : X \to X$ of a connected compact group with

$$\frac{\log \mathsf{F}_T(n)}{\log \theta_n} \longrightarrow 1$$

as $n \to \infty$.

With more effort, taking advantage of known estimates for the size of the primitive part of linear recurrence sequence, it is likely that this result can be improved, but it is unlikely that any such approach will lead to an (unexpected) positive answer to the following.

PROBLEM 10. Does Theorem 28 still hold if the compact groups are required to be connected?

It is expected that this reduces directly to Lehmer's problem, and is as a result equally intractable. A refined version (in light of the expected negative answer) is the following; background and partial results in this direction may be found in some work of the third author [66], [67], exposing in particular some of the Diophantine problems that arise.

PROBLEM 11. Characterize the set

$$\left\{ \limsup_{n \to \infty} \frac{1}{n} \log \mathsf{F}_T(n) \mid (G, T) \in \mathcal{G} \text{ and } G \text{ is connected} \right\}$$

of real numbers.

As we have seen in Proposition 16, orbit-growth questions on the space $\mathcal{G}_1$ essentially amount to questions about rational numbers and sets of primes. In this section we will give an overview of what is known in this setting specifically for the map dual to $x \mapsto 2x$ on subrings of $\mathbb{Q}$. Thus the systems are parameterised by subsets of $\mathbb{P}$, and Figure 2 gives a guide to the results assembled here on the Hasse diagram of subsets of $\mathbb{P}$, or equivalently of subrings of $\mathbb{Q}$. In the diagram the standard notation for localization of rings is used for convenience, so for example $\mathbb{Z}[\frac{1}{p} \mid p \neq 2, 3]$ is written $\mathbb{Z}_{(2,3)}$ and so on.

At the bottom of Figure 2 the compact group is the circle, and strictly speaking this has no ergodic automorphisms. However, as seen in Example 15, if we permit the endomorphisms $x \mapsto ax$ (or invert only those primes appearing in a) then the zeta functions arising are rational.

The first observation one can make about the lower part of Figure 2 (subrings of $\mathbb{Q}$ in which $k_p = \infty$ for only finitely many primes p) concerns the exponential growth rate of periodic points. If $S \subset \mathbb{P}$ has $|S| < \infty$, then there are constants $A, B > 0$ with

$$(6.5) \qquad \frac{A}{n^B} \leqslant \prod_{p \in S} |a^n - b^n|_p \leqslant 1$$

for all $n \geqslant 1$ (see [15, Th 6.1] for the details). A simple consequence of Proposition (16) and the estimate (6.5) gives the construction needed for Theorem 29.

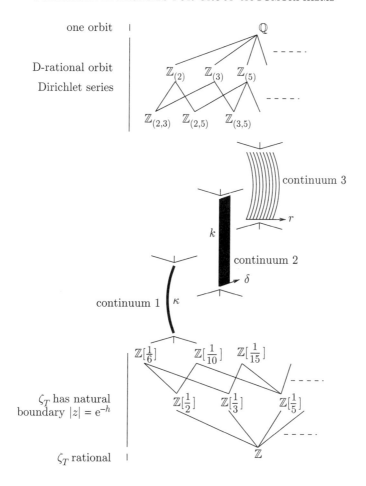

FIGURE 2. The Hasse diagram of subrings of $\mathbb{Q}$.

PROOF OF THEOREM 29. Define a sequence (S_k) of sets of primes by

$$S_k = \{p \in \mathbb{P} \mid p \mid 2^n - 1 \text{ for some } n < k\} \cup \{2\}.$$

Thus $S_1 = \{2\}$, $S_2 = \{2,3\}$, $S_3 = \{2,3,7\}$, and so on, and let $T_k : X_k \to X_k$ be the automorphism dual to $x \mapsto 2x$ on $\mathbb{Z}[\frac{1}{p} : p \in S_k]$. By Proposition 16 and (6.5) it follows that $\mathsf{F}_{T_k}(n) = 1$ for $n < k$ and $\lim_{n \to \infty} (1/n) \log \mathsf{F}_{T_k}(n) = \log 2$. It may be helpful to have in mind that the sequence of fixed point counts for these systems begins as follows:

$$\mathsf{F}_{T_1} = (1,3,7,15,31,\dots)$$
$$\mathsf{F}_{T_2} = (1,1,7,5,31,\dots)$$
$$\mathsf{F}_{T_3} = (1,1,1,5,31,\dots)$$
$$\mathsf{F}_{T_4} = (1,1,1,1,31,\dots),$$

and so on. We now inductively construct an infinite-dimensional solenoid X together with an automorphism T as follows, starting at 2 for convenience.

At the first stage, let $n_2 = \lceil \log_3 \theta_2 \rceil$ and let $(X^{(2)}, T^{(2)})$ be the n_2-fold Cartesian product of copies of (X_1, T_1). By construction,

$$(6.6) \qquad\qquad 1 \leqslant \frac{\mathsf{F}_{T^{(2)}}(2)}{\theta_2} \leqslant 3 = \mathsf{F}_{T_1}(2).$$

At the next stage, let $n_3 = \lceil \log_7 \theta_3 \rceil - m_2$ (and assume, as we may, that this is non-negative by amending the early terms of the sequence θ), and let $(X^{(3)}, T^{(3)})$ be the product of $(X^{(2)}, T^{(2)})$ and the n_3-fold Cartesian product of copies of (X_2, T_2), where m_2 is chosen to have

$$\mathsf{F}_{T^{(3)}}(3) \geqslant \theta_3$$

and to minimize

$$\mathsf{F}_{T^{(3)}}(3) - \theta_3.$$

By construction, we retain (6.6) but also have

$$1 \leqslant \frac{\mathsf{F}_{T^{(3)}}(3)}{\theta_3} \leqslant 7 = \mathsf{F}_{T_2}(3).$$

We continue in this way, multiplying at the nth stage by a number of copies of (X_k, T_k) to approximate as well as possible the number of points of period n sought, eventually producing an infinite product of one-dimensional solenoids T with an automorphism T so that

$$1 \leqslant \frac{\mathsf{F}_T(n)}{\theta_n} \leqslant \mathsf{F}_{T_{n-1}}(n) \leqslant 2^n.$$

Thus

$$1 \leqslant \frac{\log \mathsf{F}_T(n)}{\log \theta_n} \leqslant 1 + \frac{n \log 2}{\log \theta_n},$$

giving the result by the growth-rate assumption on θ. $\square$

EXAMPLE 30. The map T dual to $x \mapsto 2x$ on $\mathbb{Z}[\frac{1}{6}]$ has

$$\zeta_T(z) = \exp \sum_{n \geqslant 1} \frac{|2^n - 1||2^n - 1|_3}{n} z^n.$$

The radius of convergence of ζ_T is $\frac{1}{2}$ by (6.5), and a simple argument from [21, Lem. 4.1] shows that the series defining $\zeta_T(z)$ has a zero at all points of the form $\frac{1}{2} e^{2\pi i j/3^r}$ for $j \in \mathbb{Z}$ and $r \geqslant 1$, so $|z| = \frac{1}{2}$ is a natural boundary.

The appearance of a natural boundary for the zeta function at $\frac{1}{2} = \exp(-h(T))$ prevents an analysis of π_T and M_T using analytic methods. On the other hand, the dense set of singular points on the circle $|z| = \frac{1}{2}$ form a discrete set in $\mathbb{C}_3$ (the completion of the algebraic closure of $\mathbb{Q}_3$), raising the following question.

PROBLEM 12. For the map in Example 30 develop a notion of a dynamical zeta function $\zeta_T : \mathbb{C}_3 \to \mathbb{C}_3$ in such a way that it has a meromorphic extension beyond its radius of convergence, and exploit this to give analytic or Tauberian proofs of statements like

$$\mathsf{M}_T(X) = \tfrac{5}{8} \log N + C_T + \mathrm{O}(1/N).$$

There are several difficulties involved here, starting with the fact that the formal power series $\exp$ on $\mathbb{Q}_3$ has radius of convergence $3^{-1/2} < 1$; we refer to Koblitz [30] for the background needed to address this problem.

PROBLEM 13. For a system in $\mathcal{G}_1$ for which the set of primes of infinite height is finite, show that if the zeta function is irrational, then it has natural boundary at e^{-h}.

More generally, there is some reason to suspect a stronger result (see [9] for more on the methods involved here, and some partial results). In the disconnected case it is known that the zeta function is typically non-algebraic [69].

PROBLEM 14 (Bell, Miles and Ward [9]). Is there a Pólya–Carlson dichotomy for zeta functions of automorphisms of compact groups? That is, is it true that such a zeta function is either rational or has a natural boundary at its circle of convergence?

Results of [19] give precise information about the systems at the lower end of Figure 2, and in the special case of one-solenoids they take the following form.

THEOREM 31 (Everest, Miles, Stevens and Ward [19]). *If T is an automorphism in $\mathcal{G}_1$ with a finite set of primes of infinite height, then there are constants k_T in $\mathbb{Q}$ and $C_T > 0$ with*

$$\mathsf{M}_T(n) = k_T \log N + C_T + \mathrm{O}(1/N).$$

EXAMPLE 32. If T is the automorphism dual to $x \mapsto 2x$ on $\mathbb{Z}[\frac{1}{21}]$ then

$$\mathsf{M}_T(N) = \tfrac{269}{576} \log N + C + \mathrm{O}(1/N).$$

Problem 12 also arises in this setting, taking the following form.

PROBLEM 15. For systems in $\mathcal{G}_1$ for which the set of primes of infinite height is finite, develop a notion of dynamical zeta function on the space $\prod_{p:k_p=\infty} \mathbb{C}_p$ so that it has a meromorphic extension beyond its radius of convergence, and exploit this to give an analytic proof of Theorem 31.

At the opposite extreme, the system corresponding to the subring $\mathbb{Q}$ has one closed orbit of length 1 by the product formula. If the set of primes of infinite height is co-finite, then the same basic p-adic estimate (6.5) shows that

$$(6.7) \qquad\qquad \mathsf{F}_T(n) \leqslant n^A$$

for some $A > 0$. This polynomial bound makes it natural to study the orbit growth using the orbit Dirichlet series

$$\mathsf{d}_T(s) = \sum_{n \geqslant 1} \frac{\mathsf{O}_T(n)}{n^s},$$

and in [20] this is done for a large class of automorphisms of connected groups where (6.7) is guaranteed. At the level of generality adopted in [20], estimates for the growth of $\pi_T(N)$ depend on the abscissa of convergence of $\mathsf{d}_T(s)$. However, if T is assumed to be an automorphism in $\mathcal{G}_1$, then $\mathsf{d}_T(s)$ is found to have abscissa of convergence $\sigma = 0$ (see [20, Rmk. 3.7(1)]), and the following applies.

THEOREM 33 (Everest, Miles, Stevens and Ward [20]). *If T is an automorphism in $\mathcal{G}_1$ with a co-finite set of primes of infinite height, then there is a finite set $M \subset \mathbb{N}$ for which $\mathsf{d}_T(s)$ is a rational function of the variables $\{s^{-m} \mid m \in M\}$, and there is a constant $C > 0$ such that*

$$\pi_T(N) = C \left(\log N\right)^K + \mathrm{O}\left((\log N)^{K-1}\right),$$

where K is the order of the pole at $s = 0$.

The special structure of the orbit Dirichlet series arising in the co-finite case, which might be called Dirichlet rationality, raises the possibility of proving some or all of Theorem 33 using Tauberian methods. Unfortunately, as the next example shows, Theorem 33 does not seem to be provable using Tauberian methods for a reason related to the natural boundary phenomenon in Example 30: in all but the simplest of settings, the analytic behavior on the critical line is poor. The following example, which is an easy calculation using Proposition 16, is taken from Everest, Miles, Stevens *et al.* [**20**, Ex. 4.2].

EXAMPLE 34. Let $T : G \to G$ be the automorphism dual to the map $x \mapsto 2x$ on the subring $\mathbb{Z}_{(3)} \cap \mathbb{Z}_{(5)} \subset \mathbb{Q}$. Then

$$\mathsf{d}_T(z) = 1 - \frac{1}{2^{z+1}} + \frac{3}{2^{z+1}} \left(1 - \frac{1}{3^{z+1}} - \frac{1}{2^{z+1}} + \frac{1}{6^{z+1}} \right) \frac{1}{1 - 3^{-z}}$$

$$+ \frac{15}{4^{z+1}} \left(1 - \frac{1}{3^{z+1}} - \frac{1}{5^{z+1}} + \frac{1}{15^{z+1}} \right) \frac{1}{(1 - 3^{-z})(1 - 5^{-z})}.$$

The obstacle to using Agmon's Tauberian theorem [**3**] or even Perron's Theorem [**23**, Th. 13] is illustrated in Example 34: not only does d_T have infinitely many singularities on the critical line, it has singularities that are arbitrarily close together. On the other hand, they are located at arithmetically special points, raising the analogue of Problem 15.

PROBLEM 16. Develop a notion of orbit Dirichlet series with variables in the space $\prod_{p:k_p=0} \mathbb{C}_p$, and develop p-adic versions of Tauberian theorems to find a meromorphic extension of the orbit Dirichlet series arising in the setting of Theorem 33, and use this to give analytic proofs of Theorem 33(1) and (3).

Theorems 31 and 33 give some insight into the range of orbit-growth invariants across those groups with finite or co-finite set of primes of infinite height. Of course most elements of $\mathcal{G}_1$ fall into neither the co-finite nor the finite camp, and new phenomena appear in this large middle ground. Using delicate constructions of thin sets of primes, Baier, Jaidee, Stevens *et al.* [**7**] find several different continua of different orbit-growth invariants inside $\mathcal{G}_1$.

THEOREM 35 (Baier, Jaidee, Stevens and Ward [**7**]). *As illustrated in Figure 2, three different continua of orbit-growth rates exist in $\mathcal{G}_1$ as follows.*

  (1) *For any $\kappa \in (0,1)$ there is a system $(G,T) \in \mathcal{G}_1$ with*

$$\mathsf{M}_T(N) \sim \kappa \log N$$

  (see Continuum 1*).*
  (2) *For any $\delta \in (0,1)$ and $k > 0$ there is a system $(G,T) \in \mathcal{G}_1$ with*

$$\mathsf{M}_T(N) \sim k(\log N)^\delta$$

  (see Continuum 2*).*
  (3) *For any $r \in \mathbb{N}$ and $k > 0$ there is a system $(G,T) \in \mathcal{G}_1$ with*

$$\mathsf{M}_T(N) \sim k(\log \log N)^r$$

  (see Continuum 3*).*

Acknowledgements. We are grateful to Anthony Flatters for the calculation in Example 21(2), to Alex Clark for pointing out the paper of Markus and Meyer [**44**], and to an anonymous referee for raising Problem 8.

References

[1] L. M. Abramov, *The entropy of an automorphism of a solenoidal group* (Russian, with English summary), Teor. Veroyatnost. i Primenen **4** (1959), 249–254. MR0117322 (22 #8103)

[2] R. L. Adler and R. Palais, *Homeomorphic conjugacy of automorphisms on the torus*, Proc. Amer. Math. Soc. **16** (1965), 1222–1225. MR0193181 (33 #1402)

[3] S. Agmon, *Complex variable Tauberians*, Trans. Amer. Math. Soc. **74** (1953), 444–481. MR0054079 (14,869a)

[4] N. Aoki, *A simple proof of the Bernoullicity of ergodic automorphisms on compact abelian groups*, Israel J. Math. **38** (1981), no. 3, 189–198, DOI 10.1007/BF02760804. MR605377 (82d:22010)

[5] M. Baake, E. Lau, and V. Paskunas, *A note on the dynamical zeta function of general toral endomorphisms*, Monatsh. Math. **161** (2010), no. 1, 33–42, DOI 10.1007/s00605-009-0118-y. MR2670229 (2011j:37036)

[6] R. Baer, *Abelian groups without elements of finite order*, Duke Math. J. **3** (1937), no. 1, 68–122, DOI 10.1215/S0012-7094-37-00308-9. MR1545974

[7] S. Baier, S. Jaidee, S. Stevens, and T. Ward, *Automorphisms with exotic orbit growth*, Acta Arith. **158** (2013), no. 2, 173–197, DOI 10.4064/aa158-2-5. MR3033948

[8] R. A. Beaumont and H. S. Zuckerman, *A characterization of the subgroups of the additive rationals*, Pacific J. Math. **1** (1951), 169–177. MR0044522 (13,431c)

[9] J. Bell, R. Miles, and T. Ward, *Towards a Pólya–Carlson dichotomy for algebraic dynamics*, Indag. Math. (N. S.) **25** (2014), no. 4, 652–668.

[10] S. Bhattacharya and T. Ward, *Finite entropy characterizes topological rigidity on connected groups*, Ergodic Theory Dynam. Systems **25** (2005), no. 2, 365–373, DOI 10.1017/S0143385704000501. MR2129101 (2006g:37003)

[11] R. Miles and M. Björklund, *Entropy range problems and actions of locally normal groups*, Discrete Contin. Dyn. Syst. **25** (2009), no. 3, 981–989, DOI 10.3934/dcds.2009.25.981. MR2533986 (2010g:37010)

[12] R. Bowen, *Entropy for group endomorphisms and homogeneous spaces*, Trans. Amer. Math. Soc. **153** (1971), 401–414. MR0274707 (43 #469)

[13] D. W. Boyd, *Speculations concerning the range of Mahler's measure*, Canad. Math. Bull. **24** (1981), no. 4, 453–469, DOI 10.4153/CMB-1981-069-5. MR644535 (83h:12002)

[14] D. W. Boyd, *Inverse problems for Mahler's measure*, Diophantine analysis (Kensington, 1985), London Math. Soc. Lecture Note Ser., vol. 109, Cambridge Univ. Press, Cambridge, 1986, pp. 147–158. MR874125 (88b:11051)

[15] V. Chothi, G. Everest, and T. Ward, *S-integer dynamical systems: periodic points*, J. Reine Angew. Math. **489** (1997), 99–132. MR1461206 (99b:11089)

[16] A. Clark and R. Fokkink, *Topological rigidity of semigroups of affine maps*, Dyn. Syst. **22** (2007), no. 1, 3–10, DOI 10.1080/14689360601028043. MR2308207 (2008a:37030)

[17] M. Einsiedler, E. Lindenstrauss, and T. Ward, *Entropy in dynamics* (to appear). http://maths.dur.ac.uk/~tpcc68/entropy/welcome.html.

[18] J. W. England and R. L. Smith, *The zeta function of automorphisms of solenoid groups*, J. Math. Anal. Appl. **39** (1972), 112–121. MR0307280 (46 #6400)

[19] G. Everest, R. Miles, S. Stevens, and T. Ward, *Orbit-counting in non-hyperbolic dynamical systems*, J. Reine Angew. Math. **608** (2007), 155–182, DOI 10.1515/CRELLE.2007.056. MR2339472 (2008k:37042)

[20] G. Everest, R. Miles, S. Stevens, and T. Ward, *Dirichlet series for finite combinatorial rank dynamics*, Trans. Amer. Math. Soc. **362** (2010), no. 1, 199–227, DOI 10.1090/S0002-9947-09-04962-9. MR2550149 (2011c:37038)

[21] G. Everest, V. Stangoe, and T. Ward, *Orbit counting with an isometric direction*, Algebraic and topological dynamics, Contemp. Math., vol. 385, Amer. Math. Soc., Providence, RI, 2005, pp. 293–302, DOI 10.1090/conm/385/07202. MR2180241 (2006k:37046)

[22] F. Grunewald and D. Segal, *Decision problems concerning S-arithmetic groups*, J. Symbolic Logic **50** (1985), no. 3, 743–772, DOI 10.2307/2274327. MR805682 (87a:22023)

[23] G. H. Hardy and M. Riesz, *The general theory of Dirichlet's series*, Cambridge Tracts in Mathematics and Mathematical Physics, No. 18, Stechert-Hafner, Inc., New York, 1964. MR0185094 (32 #2564)

[24] H. M. Hastings, *On expansive homeomorphisms of the infinite torus*, The structure of attractors in dynamical systems (Proc. Conf., North Dakota State Univ., Fargo, N.D., 1977), Lecture Notes in Math., vol. 668, Springer, Berlin, 1978, pp. 142–149. MR518555 (80d:58051)

[25] A. Haynes and C. J. White, *Group automorphisms with prescribed growth of periodic points, and small primes in arithmetic progressions in intervals*, Adv. Math. **252** (2014), 572–585, DOI 10.1016/j.aim.2013.11.014. MR3144241

[26] E. Hewitt and K. A. Ross, *Abstract harmonic analysis. Vol. I*, 2nd ed., Grundlehren der Mathematischen Wissenschaften [Fundamental Principles of Mathematical Sciences], vol. 115, Springer-Verlag, Berlin-New York, 1979. Structure of topological groups, integration theory, group representations. MR551496 (81k:43001)

[27] Y. Katznelson, *Ergodic automorphisms of $\mathbb{T}^n$ are Bernoulli shifts*, Israel J. Math. **10** (1971), 186–195. MR0294602 (45 #3672)

[28] B. Kitchens and K. Schmidt, *Automorphisms of compact groups*, Ergodic Theory Dynam. Systems **9** (1989), no. 4, 691–735, DOI 10.1017/S0143385700005290. MR1036904 (91g:22008)

[29] B. P. Kitchens, *Expansive dynamics on zero-dimensional groups*, Ergodic Theory Dynam. Systems **7** (1987), no. 2, 249–261, DOI 10.1017/S0143385700003989. MR896796 (88i:28039)

[30] N. Koblitz, *p-adic numbers, p-adic analysis, and zeta-functions*, 2nd ed., Graduate Texts in Mathematics, vol. 58, Springer-Verlag, New York, 1984. MR754003 (86c:11086)

[31] M. Kontsevich and D. Zagier, *Periods*, Mathematics unlimited—2001 and beyond, Springer, Berlin, 2001, pp. 771–808. MR1852188 (2002i:11002)

[32] A. Kurosch, *Primitive torsionsfreie Abelsche Gruppen vom endlichen Range* (German), Ann. of Math. (2) **38** (1937), no. 1, 175–203, DOI 10.2307/1968518. MR1503333

[33] C. G. Latimer and C. C. MacDuffee, *A correspondence between classes of ideals and classes of matrices*, Ann. of Math. (2) **34** (1933), no. 2, 313–316, DOI 10.2307/1968204. MR1503108

[34] W. Lawton, *Expansive Transformation Groups* (Ph.D. thesis, Wesleyan University, 1972).

[35] W. Lawton, *The structure of compact connected groups which admit an expansive automorphism*, Recent advances in topological dynamics (Proc. Conf., Yale Univ., New Haven, Conn., 1972; in honor of Gustav Arnold Hedlund), Springer, Berlin, 1973, pp. 182–196. Lecture Notes in Math., Vol. 318. MR0391051 (52 #11873)

[36] D. H. Lehmer, *Factorization of certain cyclotomic functions*, Ann. of Math. (2) **34** (1933), no. 3, 461–479, DOI 10.2307/1968172. MR1503118

[37] D. Lind, K. Schmidt, and T. Ward, *Mahler measure and entropy for commuting automorphisms of compact groups*, Invent. Math. **101** (1990), no. 3, 593–629, DOI 10.1007/BF01231517. MR1062797 (92j:22013)

[38] D. A. Lind, *Ergodic automorphisms of the infinite torus are Bernoulli*, Israel J. Math. **17** (1974), 162–168. MR0346130 (49 #10856)

[39] D. A. Lind, *The structure of skew products with ergodic group automorphisms*, Israel J. Math. **28** (1977), no. 3, 205–248. MR0460593 (57 #586)

[40] D. A. Lind, *Ergodic group automorphisms are exponentially recurrent*, Israel J. Math. **41** (1982), no. 4, 313–320, DOI 10.1007/BF02760537. MR657863 (83i:28022)

[41] D. A. Lind, *A zeta function for $\mathbb{Z}^d$-actions*, Ergodic theory of $\mathbb{Z}^d$ actions (Warwick, 1993), London Math. Soc. Lecture Note Ser., vol. 228, Cambridge Univ. Press, Cambridge, 1996, pp. 433–450, DOI 10.1017/CBO9780511662812.019. MR1411232 (97e:58185)

[42] D. A. Lind and T. Ward, *Automorphisms of solenoids and p-adic entropy*, Ergodic Theory Dynam. Systems **8** (1988), no. 3, 411–419, DOI 10.1017/S0143385700004545. MR961739 (90a:28031)

[43] A. Malcev, 'Torsionsfreie Abelsche Gruppen vom endlichen Range', Rec. Math. Moscou **4** (1938), 45–68.

[44] L. Markus and K. R. Meyer, *Periodic orbits and solenoids in generic Hamiltonian dynamical systems*, Amer. J. Math. **102** (1980), no. 1, 25–92, DOI 10.2307/2374171. MR556887 (81g:58013)

[45] G. Miles and R. K. Thomas, *The breakdown of automorphisms of compact topological groups*, Studies in probability and ergodic theory, Adv. in Math. Suppl. Stud., vol. 2, Academic Press, New York-London, 1978, pp. 207–218. MR517262 (80c:22007)

[46] R. Miles, *Periodic points of endomorphisms on solenoids and related groups*, Bull. Lond. Math. Soc. **40** (2008), no. 4, 696–704, DOI 10.1112/blms/bdn052. MR2441142 (2009e:37015)

[47] R. Miles and T. Ward, *Orbit-counting for nilpotent group shifts*, Proc. Amer. Math. Soc. **137** (2009), no. 4, 1499–1507, DOI 10.1090/S0002-9939-08-09649-4. MR2465676 (2010b:37011)

[48] R. Miles and T. Ward, *A dichotomy in orbit growth for commuting automorphisms*, J. Lond. Math. Soc. (2) **81** (2010), no. 3, 715–726, DOI 10.1112/jlms/jdq010. MR2650793 (2011j:22012)

[49] P. Moss, *The arithmetic of realizable sequences* (Ph.D. thesis, Univ. of East Anglia, 2003).

[50] D. Ornstein, *Bernoulli shifts with the same entropy are isomorphic*, Advances in Math. **4** (1970), 337–352 (1970). MR0257322 (41 #1973)

[51] D. S. Ornstein, *Ergodic theory, randomness, and dynamical systems*, Yale University Press, New Haven, Conn.-London, 1974. James K. Whittemore Lectures in Mathematics given at Yale University; Yale Mathematical Monographs, No. 5. MR0447525 (56 #5836)

[52] Y. Puri and T. Ward, *Arithmetic and growth of periodic orbits*, J. Integer Seq. **4** (2001), no. 2, Article 01.2.1, 18. MR1873399 (2002i:11026)

[53] D. Ramakrishnan and R. J. Valenza, *Fourier analysis on number fields*, Graduate Texts in Mathematics, vol. 186, Springer-Verlag, New York, 1999. MR1680912 (2000d:11002)

[54] C. G. J. Roettger, *Periodic points classify a family of Markov shifts*, J. Number Theory **113** (2005), no. 1, 69–83, DOI 10.1016/j.jnt.2004.11.012. MR2141759 (2006a:37015)

[55] D. J. Rudolph, *A characterization of those processes finitarily isomorphic to a Bernoulli shift*, Ergodic theory and dynamical systems, I (College Park, Md., 1979–80), Progr. Math., vol. 10, Birkhäuser, Boston, Mass., 1981, pp. 1–64. MR633760 (83c:28014)

[56] K. Schmidt, *Measurable rigidity of algebraic $\mathbb{Z}^d$-actions*, Smooth ergodic theory and its applications (Seattle, WA, 1999), Proc. Sympos. Pure Math., vol. 69, Amer. Math. Soc., Providence, RI, 2001, pp. 661–676, DOI 10.1090/pspum/069/1858549. MR1858549 (2002f:37007)

[57] T. L. Seethoff, *Zero-Entropy automorphisms of a compact abelian group*, ProQuest LLC, Ann Arbor, MI, 1969. Thesis (Ph.D.)–Oregon State University. MR2617673

[58] S. Smale, *Differentiable dynamical systems*, Bull. Amer. Math. Soc. **73** (1967), 747–817. MR0228014 (37 #3598)

[59] M. Staines, *On the inverse problem for Mahler measure* (Ph.D. thesis, Univ. East Anglia, 2013).

[60] J. T. Tate, *Fourier analysis in number fields, and Hecke's zeta-functions*, Algebraic Number Theory (Proc. Instructional Conf., Brighton, 1965), Thompson, Washington, D.C., 1967, pp. 305–347. MR0217026 (36 #121)

[61] J. T. Tate Jr., *Fourier analysis in number fields and Hecke's zeta-functions*, ProQuest LLC, Ann Arbor, MI, 1950. Thesis (Ph.D.)–Princeton University. MR2612222

[62] S. Thomas, *The classification problem for torsion-free abelian groups of finite rank*, J. Amer. Math. Soc. **16** (2003), no. 1, 233–258, DOI 10.1090/S0894-0347-02-00409-5. MR1937205 (2004b:20081)

[63] S. Thomas, *Borel superrigidity and the classification problem for the torsion-free abelian groups of finite rank*, International Congress of Mathematicians. Vol. II, Eur. Math. Soc., Zürich, 2006, pp. 93–116. MR2275590 (2007m:20085)

[64] T. Ward, *The entropy of automorphisms of solenoidal groups* (Master's thesis, University of Warwick, 1986).

[65] T. Ward, *Automorphisms of $\mathbb{Z}^d$-subshifts of finite type*, Indag. Math. (N.S.) **5** (1994), no. 4, 495–504, DOI 10.1016/0019-3577(94)90020-5. MR1307966 (97a:28014)

[66] T. Ward, *An uncountable family of group automorphisms, and a typical member*, Bull. London Math. Soc. **29** (1997), no. 5, 577–584, DOI 10.1112/S0024609397003330. MR1458718 (98k:22028)

[67] T. B. Ward, *Almost all S-integer dynamical systems have many periodic points*, Ergodic Theory Dynam. Systems **18** (1998), no. 2, 471–486, DOI 10.1017/S0143385798113378. MR1619569 (99k:58152)

[68] T. Ward, *A family of Markov shifts (almost) classified by periodic points*, J. Number Theory **71** (1998), no. 1, 1–11, DOI 10.1006/jnth.1998.2242. MR1631042 (99j:11089)

[69] T. Ward, *Dynamical zeta functions for typical extensions of full shifts*, Finite Fields Appl. **5** (1999), no. 3, 232–239, DOI 10.1006/ffta.1999.0250. MR1702897 (2000m:11067)

[70] T. Ward, *Group automorphisms with few and with many periodic points*, Proc. Amer. Math. Soc. **133** (2005), no. 1, 91–96 (electronic), DOI 10.1090/S0002-9939-04-07626-9. MR2085157 (2005f:37054)

[71] A. Weil, *Basic number theory*, Die Grundlehren der mathematischen Wissenschaften, Band 144, Springer-Verlag New York, Inc., New York, 1967. MR0234930 (38 #3244)

[72] A. J. Windsor, *Smoothness is not an obstruction to realizability*, Ergodic Theory Dynam. Systems **28** (2008), no. 3, 1037–1041, DOI 10.1017/S0143385707000715. MR2422026 (2009a:37042)

[73] S. A. Juzvinskiĭ, *Calculation of the entropy of a group-endomorphism* (Russian), Sibirsk. Mat. Ž. **8** (1967), 230–239. MR0214726 (35 #5575)

[RM] SCHOOL OF MATHEMATICS, UNIVERSITY OF EAST ANGLIA, NORWICH NR4 7TJ, UK

[MS] SCHOOL OF MATHEMATICS, UNIVERSITY OF EAST ANGLIA, NORWICH NR4 7TJ, UK

[TW] DEPARTMENT OF MATHEMATICAL SCIENCES, DURHAM UNIVERSITY, DH1 3LE, UK
E-mail address: t.b.ward@durham.ac.uk

Selected Published Titles in This Series

For a complete list of titles in this series, visit the
AMS Bookstore at **www.ams.org/bookstore/conmseries/**.